NONDESTRUCTIVE EVALUATION
A Tool for Design, Manufacturing, and Service

McGraw-Hill Series in Mechanical Engineering

Jack P. Holman, Southern Methodist University
Consulting Editor

Also available from McGraw-Hill

Schaum's Outline Series in Mechanical and Industrial Engineering

Each outline includes basic theory, definitions, and hundreds of solved problems and supplementary problems with answers.

Current List Includes

Acoustics
Basic Equations of Engineering
Continuum Mechanics
Engineering Economics
Engineering Mechanics, 4th edition
Fluid Dynamics
Fluid Mechanics & Hydraulics
Heat Transfer
Introduction to Engineering Calculations
Lagrangian Dynamics
Machine Design
Mechanical Vibrations
Operations Research
Strength of Materials, 2d edition
Theoretical Mechanics
Thermodynamics

Available at Your College Bookstore

NONDESTRUCTIVE EVALUATION
A Tool for Design, Manufacturing, and Service

Don E. Bray, Ph.D., P.E.

Department of Mechanical Engineering
Texas A & M University

Roderic K. Stanley, Ph.D.

International Pipe Inspectors Association

McGraw-Hill Book Company

New York   St. Louis   San Francisco   Auckland   Bogotá   Caracas
Colorado Springs   Hamburg   Lisbon   London   Madrid   Mexico   Milan
Montreal   New Delhi   Oklahoma City   Panama   Paris   San Juan
São Paulo   Singapore   Sydney   Tokyo   Toronto

This book was set in Times Roman by Science Typographers, Inc.
The editors were John Corrigan and John M. Morriss;
the designer was Rafael Hernandez;
the production supervisor was Janelle S. Travers.
Project supervision was done by Science Typographers, Inc.
R. R. Donnelley & Sons Company was printer and binder.

NONDESTRUCTIVE EVALUATION
A Tool for Design, Manufacturing, and Service

234567890 DOC DOC 89432109

ISBN 0-07-007351-1

Library of Congress Cataloging-in-Publication Data

Bray, Don E.
    Nondestructive evaluation: a tool for design,
        manufacturing, and service/Don E. Bray, Roderic K. Stanley.
    p. cm.--(McGraw-Hill series in mechanical
        engineering)
    Includes bibliographies and index.
    ISBN 0-07-007351-1
    1. Non-destructive testing.   I. Stanley, Roderic K.
    II. Title.   III. Title: Non-destructive evaluation.   IV. Series.
TA417.2.B63 1989
620.1′127--dc19   88-19015

ABOUT THE AUTHORS

Dr. Don E. Bray is Associate Professor of Mechanical Engineering at Texas
A & M University. Dr. Bray's experience in Nondestructive Evaluation (NDE)
spans over 20 years of work in industry, government, and academic teaching and
research. Most recent research activities include inspection with critically refracted waves such as P, SV, and higher-order Rayleigh waves, ultrasonic angle-
beam inspection in anisotropic media, and ultrasonic measurement of stress. He
has authored several technical papers and reports on NDE topics and teaches
undergraduate and graduate elective courses in NDE at Texas A & M University.
His education includes a B.S.M.E. from Southern Methodist University, a
M.S.M.E. from the University of Houston, and a Ph.D. from the University of
Oklahoma. Dr. Bray is active in the NDE Engineering Dvision of the American
Society of Mechanical Engineers (ASME). He has been certified at Level III in
ultrasonics by ASNT examination and is a registered professional engineer in
Texas and Oklahoma. In 1980, he jointly received the Achievement Award from
The American Society for Nondestructive Testing (ASNT) for the outstanding
publication in *Materials Evaluation*. ASNT awarded him the rank of Fellow in
1988. He has traveled in Europe, Japan, and the Soviet Union on NDE assignments.

Dr. Roderic K. Stanley is Executive Director of the International Pipe
Inspectors Association. Dr. Stanley obtained his Ph.D. degree in solid state
physics from Florida State University after teaching for several years in East
Africa. His industrial experience includes seven years in the research group at
Baker Tubular Service, where he assisted in the development of new tools for the
inspection of oil field tubular products, and one year at NL McCullough, where
he began development of down-hole inspection devices utilizing multifrequency

eddy-current methods. He also consults in inspection matters that utilize magnetic and electromagnetic techniques.

He has authored several papers on magnetic flux leakage inspection, and written one book on the subject for oil field inspectors. As an active member of ASNT, he chairs the Electromagnetics Committee, which currently oversees the production of Volume IV (Electromagnetic Tests) and Volume VII (Magnetic Testing) of the ASNT Handbook. Other professional societies include the Society of Petroleum Engineers and the American Physical Society.

"Magnetic testing techniques matured with inspection in the steel industry and this author seems unable to avoid it. Part III of this text is dedicated to my father, who worked all his life in the steel mills of Sheffield, England, rolling various types of plain carbon and high alloy steels with that great pride which is found in the craftsmen of his generation. Both of his sons are proud to work in the steel industry. It is my regret that he will not read this material."

RKS

CONTENTS

12 Macroscopic Field Relations   217

13 Magnetic Circuit   230

14 Magnetization in the Circumferential Direction — 254

15 Field Levels In Magnetic NDE — 268

16 Magnetic Flux Leakage From Tight Flaws — 276

PREFACE

In this book, nondestructive evaluation (NDE) topics are described in depth in order to provide sufficient technical background for the engineers and scientists who have responsibility for NDE activities. With this background, NDE inspection schemes and results may be proposed or evaluated with confidence. The primary techniques, i.e., ultrasonics, magnetics, radiography, penetrants, and eddy current are introduced from the fundamental laws of physics. Partial differential equations are used where needed to develop these fundamentals. In each part of the book, examples are given that show typical NDE applications of the physical principles. Research and future applications are also covered.

The material in this book is oriented toward fourth year undergraduate or first year graduate level engineering and science students who are familiar with calculus and engineering materials. The first part introduces the concept of Nondestructive Evaluation (NDE) and provides a review of probability, NDE in design, and NDE in manufacturing and maintenance. The probability topic is included simply to introduce some humility into the thought process. As much as we would like to design fail-free machines, we do not. As much as we would like to detect all cracks, we miss some. With the correct application of NDE, however, failures can be reduced and machinery and systems can be operated with confidence in an extended life mode. Thus, there are economic justifications for the growing interest in NDE.

There are several NDE techniques that are not discussed, e.g., thermal, acoustic emission, noise analysis, etc. Including these topics at a level of presentation equal to that given for the present material would involve considerably more time and would make the size of the book quite large. Those and other emerging NDE topics will be saved for a later volume.

Because nondestructive evaluation is a rapidly changing field, it is impossible to produce a textbook that covers the most recently released research. To stay up-to-date, the reader is directed to the most recent issues of the engineering and science journals and the regularly issued series of NDE reference books that are cited in each topical part of the text.

In the process of writing this book, it has been our pleasure to receive comments from several individuals. These comments have helped to assure that the material is more clearly presented and more oriented toward our audience. Dr. Alan Wolfenden, Dr. Sherif Noah, and Dr. Omer Jenkins of Texas A&M University and the late Merritt Goff are particularly recognized for their comments and advice. Others who gave useful comments on the material include Dr. Frank Iddings of Southwest Research Institute and Frank Malek, NDE consultant. Several individuals who have furnished material that has been used in this book have been cited in the text. Graduate students T. Leon-Salamanca and C. Salkowski have assisted with calculations and suggestions. We appreciate these contributions as well as the contributions of the following reviewers: David M. Egle, University of Oklahoma; Jerald E. Jones, Colorado School of Mines; Phillip L. Jones, Duke University; J. Bruce Nestleroth, Batelle Columbus Laboratories; Ramisamy Palanisamy, The Timken Company; and Stuart Stock, Georgia Institute of Technology.

Additional recognition goes to the students who have used the book and who offered comments that have been useful in correcting errors and improving the presentation. Further, our numerous NDE colleagues who through the years have shared thoughts and ideas on the subject and how to teach it to others should be acknowledged. And we acknowledge our families, who have tolerated the added endeavors that go along with the writing of a textbook. To all of these, we say thank you.

Don E. Bray
Roderic K. Stanley

NONDESTRUCTIVE EVALUATION
A Tool for Design, Manufacturing, and Service

PART
I

PROBABILITY, DESIGN, AND MANAGEMENT IN NONDESTRUCTIVE EVALUATION

CHAPTER
1

INTRODUCTION

Since people first realized the fallibility of man and, thus, his machines, they have recognized a need to inspect these machines in order to prevent failures. A wide variety of test schemes exist, some destructive and some nondestructive. The practical benefits of nondestructive inspection are obvious, as long as the results are reliable and the inspection is cost-effective.

In general, the various nondestructive evaluation (NDE) techniques can be placed into two categories: active and passive. The active techniques are those where something is introduced into or onto the specimen and a response is expected if a defect is present. Magnetics, ultrasonics, and radiography fall into this category. Passive techniques, on the other hand, are those that monitor or observe the item in question during either a typical load environment or a proof cycle and attempt to determine the presence of a defect through some reaction of the specimen. Acoustic emission, noise analysis, leak testing, visual examination, and some residual magnetic techniques are in this classification. Most of the techniques that will be described herein are active types.

The beneficial effects of nondestructive evaluation can be felt in engineering design. For example, in mechanical design, a factor-of-safety is introduced in order to allow for a variety of uncertainties. The nature and the often catastrophic results of these uncertainties have been well described in the literature dealing with fracture and material failure. One of the principal uncertainties is the performance of the components used in the construction of a mechanical system. Manufacturing irregularities, such as voids, inclusions, unfavorable patterns, and hardness, affect the performance of the final part. It is no longer satisfactory for the engineer to simply specify that the material shall be free of defects. There must be more assurance that this is the case. The use of nondestructive evaluation in the quality control of manufactured parts can provide this assurance and thus

3

increase the certainty that an item will perform as intended. With this, then, a lower factor-of-safety may be possible with a resulting overall saving in weight and cost of an item.

Nondestructive testing can also be beneficial in reducing the frequency of unscheduled maintenance, which usually is more expensive than regularly scheduled maintenance. Often, NDE can be used to inspect questionable parts in-place on the equipment, thereby preventing an unscheduled and unnecessary shutdown if the part is in fact defect free. With assurance that there is no defect present, the equipment may continue operating without fear of failure.

Additionally, scheduled maintenance periods may be lengthened with the proper use of NDE in the maintenance cycle. Knowing from an inspection that crucial parts are not approaching failure may allow the machine to operate safely for a longer period of time. Less frequent maintenance may be cost-effective provided that the cost of operation is not increased due to an unexpected failure.

Considerable dedication is required in order to establish a nondestructive testing program that will yield sufficient confidence as to where a factor-of-safety or maintenance cycle can be changed. To do this requires positive communication and thorough understanding of the technical principles involved. One "missed defect" can cause a failure that can erase the savings accrued from years of expensive testing.

The reader will note the frequent use here and elsewhere of the terms nondestructive evaluation (NDE), nondestructive inspection (NDI), and nondestructive testing (NDT). Sometimes they are equal and interchangeable, while on other occasions they are not. While there are obvious inconsistencies in using only one term for all discussion in this text, the term NDE will be adopted because it generally represents the broadest range in the definitions.

CHAPTER 2

PROBABILITY APPLICATIONS IN NONDESTRUCTIVE EVALUATION

2-1 INTRODUCTION

By definition, the concept of probability implies uncertainty. At the outset, this concept of uncertainty may appear to have no place in a field like nondestructive evaluation. Where safety and large maintenance costs hang in the balance, the need is for assurance rather than uncertainty. The presence of probability and uncertainty in engineering is, unfortunately, a matter of fact rather than speculation and it is best to deal with them in a serious manner instead of ignoring them. A discussion of probability necessitates the use of a variety of statistical concepts; some of the more fundamental concepts are described in more detail in Appendix A.

2-2 PROBABILITY IN NONDESTRUCTIVE EVALUATION

In the context of nondestructive evaluation, the first probability concept that comes to mind is the probability of the existence of a defect in a particular item. If we knew that a defect existed in the part, there would be little need for inspection. Several other probability expressions listed in Table 2-1 also relate to NDE. Obviously there are an infinite number of extensions of these expressions. Some of these will be discussed in more detail in the material to follow. The

TABLE 2-1
Typical probability expressions in nondestructive evaluation

---

1. What is the probability of existence of a defect?
2. Given the existence of a defect, what is the probability that its size is in a specified range?
3. Given the existence of a defect in a specified size range, what is the probability of detection for a particular NDE system?
4. What is the probability of a defect of a certain size propagating to failure prior to the next inspection?

---

reader is directed to the several references for more information on probability and its use in NDE.

2-3 PROBABILITY EVENTS AND COMBINATIONS

Probability is a study of the likelihood of occurrence of specific events. A simple yet useful application of probability analysis is in studying the results from a toss of a single coin. Assuming that the coin is "balanced," the probability of it showing heads or tails on a single toss is even. For example, for one toss, the probability of heads $P(H)$ is

$$P(H) = \frac{1 \text{ event}}{\text{number of equally possible events}}$$

$$= \tfrac{1}{2} = 0.5 \tag{2-1}$$

Similarly, the probability of tails $P(T)$ is

$$P(T) = \tfrac{1}{2} = 0.5 \tag{2-2}$$

For the roll of one six-sided balanced die, there are six equally possible events (1,2,3,4,5,6) and the probability of a single event (e.g., rolling a 4) is

$$P(4) = \tfrac{1}{6} = 0.167 \tag{2-3}$$

If a pair of dice is thrown, the probability of rolling a pair of 4s is

$$P(4,4) = \left(\tfrac{1}{6}\right)\left(\tfrac{1}{6}\right) = 0.028 \tag{2-4}$$

This indicates that the probability of two single independent events occurring is the product of their individual event probabilities. The multiple event probability, therefore, will be lower than any single event probability.

Obviously, then, there is much greater assurance that a person could call the flip of a coin as compared to a roll of the dice. It is important to stress, however, that neither event is impossible and neither event is absolute. There is an element of chance, or probability, in each.

A probability of 1 is used for absolute assurance that an event will occur. For example, for a common six-sided die, the probability that a thrown die will land with any side facing upward is

$$P(6) = \tfrac{6}{6} = 1 \tag{2-5}$$

The example of drawing marbles from a box introduces some greater variations into the concepts just introduced. For a box of N marbles, there are n_1 red marbles and n_2 white marbles, where $n_1 + n_2 = N$. If one draw is made from the box, the probability of drawing a red marble is

$$P(R) = \frac{n_1}{N} \tag{2-6}$$

Similarly, the probability of drawing a white marble is

$$P(W) = \frac{n_2}{N} \tag{2-7}$$

Also, since the number of white marbles is related to the number of red ones, the previous probabilities can be written as

$$P(R) = 1 - P(W)$$
$$= 1 - \frac{n_2}{N} = \frac{n_1}{N} \tag{2-8}$$

A further modification of this problem occurs when a second draw is made from the box. Two possibilities exist. If the original marble is replaced, probabilities for the second draw are identical to the first. This is drawing with replacement. If, on the other hand, the marble first drawn is set aside and not replaced, this is defined as sampling without replacement. If a red marble is first drawn without replacement, on the second draw the probability of a red marble is

$$P(R) = \frac{n_1 - 1}{N - 1} \tag{2-9}$$

which is clearly smaller than $P(R)$ for the first draw.

Ordered samples are also a topic of some importance. For this group, assume again a box of N marbles. In this case, however, each is different. The question then arises: How many ordered samples are possible for r draws? Here again, the with and without replacement distinction may apply. With replacement, the number of ordered samples possible for r draws is

$$N^*(N, r) = N_1 N_2 N_3 \cdots N_r = N^r \tag{2-10}$$

where the subscripts indicate the number of the draw.

Drawing for the game of bingo is an example of how an ordered sample is drawn without replacement. Each bingo ball is unique and it is placed on a board and not returned to the cage holding the remaining balls. For this, the number of different permutations or ordered samples for r draws from a lot of size N is

$$N^{**}(N,r) = N(N - 1)(N - 2) \cdots (N - r + 1) = \frac{N!}{(N - r)!} \tag{2-11}$$

The scope of the material on probabilities has been necessarily limited since the intent has been to introduce only statistical concepts without delving too deeply into the myriad of often-required considerations of other factors. Two examples of the effects of other considerations will serve to illustrate the point. First, Eq. (2-1) tends to imply that where there are only two possible outcomes, the probability of either event is 0.5. That this is not the case can be realized by noting that there are basically two possible ways to terminate an airplane flight, the plane can land normally or it can crash. Even though we consider only two possible outcomes, we hope that the probability of the former far exceeds that of the latter. The reason is that the outcome is heavily weighted in favor of the successful termination, which is equivalent to stating that the probability of the event is conditioned by other factors. Similarly, it also can be shown that the results expected from Eq. (2-9) are conditioned by the results of the previous draw.

Example 2-1. A box contains 25 red balls and 75 white balls. For one draw, what is the probability of drawing a red ball? Also, what is the probability of drawing a white ball?

$$P(R) = \frac{25}{100} = 0.25 \qquad P(W) = \frac{75}{100} = 0.75$$

Also

$$P(W) = 1 - P(R) = 1 - 0.25 = 0.75$$

Example 2-2. For the lot in Example 2-1, if a red ball is taken on the first draw and not replaced, what is the probability of drawing a red ball on the second draw? For the same conditions, what is the probability of drawing a white ball on the second draw? Compare these results with those in Example 2-1.

$$P(R) = \frac{24}{99} = 0.242$$

$$P(W) = \frac{75}{99} = 0.758$$

Example 2-3. For a series of 5 draws from a box of 40 different balls, how many ordered samples are possible, assuming that the balls are replaced after each draw?

$$N^* = (40)^5 = 102.4 \times 10^6$$

Example 2-4. For Example 2-3, how many possible samples are there for the without-replacement condition?

$$N^{**} = \frac{40!}{35!} = 79 \times 10^6$$

Example 2-5. Three balls in a box are labeled A, B, and C. How many ordered samples are possible from this box with two draws and with replacement?

$$N^* = N^r + 3^2 = 9 \qquad \{AB \quad AC \quad BC \quad BA \quad CA \quad CB \quad AA \quad BB \quad CC\}$$

How many would be possible under without-replacement conditions?

$$N^{**} = \frac{N!}{(N-r)!} = \frac{3!}{1!} = 6 \qquad \{AB \quad AC \quad BC \quad BA \quad CA \quad CB\}$$

2-4   PROBABILITY SAMPLE SPACES AND FLAW SIZE DISTRIBUTIONS

The categorization of possible events for a particular situation is defined as the sample space. This sample space can be both finite and infinite, as will be described.

A finite sample space applies when there is a limit on the number of possible events. Consider three coins to be tossed simultaneously in the air. There are a finite number of possible head and tail combinations, as shown in Fig. 2-1.

The following probabilities may be calculated for the indicated number of heads in one toss.

$$P(0) = P\{TTT\} = \tfrac{1}{2} \cdot \tfrac{1}{2} \cdot \tfrac{1}{2} = \tfrac{1}{8} \tag{2-12}$$

$$P(1) = P\{HTT\} + P\{THT\} + P\{TTH\}$$
$$= \tfrac{1}{2} \cdot \tfrac{1}{2} \cdot \tfrac{1}{2} + \tfrac{1}{2} \cdot \tfrac{1}{2} \cdot \tfrac{1}{2} + \tfrac{1}{2} \cdot \tfrac{1}{2} \cdot \tfrac{1}{2} = \tfrac{3}{8} \tag{2-13}$$

$$P(2) = P\{HHT\} + P\{THH\} + P\{HTH\} = \tfrac{3}{8} \tag{2-14}$$

$$P(3) = P\{HHH\} = \tfrac{1}{2} \cdot \tfrac{1}{2} \cdot \tfrac{1}{2} = \tfrac{1}{8} \tag{2-15}$$

These calculations may be verified with the sample space in Fig. 2-1.

An infinite sample space is often more practically useful in fatigue, failure, and defect detection than the previously described finite space. An infinite sample space is demonstrated by the concept of tossing a coin until heads appears where there is no limit on the number of tosses that may be required to obtain a heads.

The probability of requiring n tosses to achieve a head is defined by the ratio of the number of events of interest divided by the number of possible events raised to the nth power. For the coin, there is one event of interest (a heads) and two possible events (a head and a tail), and the required probability is

$$P(n) = \left(\tfrac{1}{2}\right)^n \tag{2-16}$$

The probability of requiring one, two, or an infinite number of tosses to obtain a heads is

$$P(1) = \tfrac{1}{2} = 0.5 \tag{2-17}$$

$$P(2) = \tfrac{1}{2} \cdot \tfrac{1}{2} = \tfrac{1}{4} = 0.25 \tag{2-18}$$

$$P(\infty) = \left(\tfrac{1}{2}\right)^{\infty} \simeq 0 \tag{2-19}$$

The probability of requiring a large number of tosses to get one head is clearly a diminishing number.

T	T	T	H	H	T
H	T	T	T	H	H
T	H	T	H	T	H
T	T	H	H	H	H

FIGURE 2-1
Sample space for tossing three coins in the air.

TABLE 2-2
Distribution of typical crack depth

Dimensions (mm)	Number of occurrences	P(flaw)
0 < x < 2	2	0.051
2 ≤ x < 4	6	0.154
4 ≤ x < 6	10	0.256
6 ≤ x < 8	15	0.385
8 ≤ x < 10	4	0.103
x ≤ 10	2	0.051
	39	1.000

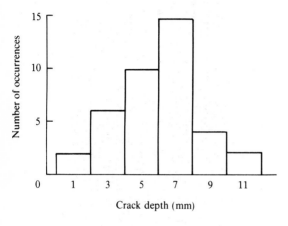

Crack depth (mm)

FIGURE 2-2
Distribution of crack depths in turbine rotors (mm).

The dimensions of a fatigue crack can fill an infinite sample space since there are an infinite number of possibilities. Usual practice, however, is to categorize the data, i.e., define numerical categories and fit the fatigue crack data within those limits. For example, a survey of a group of turbine rotors could result in the data shown in Table 2-2 and Fig. 2-2 for existing fatigue cracks. If circumstances required that the crack depth be catalogued on a smaller interval, more size categories would be required. In the limit, for an extremely small individual dimension range, an extremely large number of categories would be required. The number of categories is the sample space and this could, in fact, be carried to infinity. Considering only the shafts containing flaws within each defined category, the probability of a flaw P(flaw) in the specified category is equal to the number of flaws in the category divided by the total number of known flaws.

2-5   CONDITIONAL PROBABILITY

A conditional probability is an expression of the likelihood of one event occurring given that another event occurs. For example, the probability of a person getting wet when walking outdoors is dependent on how frequently he is outdoors and the probability that it is raining. This is the question that conditional probability addresses.

A more definitive example of conditional probability can be derived from the seat assignment process on a typical airliner. Figure 2-3 shows a seating arrangement of 20 rows of 5 seats each. For this, define event A as the probability of being assigned a seat in rows 1–7 and event B as the probability of being assigned a seat in rows 6–15, assuming that there is an equal likelihood of being assigned any seat. The following events are defined:

$$A = \text{assigned a seat in rows } 1\text{–}7$$

$$B = \text{assigned a seat in rows } 6\text{–}15$$

First, the probabilities of each event as previously defined are

$$P(A) = \frac{7 - 0}{20} = 0.35 \tag{2-20}$$

$$P(B) = \frac{15 - 5}{20} = 0.50 \tag{2-21}$$

For conditional probability, the general sample spaces shown in Fig. 2-4 can be utilized. Here E_1 and E_2 represent events 1 and 2, respectively. Where they overlap, as in Fig. 2-4(a), it is possible for an event to occur in both spaces, as shown by the cross-hatched area. If there is no possibility of an overlap of the two events, the sample spaces are said to be mutually exclusive as shown in Fig. 2-4(b).

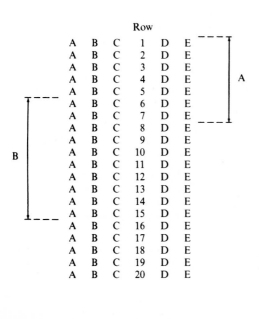

FIGURE 2-3
Typical aircraft seating arrangement.

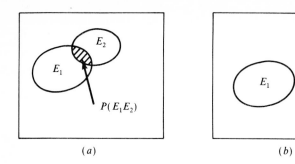

FIGURE 2-4
(*a*) Conditional sample spaces. (*b*) Mutually exclusive sample spaces.

The expression for conditional probabilities for two samples spaces is

$$P(E_1|E_2) = \frac{P(E_1 E_2)}{P(E_2)} \tag{2-22}$$

This expression says that the probability of both events occurring, given that E_2 must occur first, is given by the overlap area $P(E_1 E_2)$ divided by the sample space for E_2, $P(E_2)$. The effect of the conditional probability as seen in Fig. 2-4(*a*) and (*b*) is to shrink the sample space. For example, in Fig. 2-4(*a*) the events in E_1 that are stated by Eq. (2-22) to be of interest are limited to those that are also in space E_2.

For the preceding seating assignment problem, the probability of A occurring given that B must occur first is

$$P(A|B) = \frac{P(AB)}{P(B)} = \frac{(2/20)}{(10/20)} = 0.2 \tag{2-23}$$

2-6 EFFECT OF INSPECTION ON FLAW SIZE DISTRIBUTION

In general, nondestructive inspection systems are more likely to detect large defects than small ones. Empirical tests as reported by Packman [7] and more recently reviewed by Williams and Mudge [8] typically are used to establish the probability of detection as a function of flaw size for a particular system $P(D|x)$. Figure 2-5 shows a hypothetical curve for $P(D|x)$. The inspection system depicted in Fig. 2-5 applied to the flaw size distribution shown in Fig. 2-2 would clearly result in the largest flaws being detected with some of the smaller defects left in the rotors. The outcome, then, is a new distribution for the remaining flaws with the mean flaw size of the new set decreased in comparison to the previous, uninspected set.

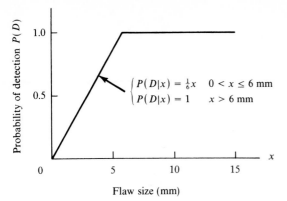

$$P(D|x) = \tfrac{1}{6}x \quad 0 < x \le 6 \text{ mm}$$
$$P(D|x) = 1 \quad x > 6 \text{ mm}$$

FIGURE 2-5
Probability of detection and flaw depth.

An example of the effect of inspection on the size distribution of the undetected flaws has been given by Ang and Tang [3] and Tang [9]. Considering a set of welds, they present a distribution of flaw depths prior to inspection, $f_x(x)$, to be represented by

$$f_x(x) = \begin{cases} 208.3x & 0 < x \le 0.06 \text{ in} \\ 20 - 125x & 0.06 < x \le 0.16 \text{ in} \\ 0 & x > 0.16 \text{ in} \end{cases}$$

(note: 0.06 in = 1.5 mm, 0.16 in = 4.1 mm) and shown by the triangular curve in Fig. 2-6. All dimensions are in inches. Additionally, they assume that the nondestructive inspection system has the detection capabilities

$$P(D|x) = \begin{cases} 0 & x \le 0 \text{ in} \\ 8x & 0 < x \le 0.125 \text{ in} \\ 1.0 & x > 0.125 \text{ in} \end{cases}$$

(note: 0.125 in = 3.2 mm) where $P(D|x)$ is the probability of detection of a defect of size x. Since all flaws of 0.125 in (3.2 mm) or greater would be detected by this system, a shift of the mean flaw size to a smaller value would be expected for the missed defects. The expression that is sought is the posterior distribution function $f_x(x|\overline{D})$ where the $\overline{D}$ indicates a missed flaw. The density function for the size of the missed flaws is given by

$$f_x(x|\overline{D}) = kP(\overline{D}|x)f_x(x) \tag{2-24}$$

where k = normalization constant
$P(\overline{D}|x)$ = probability of missing a defect of size x
$f_x(x)$ = the size distribution prior to inspection

Using standard techniques, the results show that the density function for the missed defects is

$$f_x(x|\overline{D}) = \begin{cases} 0 & x \le 0 \text{ in} \\ 495x - 3964x^2 & 0 < x \le 0.06 \text{ in} \\ 47.6 - 678x + 2379x^2 & 0.06 < x \le 0.125 \text{ in} \\ 0 & x > 0.125 \text{ in} \end{cases}$$

which is skewed to the left and has a mean value less than that of the previous, uninspected peak.

The exact nature of the remaining flaw set has been shown to be directly related to the inspection characteristics of the NDE system. For example, a relatively insensitive and inexpensive technique could be used to detect only the larger cracks. For this case, the shift of the peak to the left would not be nearly as great as would be achieved with the use of a far more sensitive inspection system. The cost of an inspection system, however, is often directly related to the size sensitivity and this must be considered in the overall planning.

Example 2-6. Prior studies have shown that a flaw distribution as shown in Table 2-2 may be expected for a particular rotor design used in gas turbine engines. One hundred rotors received from a supplier are to be inspected using an NDE system having flaw detection capabilities as shown in Fig. 2-5. If the sample of 100 rotors actually contains 5 defects, what is the probability of detecting a defect in the inspection of a single rotor selected randomly from the sample? If all of the shafts are inspected without replacement, how many defects will be located and how many will be missed?

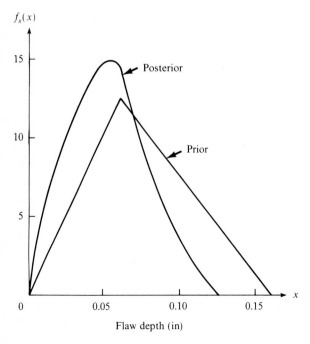

FIGURE 2-6
Flaw depth distribution in welds before and after inspection. (*From Ang and Tang* [3]. *Courtesy John Wiley.*)

The following probabilities are defined:

$$P(D) = \text{probability of detection of a defect}$$

$$P(D/a_i < X \leq b_i) = \text{probability of detecting a defect in the size range} \\ (a_i < X \leq b_i) \text{ (from Fig. 2-5)}$$

$$P(a_i < X \leq b_i) = \text{probability of existence of a defect in size range} \\ (a_i < X \leq b_i) \text{ (from Table 2-2)}$$

$$P(X) = \text{probability of existence of any defect}$$

$$a, b = \text{lower and upper bounds of each interval } i$$

For these defined terms and in the individual size groups, the probability of detecting a defect may be written from Eq. (2-22):

$$P(Da_i < X \leq b_i) = P(D|a_i < X \leq b_i) P(a_i < X \leq b_i)$$

Summing over the interval containing the flaw distribution results in

$$P(D) = \left[\sum_x P(D|x) \right] P(X)$$

$$\sum_x P(DX) = \sum [P(D|a_i < X \leq b_i) P(a_i X \leq b_i)]$$

$$\sum_x P(DX) = \underset{0-2}{(0.167 \times 0.051)} + \underset{2-4}{(0.5 \times 0.154)} + \underset{4\text{-}6\text{-}6}{(0.83 \times 0.256)}$$
$$+ \underset{6-8}{(1 \times 0.385)} + \underset{8-10}{(1 \times 0.103)} + \underset{> 10}{(1 \times 0.051)}$$
$$= 0.837$$

Thus, the probability of locating a defect within the distribution given is 0.837. Because, for the present case, the probability of any rotor having a defect is 0.05, the probability of finding a defect in the rotors is $P(D) = 0.837 \times 0.05 = 0.042$. For this example, then, four defects would be found and one would be missed. In a series of inspections on groups of shafts using these same conditions, actual results would vary on either side of this average of 4.2 defects detected.

2-7 DATA SET CHARACTERISTICS

It is useful at this point to emphasize the character of the information that might emerge from a small data set as opposed to a large one and the difference within these data sets between the probability of the existence of a defect and the distribution of defects within the sample set. If the data in Table 2-2 were obtained from an inspection of 10,000 shafts, the probability of a defect in these shafts would clearly be 0.0039. Moreover, the data set appears to be reasonably normal and an analysis will show a well defined mean and standard deviation. What could be expected to emerge, however, if only 100 shafts were inspected? Although a mean flaw size as well as a standard deviation could be calculated for the smaller data set, how well would these results represent the overall population of these shafts?

Evolutionary variations that occur in empirical data may be observed using a chronological record of rocket launches as provided by the National Aeronautics and Space Administration [10, 11]. The Delta rocket was first launched in 1960 with a capability of lifting about 91 kg (200 lb) to the geosynchronous transfer orbit. The 3900 series of the Delta was capable of lifting approximately 1270 kg (2800 lb) into the same orbit. In the 26 year period from 1960 to mid-1986, 178 launches were attempted.

The probability of success for a specific rocket launch is strictly a function of the conditions existing in the rocket, the support facilities, and the environment as well as other factors that may affect the launch. Certainly, evolutionary design changes in the rocket, support systems, etc., improve the probability of success of each succeeding launch. Nonetheless, it might be reasonable to assume that at the time of a particular launch, prior launch data also may be useful in predicting the probability of success $P(S)$ for that launch.

With the assumption that prior launch data might be indicative of the probability of success of an individual launch, it should be noted that the first attempt at launching the Delta failed, yielding $P(S) = 0$ for the next attempt. Clearly, this was not realistic since the second launch was successful. After early fluctuations, the value of $P(S)$ from 1964 to mid-1986 ranged from 0.92–0.96 and in the last nine years, ranged from 0.93–0.94. Thus, where one single failure in 1960 greatly affected $P(S)$, the failure of launch number 178 on 3 May 1986 did not significantly affect the overall probability of success for the Delta rocket.

In answering the question on how well small amounts of data may represent a full population of data, the discussion on Delta rocket failures shows that a sample of data obtained from only the early launches would be not at all representative of the overall data set. In contrast, small amounts of data taken from later launches would yield success estimates more closely resembling the overall $P(S)$. Thus, different conclusions might be drawn, depending on which data set was used.

The previous discussion introduces in a qualitative manner the concept of statistical confidence. It is easy to recognize that we should be wary that small amounts of data can truly represent the overall data set. More detailed information on statistical confidence is contained in Appendix A.

2-8 SUMMARY

The close relationship of NDE and probability has been presented in this chapter with a primary emphasis on the probability of existence of a defect and the size effect on the probability of detection. While the probability of existence of a defect is a factor over which the NDE engineer has no control, the probability of detection is a factor that is very much in the control of the NDE engineer. More sensitive systems, with a higher probability of detection, most likely will be more expensive. Do the risks involved in potential failures justify the added expenditure? Often times they do; sometimes they do not. Other important factors that are involved in this decision are the material characteristics and costs of failures. These topics will be discussed in more detail in the chapters to follow.

3

NONDESTRUCTIVE
EVALUATION
IN DESIGN

3-1 INTRODUCTION

The engineer is under increasingly severe pressure to produce reliable and safe designs for structures and equipment with the added constraint that they must compete economically in the world market. To accomplish this, a designer must have confidence in the materials that are used and full knowledge of the likely service environment that will be encountered. In the final analysis, however, the design engineer has come to "expect the unexpected." Nondestructive evaluation plays a key role in minimizing the effects of the unexpected load condition or material property.

Figure 3-1 illustrates in a simplistic manner the interrelationship among material properties, load environment, and nondestructive evaluation. The application of NDE to the inspection of parts in the early stages of manufacture can minimize the chance of defective items being placed into service. Additionally, ordinary items can fail in fatigue, often times due to a change in load environment. NDE can often detect fatigue cracks in the early stage before catastrophic failure occurs. The related field of fracture mechanics has placed more responsibility on those in nondestructive evaluation. When the fracture mechanics specialist states that a component possibly valued at several million dollars can remain in service provided that there are no defects present greater than a certain length, the nondestructive evaluation specialist must know the capabilities of his system to detect defects of that critical length. Thus, NDE is a valued contributor to the safe and economical performance of equipment and structures.

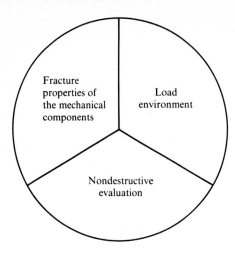

FIGURE 3-1
Total failure prevention program.

3-2 NDE IN DESIGN

Nondestructive evaluation can play a significant role in obtaining an efficient, long-lived design of components and operating machines. In a review, Mordfin [13] argued that the combined advances of NDE and fracture mechanics in recent years have radically affected the approach to mechanical design. Several axioms describing this new approach to design are given. Axiom 1 states "All materials contain flaws." This is in contrast to the earlier philosophy where it was felt to be sufficient for the design engineer to simply state in his specifications that the material used shall be free of defects. With the new thought that all materials contain flaws, it is now the responsibility of the design engineer to first know the fracture mechanics characteristics of the material that is being used and, second, to assure that NDE has been used to prevent the occurrence of potentially hazardous defects. The beneficial role that NDE is able to play in the initial design process may be presented using the typical fatigue ($\sigma - N$) curve as shown in Fig. 3-2. These curves are often used to evaluate the fatigue characteristics of components. The shape of the curve and, hence, the fatigue characteristics are functions of both the material being used and the design of the component being tested.

The general conclusion from these diagrams is that a part operating at a stress level σ_2 would be expected to have a longer fatigue life than one operating at a higher stress level σ_1. Largely due to material and manufacturing variations, however, actual practice has shown that in the vicinity of the fatigue failure line there is considerable uncertainty. In order to produce a more conservative design, the operating stress level must be removed from the uncertain area. This is accomplished with the *factor-of-safety*, typically defined as the ratio of the design stress level over the expected stress level. Thus, for a part expected to operate for N_2 cycles at a stress level of σ_2, the load factor-of-safety is σ_2/σ_2'.

It is easy to see that changes in the operating conditions after the component is placed in service will result in a new factor-of-safety. For example, either an increase in the load stress to σ_2'' or an extended-life operation to N_3 would result in a decrease in the factor-of-safety and a corresponding increase in the likelihood of failure.

Recognizing the uncertainty in the area near to the fatigue line, one might expect failure data arising from actual operations to indicate the true factor-of-safety of the various components. When historical data indicate that a particular component has not shown any sensitivity to changes in loading, it is reasonable to assume that the true stress level is safely removed from the fatigue area. On the other hand, a noticeable change in failure rate with a change in operating conditions should suggest that operation has been moved into the sensitive area. More detailed discussions on the fatigue factor-of-safety are contained in a number of sources, including Refs. 6 and 14.

NDE can be a significant contributor in a decision on increasing the load or life cycle of an existing component. Frequent monitoring of the equipment for a suitable period could produce indications of whether or not the operating conditions had been placed too near the failure line. There may be sufficient economic justification for increasing the load or extending the life, provided that it can be done without catastrophic consequences.

The uncertainty around the fatigue failure line can be decreased using NDE. Much of this scatter is due to material property variations such as inclusions, forging bursts, improper heat treatment, or even incorrect chemical composition. Stricter quality control at the manufacturing plant can decrease the variations in material properties with a resulting decrease in the scatter at the fatigue line. NDE, of course, is a crucial contributor to manufacturing quality control. Better quality control at all stages of manufacture can give designers sufficient confidence to be able to design with a lower factor-of-safety without

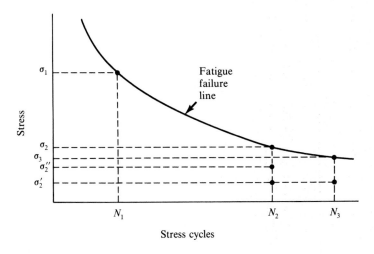

FIGURE 3-2
Effect of changes in loading on factor-of-safety [14].

any sacrifice in safety. One additional benefit of the lower factor-of-safety is that less material would be required to construct a particular machine.

3-3 FAILURE RATE, RELIABILITY, AND NDE

Failure rate information is frequently useful in planning a nondestructive evaluation program. The mortality or "bathtub" curve shown in Fig. 3-3 is typically used for describing failure rates [4]. Each component will have its own unique failure rate curve, affected by a number of things such as material condition, operating environment, and source of manufacture, to name a few.

For the hypothetical component whose failure rate characteristics are shown in Fig. 3-3, the "burn-in" region from time 0 to T_E includes the very early failures generally associated with either manufacturing defects or severe overload. Too loose or too tight a fit, a defective weld, material inclusions, and too hard or too soft material are among typical conditions that cause these early failures. Severe overload can originate from either a load going higher than the designer anticipated or when an underrated part is erroneously installed. There is little that NDE can do to reduce these failures, once the parts are installed on the machine. The failures simply occur too swiftly for most techniques to have any effect. NDE utilized during construction to locate weak or nonspecification parts or materials is the best defense against failures of this type.

The middle portion of the curve from T_E to T_w is called the useful life or constant failure region. This mostly flat portion covers a very large time, and failures that occur here are generally random in nature. Some very late burn-in failures and early fatigue failures can occur in this region. This is the region where NDE is most difficult to utilize. Due to the long time span covered by this region and the random nature of the failures, periodic monitoring is an ineffective and costly defense. Constant monitoring techniques are most effective here, but, due to the costs, can only be justified where the consequences of failure are severely catastrophic. Due to the rather small number of failures in this region, these events are often viewed as an acceptable business cost.

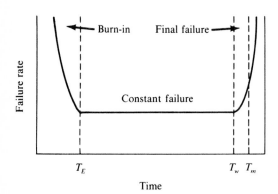

FIGURE 3-3
Typical failure rate curve.

The final failure region, which begins at T_w, is generally dominated by fatigue failures. At time T_m, the mean life, approximately one-half of the components will have failed. Typically, the failure rate will show only a slight increase at T_w followed by an increasingly steep curve. In some situtations, the total time in the final failure region could be as long as for the useful life portion, while in other cases the failure rate could increase quite rapidly after T_w. The relationship of T_w and T_m may be associated with a number of factors such as load environment and material characteristics, as will be discussed later. Effective nondestructive evaluation can assist in locating the start of the final failure portion T_w and in the removal of potentially hazardous parts before time T_m is reached.

Often the characteristics of the final failure portion of the curve may be estimated using existing failure data and mathematical analyses. The failure rate curve shown in Fig. 3-4 represents data from in-service failures of rotary couplers used in rotary-dump hopper cars hauling coal to Texas power plants [15]. The

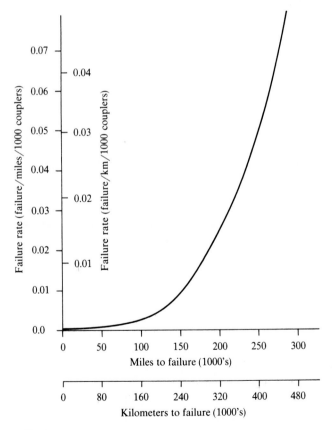

FIGURE 3-4
Failure rate for railroad coal train couplers that failed in service [15].

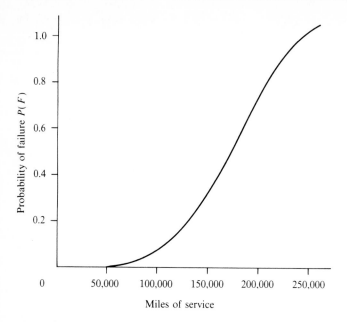

FIGURE 3-5
Probability of failure for railroad coal car couplers.

coupler is the mechanical connection between the cars, transmitting the full train forces. A Weibull analysis of the actual failure data produced this final failure portion of the failure rate curve. No nondestructive inspection was used in this instance, resulting in a data set where every failed component was allowed to remain in service until that event occurred. The failure rate curve could be used to establish a proper NDE interval, once an acceptable threshold failure rate was chosen. This is discussed in more detail in what follows. The probability of failure $P(F)$ of operating systems also may be established with the Weibull analysis previously used for the coupler failures. Figure 3-5 shows $P(F)$ as a function of miles of service for the rotary couplers. While the probability of failure at 100,000 miles is quite low, the rate of increase becomes quite sharp after that. At 180,000 miles, approximately half of the couplers would have failed. Virtually all couplers would have failed at 250,000 miles.

3-4 FLAW CHARACTERISTICS AND NDE

Axiom 3 from Mordfin [13] states that "The detectibility of a flaw generally increases with its size." It is also true, however, that the probability of failure generally increases with flaw size, creating a challenge for the NDE engineer to assure inspection, detection, and characterization before failure occurs.

Propagating fatigue flaws may originate from a number of sources, notably material imperfections and damage occurring in service. In order for NDE to be effective against fatigue failures, a stable crack that is of sufficient size to be detected must exist at the time of inspection. A typical fatigue failure pattern is shown in Fig. 3-6. Initially, the crack may start at a single location and grow rather slowly for quite some time. At some stage in its growth, the propagation rate increases rapidly until final failure occurs. The crack growth lines and the final, sudden rupture portions of the failure surface are shown in the figure. Where a ductile material may have a rather large crack growth area before final failure occurs, a brittle material may have an extremely small fatigue crack region. It follows, therefore, that NDE is more effective in preventing ductile fatigue failures than brittle failures. Brittle failures caused by initial material imperfections are best prevented by stringent quality control in the manufacturing process. NDE, of course, is an important component in a quality control program. The discussion that follows gives more detail on the use of fracture mechanics in describing material fracture characteristics.

Most NDE techniques can be used to obtain some information on the size, shape, and orientation of the defect. It is useful, then, to have some knowledge on the more typically encountered defects. There are a large number of reference sources that describe material failures, analyze the cause of the failure, and provide photographs or sketches of the fracture surfaces. References 16–21 contain a wide range of information on material failures.

The appearance of the fracture surfaces for several types of loadings of a circular shaft are shown in Fig. 3-7 [21]. Knowledge of the expected fracture surface of an item being inspected can sometimes aid the NDE engineer in differentiating a crack from a spurious indication. Figure 3-7 shows two major columns: high nominal stress on the left and low nominal stress on the right. Clearly, a part in a relatively low stress field may be expected to endure a crack of considerably greater magnitude before failure than one operating in a high stress field. The effect of various loadings such as uniaxial tension–compression and several types of bendings are also shown. A shaft in rotating bending service may crack around the entire periphery while one in unidirectional or reverse bending will have the crack surface generally confined to one or two loactions. Torsional defects, as shown on the bottom, may propagate in a variety of patterns.

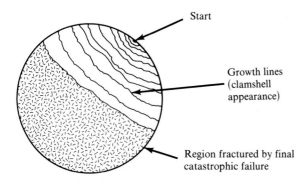

Start

Growth lines (clamshell appearance)

Region fractured by final catastrophic failure

FIGURE 3-6
Typical fatigue crack appearance in a circular shaft.

3-5 FRACTURE MECHANICS AND NDE

Axiom 2, as stated by Mordfin [13], says "Flaws in a material do not necessarily render it unfit for its intended purpose." Where a flaw is defined as a material imperfection, it is easy to recognize that a flaw that is not growing in size with time in service is unlikely to result in an ultimate component failure. Four significant factors in final failure are the number and character of the flaws, the load environment, the residual stress levels, and the mechanism of failure for the material. NDE can aid the design engineer in characterizing the flaws and establishing the residual stress level. With this, then, judgement on the suitability

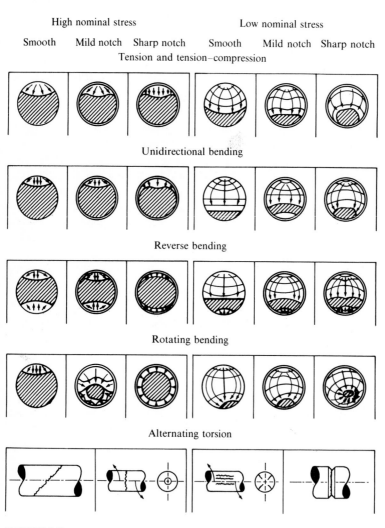

FIGURE 3-7
Appearance of fatigue fracture as affected by mode of loading, magnitude of stress, and degree of stress concentration. (*From Jacoby* [21]. *Courtesy Society for Experimental Mechanics.*)

of the component to serve its intended purpose may be made with greater assurance.

Many examples exist where, for brittle materials, a crack unexpectedly propagates from a small size to cause total failure. NDE obviously is of limited use in this instance. Fatigue cracks are assumed to travel at a slower rate, making them detectable with the proper NDE technique and a satisfactory inspection cycle.

The study of fracture mechanics has given the design engineer an added dimension in the creation of designs that will have a high likelihood of successful service. Fracture mechanics serves as an integrator of material properties, design stress, and flaw detectibility. Fracture mechanics is covered in a number of references, e.g., Hertzberg [22]. In fracture mechanics, a distinction is made between the initiation of a crack and its subsequent propagation. In calculating the critical stress levels, fracture mechanics assumes the presence of a crack when most likely one does not exist. Thus, the analysis begins with a conservative assumption. With the presence of a crack assumed, the propagation in metals begins from the onset of plastic flow in a very localized region at the crack tip. Once started, it is further assumed that the crack will travel at a speed near to the speed of sound until arrested by some condition in the material. Failure prevention, then, demands that the crack be detected before reaching the critical, final failure stage.

The initial parameter to be defined in fracture mechanics is the stress intensity factor K, which is a function of the geometry of the crack and the surrounding stress field. A very sharp crack tip is assumed and empirical estimates of K for a variety of crack shapes in various materials are available in the literature. For a flat crack of width $2a$ oriented 90° to the stress field σ and existing through the thickness of the material, K is given as

$$K = \sigma\sqrt{\pi a} \qquad (3\text{-}1)$$

In design, the important parameter is the crack size that will lead to failure. For a stress field dictated by the engineering application, the critical crack size, $2a$ in this example, is a function of the NDE system capability, while the critical stress intensity factor K_c is a function of the material. For critical conditions, Eq. (3-1) may be rearranged to

$$\sigma_c = K_c/\sqrt{\pi a} \qquad (3\text{-}2)$$

Where the stress field, i.e., the applied plus the residual stress, is less than σ_c, crack propagation will not occur. Stresses above the critical level, on the other hand, will result in crack growth. Thus, for a particular material, the availability of a NDE system capable of detecting very small cracks would allow a design at a higher stress level with a resulting savings in the amount of required material. The lack of a suitable NDE system, on the other hand, would force a design at a lower stress level. The factor-of-safety and the fatigue curve that have been previously discussed are obviously connected to the critical factors obtained through fracture mechanics analyses.

Once defects are detected, the question of the growth rate immediately arises. If there is no growth, there is no cause for alarm. If growth is slow and the crack is considerably smaller than the critical size, then it may be reasonable to leave the defective component in service and inspect it at a prudent interval. In this event, however, the estimated growth rate of the defect may be used to establish the interval.

The fatigue crack growth rate, as described by Hertzberg [22], is

$$\frac{da}{dn} = A \, \Delta K^m \tag{3-3}$$

where da/dn = fatigue crack growth rate
ΔK = stress intensity factor range ($\Delta K = K_{max} - K_{min}$)
A, $m = f$ (material variables, environment, load frequency, temperature, and stress ratio).

Thus, with sufficient empirical data, as cited by Hertzberg, an estimate of the crack growth rate may be obtained. Once the flaw is sized by some nondestructive means and the load environment is established, engineering calculations may establish the suitability of continued service and any appropriate inspection interval.

Failure mechanisms associated with wear, corrosion, and other anomalies that are identifiable with NDE are described in the various references. Similarly, typical manufacturing defects are likewise discussed.

3-6 DESIGNING FOR INSPECTABILITY

Recognizing the catastrophic potential for a fatigue related failure, it is prudent for the designer to consider future inspections in the original concept. Where considerable effort is required to gain access to inspect a critical part, the inspection will be more costly. Moreover, inspections that are required to be conducted in places that are difficult to reach are more likely to be performed in an ineffective manner. It is therefore imperative that the original designer plan for convenient inspection for critically stressed locations.

A review of items in the design process that are critical to the inspectability of the part or system is given by Mordfin [13]. Initially, he stresses the importance of selecting materials having favorable fracture properties. Certainly the ability of a material to arrest a crack rather than allowing it to propagate suddenly to full failure is conducive to successful NDE. The importance of a material having well established NDE properties is also stated. For example, ultrasonics may have a response in some cast materials that is different from similar wrought materials because of the increased scatter at the grain boundaries for the former. Fabrication processes that may inflict flaws or other anomalies into the part should be avoided. Of particular importance in this area are tensile stresses, introduced during fabrication, which can facilitate the initiation and propagation of a crack. The configuration of the part should be such that unnecessary section changes that might inhibit inspection are minimized. Further, critical areas should be easily accessible, either for visual inspection, NDE, or both. Finally, Mordfin recommends that the design engineer consult frequently with the NDE engineer so that an overall satisfactory design emerges.

CHAPTER
4

INSPECTION OPTIMIZATION USING PROBABILISTIC CRITERIA

4-1 INTRODUCTION

Optimization of the inspection process simply means that the maximum benefit will be achieved at the minimum cost. While the actual levels of benefit and cost may vary for typical applications, the need for optimization of the inspection process remains invariant. The NDE engineer's involvement in the optimization process includes an evaluation of parameters such as the minimum detectable flaw size, the operator or system reliability, the frequency of inspection, and the risks of defects missed. In the past, decisions on these matters may have been based on one's past experience. With the modern high-strength materials used at a low factor-of-safety and the potentially catastrophic results of a failure, the demand on the engineer is for considerably more expertise than past experience. Awareness of the probabilistic nature of much of the data used in the NDE decision process can yield greater confidence in being able to accurately predict the outcome for the several decisions to be made.

4-2 PROBABILITY OF DETECTION

The probability of detection is a parameter that is dependent on many other considerations, e.g.,

1. Operator training, alertness, and confidence
2. Correct application of the proper technique
3. Environment of the test—laboratory or field
4. Material homogeneity and isotropy
5. Crack orientation
6. Shape of the part
7. Calibration and capability of the system
8. Other factors

The relationship of operator performance and the probability of detection has been studied by Chin Quan and Scott [23]. Their conclusions have been summarized in Fig. 4-1.

The "ideal" circumstance is one in which no part with a defect size $x < x'$ will be identified as defective and all parts with defects of $x \geq x'$ will be correctly located. The other two curves represent more realistic conditions. Note that for the "good" curve, a small number of items will be detected with $x < x'$. These occurrences are noted as *false calls*. The curve for the "poor" operator shows performance that generally results in a number of "missed defects" and few, if any, false calls.

Organizations heavily committed to NDE are keenly interested in the performance of operators and the effectiveness of various techniques in particular applications. Further, organizations operating NDE installations at several re-

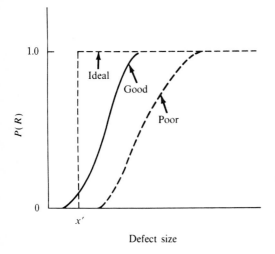

FIGURE 4-1
Probability of rejection and defect size for ideal, good, and poor operator or system performance (*From Chin Quan and Scott* [23]. *Courtesy Academic Press.*)

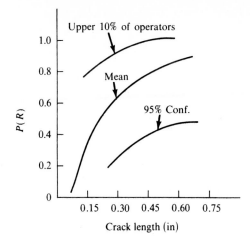

FIGURE 4-2
Probability of rejection $P(R)$ for wing box section with eddy current. (*From Lewis et al.* [*24*]. *Courtesy Nondestructive Testing Information and Analysis Center.*)

mote locations are interested in the uniformity of the inspection quality that they might expect.

The United States Air Force has surveyed the performance of NDE operators at several maintenance facilities and it has produced results showing variations in the probability of rejection (i.e., detection) as a function of flaw size, type of NDE technique, and demands of the test [24]. The results shown in Figs. 4-2 and 4-3 were obtained from a round-robin inspection of a box-wing section of an aircraft. The same part was sent to several maintenance facilities for inspection by different operators. The skin crack that was in the wing section was most amenable to detection by a surface effect NDE technique.

Figure 4-2 shows the results of eddy-current inspection, while Fig. 4-3 is for ultrasonics. The three curves in each figure are for the upper 10% of the operators, the mean for all operators, and the level of probability to be expected

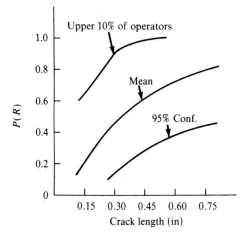

FIGURE 4-3
Probability of rejection $P(R)$ for wing box section with ultrasonics. (*From Lewis et al.* [*24*]. *Courtesy Nondestructive Testing Information and Analysis Center.*)

if 95% confidence in the results is required. In either figure, the more skilled operators are shown to have a higher probability of finding a defect. Most likely, though, this group will find more indications that prove to be false than will the less skilled group (note Fig. 4-1). While the correct judgement of an indication is quite important, the specification of a certain level of correctness may have some unexpected results. For example, where a 95% confidence is required in the correctness of the operator's judgement in locating a crack, the probability of rejection is considerably reduced in both cases.

A comparison of the effectiveness of the two techniques in finding this type of defect is also significant. In general, eddy-current techniques are more sensitive than ultrasonics to surface-breaking cracks, as is clearly indicated here. Note that for 95% confidence, the probability of finding a crack 0.3 in (7.6 mm) long is almost three times greater for eddy currents than for ultrasonics.

A survey on applications of the various NDE techniques and typical confidence levels for them has been conducted by Rich et al. for the United States Army [25]. The results showed which techniques were generally preferred for interior and surface-breaking cracks, for example, and how the techniques were ranked for pre-service, in-service, and laboratory applications.

4-3 COST AND BENEFIT ANALYSIS FOR NDE IN MANUFACTURING AND MAINTENANCE

The goal of any company obviously should be to maximize profits by obtaining the maximum service life from equipment and structures at the least possible cost. One approach to achieving this is to perform periodic maintenance at very frequent intervals so that the probability of failure is exceedingly small. With this philosophy, there will be no failures, but the maintenance costs can be exceedingly high. On the other hand is the philosophy that if a machine is not broken, there is no need to fix it. This approach is quite acceptable in some situations where the failure is inconsequential, but totally unacceptable where the potential for injury, death, or extraordinary damage is high. The combined application of fracture mechanics and NDE permits a sounder, more technically justified decision than either of the aforementioned options.

The cost of implementing an NDE program is often considerable. The initial expense of purchasing suitable NDE equipment can range from a few hundred dollars to several million. Providing facilities for housing the equipment can represent quite a large expense. Added to this is the training of personnel, the periodic purchase of supplies, and then equipment maintenance. Also of serious concern to the manager of a manufacturing plant is the lost income from the items that are rejected as faulty by the NDE system. The reliability of the system is also likely to be questioned. It is imperative that decisions on implementing a NDE program be approached first on a cost effective basis. Following that, other considerations such as public relations can be included.

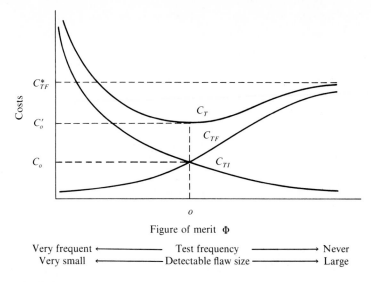

FIGURE 4-4
Effect of allowable flaw size and frequency of inspection on costs of NDE.

In order to develop a model that will quantify the benefits of NDE for a specific situation, the benefit analysis should investigate the effect on operating and inspection cost of several parameters such as inspection frequency and detectable flaw size. These factors will be studied using a model similar to the work of Johnson [26], as illustrated in Fig. 4-4.

Where NDE is included in the cost analysis, the total cost for either a manufacturing situation or an operating system such as a pipeline may be expressed as

$$C_T = C_{TF} + C_{TI} + C^* \tag{4-1}$$

where C_{TF} = the expected cost per unit time associated with failures
C_{TI} = the cost per unit time associated with inspection
C^* = other factors to be discussed

Two characteristics of a NDE inspection are the sensitivity of the test and the frequency of the test. The test sensitivity may be measured by the probability of detecting a flaw of size x and will vary among test procedures. The frequency of the test may be viewed as a combination of the cost of testing and the time between tests. It is measured in cost per unit time and again varies among test procedures.

Considering that the two characteristics of sensitivity and frequency may be combined into one overall performance figure of merit Φ, then the economic impact of NDE may be seen in Fig. 4-4. At a low level of Φ, the total cost C_T is dominated by inspection costs C_{TI} and at a high level of Φ, C_T is dominated by failure costs C_{TF}. At intervening values between the two extremes, the total C_T, as given by Eq. (4-1), is represented by the upper curve. On either side of the optimum value Φ_o, the costs would be greater than the minimum cost associated with a test C_o'.

The primary objective of a benefit analysis would be to develop a model that will permit the establishment of the location of o for a wide variety of circumstances. In many cases, the parameters needed to do this either are available, may be estimated based on available data, or may be determined from experimental data. For example, the probability of failure, and hence C_{TF}, may be estimated using probabilistic fracture mechanics techniques. The probability of detection, however, may be established using experimental data obtained from simulated inspection conditions.

As seen in Fig. 4-4, the benefit or saving B to be derived from applying nondestructive testing to a system not currently being inspected is

$$B = C_{TF}^* - C_o'. \tag{4-2}$$

Where the cost of the new or improved NDE system is C_s, the cost/benefit ratio would be

$$\frac{C}{B} = \frac{C_s}{C_{TF}^* - C_o'} \tag{4-3}$$

It must be noted that there are a number of other factors C^* that will affect the analysis. Among the factors to be considered are:

Inspection reliability. Unreliable inspection techniques will always leave some defects in service while recommending the removal of otherwise benign anomalies. This significantly affects the value of the inspection. For the preceding example the degrees of unreliability would affect how close the true inspection results could be brought to the optimum cost C_o. Unreliable inspection, for example, might achieve a cost level of C_o'.

Repair and maintenance costs. The value of an inspection may be heavily affected by the cost of repair. For example, where repair and replacement costs are extremely high, there might be a much greater demand for a highly reliable system than where the repairs may be made in an inexpensive and easy manner. These cost estimates should be available for most industrial situations.

Material effects. It is well known that the fracture resistance may vary significantly within engineering materials. Thus, the probability of detection $P(D|x)$, hence the strictness of the inspection, might need to be adjusted for more brittle materials as compared to those that are more ductile. The type of inspection system used would establish the probability of detection.

Inspection speed. In a manufacturing plant, where formed products such as extruded or forged shapes are to be inspected as they move through the forming process, and in situations such as pipeline inspection, inspection speed might be an important criteria. While a faster inspection speed might be desirable in order that more material might be inspected in a shorter period of time, there would also be an expected effect on the probability of detection that must be considered. An optimum inspection speed would be considered for the various parameters in the test system.

For a model to show the overall costs of NDE, the total cost of failure C_{TF} in the preceding model will include a number of parameters. For some circumstances, the costs associated with a failure-caused accident are quite important. For example, a failure in a gas pipeline traveling through a heavily populated area would be expected to result in a significantly larger cost than one that occurs in remote regions that are unpopulated. Further, the cost of repair would be expected to vary, depending in this example on a number of factors, including the terrain and the accessibility to the failure location. There are, of course, other cost factors that will be discovered in the full analysis.

An analysis of NDE related manufacturing costs given by Johnson [26] is as follows:

$$C_T(x) = C_M + C_I + (C_{RP} + C_i)P(R|x)(1 - P(R|x))^{-1} + C_F P(F|x) \quad (4\text{-}4)$$

where $C_T(x)$ = total cost per item
$\qquad C_M$ = base cost to manufacture the item
$\qquad C_{RP}$ = average cost to repair an item found with a defect
$\qquad C_I$ = average cost to inspect each item
$\qquad C_F$ = cost of a failure
$\qquad P(R|x)$ = probability of rejecting an item because of a defect of size x
$\qquad P(F|x)$ = probability of failure of an item after passing inspection with flaw size x

Now consider the effect on the cost of inspection of specifying a minimum allowable defect size (x). Rewrite the cost equation using

$$C_{TM}(x) = C_M + C_I + (C_{RP} + C_I)(P(R|x)/(1 - P(R|x))) \quad (4\text{-}5)$$

where $C_{TM}(x)$ is the total cost associated with manufacturing. The total cost of failure is

$$C_{TF}(x) = C_F P(F|x).$$

Therefore,

$$C_T(x) = C_{TM}(x) + C_{TF}(x) \quad (4\text{-}4a)$$

What is the effect on cost of specifying a very small allowable flaw size, i.e., $x \to 0$? The probability of rejection approaches 1; i.e., virtually all items will be judged to have a defect. Hence,

$$\frac{P(R|x)}{1 - P(R|x)} \to \infty$$

and $P(F|x) \to 0$ since any part that is passed will be unlikely to fail. Therefore, $C_T(x) \to \infty$.

To develop failure costs, consider that for a small value of x,

$$P(F|x) \to \text{small}$$

and

$$C_{TF}(x) \to \text{small}$$

For large defects,

$$P(R|x) \rightarrow \text{small}$$

and
$$C_{TM} = C_M + C_I$$

which is constant. Also,

$$P(F|x) \rightarrow 1$$

and
$$C_{TF} \rightarrow \text{constant}$$

The combination of cost to manufacture $C_{TM}(x)$ and cost of failure $C_{TF}(x)$ yields results as previously discussed for Fig. 4-4. The optimum defect size is obviously at the point of minimum total cost, while the conservative safe size would be somewhat smaller. In some instances, the typical allowable flaw size may be slightly larger than the optimum size. Interestingly, the total cost for the larger could be comparable to that seen for the smaller flaw size. The difference is in the relative values of $C_{TM}(x)$ and $C_{TF}(x)$.

The effect of inspection frequency on the total cost of inspection may be approximately the same as that presented in Fig. 4-4 for the effect of the flaw size. Using an analysis similar to the one used in that section, it may be shown that while a too frequent inspection cycle would increase the cost of inspection to an unacceptable level with only a marginal increase in the number of defects detected, a too long cycle would minimize the inspection costs while allowing the failure costs to rise to a high but constant level. Clearly, then, there is a need for an optimum interval that will minimize the combined effect of the cost of inspection and the cost of failure.

The effect of periodic maintenance and inspection can be demonstrated using transportation vehicles as examples. Once the mathematical models have

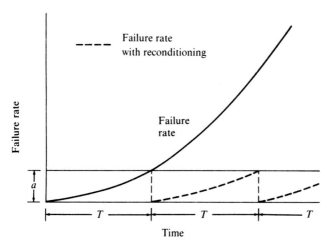

FIGURE 4-5
Effect on failure rate of reconditioning at interval T.

been established for the vehicle failure rate, as was done for the railroad coupler failures, it will be possible to estimate how the failure rate would be affected by either periodic maintenance or periodic inspection at varying intervals. Periodic maintenance and periodic inspection may affect the failure rate curves in different ways. Periodic maintenance in the form of full reconditioning could return the vehicle to "as-new" condition. In this case, the new failure rate would return to that which existed at zero time, as shown by the dotted line in Fig. 4-5.

For each reconditioning performed at an interval T when the failure rate is at threshold level a, the new failure rate would return to the as-new condition. If only noticeably needed work or obviously defective parts were replaced, the failure rate would be returned to some lower rate, but not to the as-new rate. The ability to return the vehicle to as-new condition implies that all fatigue cracks, even the smallest ones, have been removed. Clearly, NDE plays an important role in achieving the fully reconditioned failure rate.

4-4   PROBABILISTIC INSPECTION AND PROBABILISTIC FAILURE

The axioms given earlier in Chapter 3 generally emphasize the varying nature of the data sets for flaw size, flaw criticality, and inspection sensitivity. For example, where the mean flaw size from a set of failed items from an operating system may be known as the mean critical flaw size, parts that are identical to those that have failed but are still in service possibly may contain cracks larger than the mean critical flaw size for the failed items. As observed by Richardson and Fertig [27], this indicates that there are in reality two data sets commingled about the dimension of the mean critical flaw size, namely the set that will survive in service and the set that will fail. These data sets are indicated in Fig. 4-6 where $P(x|1)$ represents the surviving population and $P(x|0)$ the population that will fail.

The shaded regions designate areas where the effect of NDE is uncertain, generally because of variations in material performance and loading conditions. On either side of some decision point, e.g., the maximum allowable flaw size, the area e_0 designates the probability of a survival decision for a flaw that will in fact fail in service, while e_1 designates the probability of rejection of a flaw that will survive in further service.

If a minimum allowable flaw size x^* is designated for the NDE system being used, it is clear that even with a perfect system some defects that will fail will be passed while some that will survive will be removed because of NDE indication. For example, where x^* is moved left to a location at the tail of the e_0 region, the value of e_0 clearly goes to 0 while that of e_1 increases in value. In this case, all flaws that will fail are rejected but some that will not fail are removed from service. For the less-than-perfect inspection system or operator, the relative sizes of the areas e_0 and e_1 are affected by the additional uncertainty.

The effect of NDE system performance on the probabilistic flaw distributions has been reviewed by Richardson and Fertig [27] and is expressed using the NDE operating characteristic as shown in Fig. 4-7. A curve closer to the axes

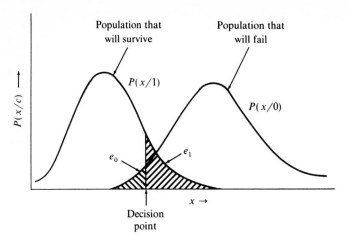

FIGURE 4-6
Distributions of failing and surviving populations with classification errors e_0 and e_1 indicated.
(*From Richardson and Fertig* [27]. *Courtesy Academic Press.*)

reflects better system performance than one away from the axes. In Fig. 4-7, technique C is inferior to the other techniques and NDE system performance improves as the curve moves further toward the axes. For a designated level of e_0, as denoted by the vertical dotted line, the false rejection probability is seen to increase for techniques A, B, and C, respectively. The better performing system, technique A, would reject a higher proportion of the rejectable anomalies, even though the survivability of these anomalies still depends on the material and loading factors as illustrated in Fig. 4-6.

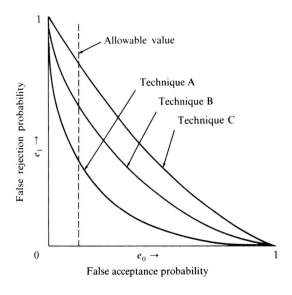

FIGURE 4-7
The NDE operating characteristic.
(*From Richardson and Fertig* [27]. *Courtesy Academic Press.*)

Additional discussion describing the relationship of NDE and probabilistic structural mechanics has been given by Balkey, Meyer, and Witt [28]. After describing probability analysis, risks, and other characteristics of probabilistic mechanics, they cite several examples where NDE has been used to enhance the performance of various engineering systems.

The *retirement-for-cause* approach, utilizing both fracture mechanics and NDE, has been applied in several industries to optimize an inspection system performance. In this analysis, it is initially recognized that a typical replacement cycle of a component in a fatigue environment generally will be based on a very early occurrence of failure. Thus, if a 1% failure rate is judged to be the acceptable limit for a particular component and all are replaced at the corresponding time, in a group of 1000 items, 990 of them would still have some useful remaining life.

In describing this situation as it applies to military gas turbines, Annis et al. [29] indicate that some turbine compressor disks have up to 80 lifetimes remaining when removed from service due to the expected failure of one turbine. For the whole group of compressor disks, they suggest that a considerable amount of unused life is removed from service without just cause. While there is sound economic argument for the retirement-for-cause approach to failure prevention, it must be emphasized that each circumstance should be approached with caution. The costs of failure and the likelihood for catastrophic events must be weighed against the economic gain.

An additional example of the integration of NDE into a maintenance program that seeks to extend the life of complex structures has been given by Hagemaier et al. [30] for large commercial aircraft. The probability of detection for various NDE systems is discussed along with damage tolerance allowance, inspection reliability, and sampling techniques as they apply to aircraft maintenance.

The application of nondestructive evaluation in the quality control process at the manufacturing facility poses a unique problem when the possible number of parts to be inspected is considerable. Sampling plans are widely used where the objective is to monitor the quality of a production process without inspecting each item. Quality control and sampling plans are covered in a number of sources, including Refs. 12 and 31.

4-5 SUMMARY

Probablistic methods are now seeing applications in virtually all aspects of engineering design, inspection, and maintenance. The NDE engineer, most of all, must be aware of the probabilistic nature of both failure and inspection. Once knowledgeable of both the risks and the assurances that accompany NDE, correct utilization of the various techniques can increase safety, increase efficiency, and optimize the design process for most engineering systems.

Often times, an engineering manager will feel that all defects must be found in order to justify an expenditure for an NDE system. Clearly, that is not possible. Moreover, where a too small allowable defect is specified, the presentation in this section has shown the costs of implementing an NDE system may be prohibitive. It has been strongly suggested that where data exist, economic judgement on whether or not to implement NDE can be quantified. In many cases, these data will not be available in clearly defined form and the NDE specialists will be required to make assumptions. The graduate of a four year engineering curriculum will be able to do this. It is fundamentally true that NDE can be a cost effective tool in many circumstances. Each application, however, must stand on its own merit. It is not true to say that an expenditure for NDE is justified in all cases, and the material in this section is intended to enable an engineer to reach the proper decision on NDE implementation.

PROBLEMS
FOR
PART I

Chapter 2

2-1. If you are standing on the corner, waiting for a bus, and you are flipping a common coin, what is the probability that you will have four straight tosses showing tails and a fifth showing heads?

2-2. A steel natural gas transmission line extends 180 km from the gas field to the industrial distribution terminal. The 50–90 km portion of the pipeline was installed with seamless pipe and the remainder was installed with welded pipe. The 70–150 km segment of the pipeline passes through rugged terrain that makes access to the line difficult. The entire line is being tested for critical corrosion that is equally likely to occur anywhere along the line. If corrosion is detected in the welded pipe, what is the probability that it also occurs in the region where access is physically difficult?

2-3. The exercise of drawing from a lot without replacement is a conditional probability. For the box of marbles in Example 2-1 and using conditional probabilities, show that the probability of drawing a red marble on the second draw is conditioned by the results of the first draw.

2-4. Use the aircraft seating arrangement shown in Fig. 2-3 to answer the following questions, assuming that all seats are open and that seat assignments are random.
(*a*) What is the probability of being assigned an aisle seat?
(*b*) What is the probability of being assigned a seat in rows 6–15 that also is in rows 1–7?

2-5. The simply supported beam AB shown in Fig. P2-5 and the load P can be used as a model for a shaft in a rotating machine subject to impulse loading at any point along its length. For a 0.15 m diameter mild steel shaft with yield stress of 270×10^6 N/m^2, assume that load P can act randomly at any point. What is the probability of failure based on a single event and yield stress for $P = 72$ and 75 kN? Classify the sample space for this data set as being finite or infinite.

2-6. Divide the length of the shaft in Problem 2-5 into 50 segments of 100 mm each. Use the loading conditions given in Problem 2-5 and again calculate the probability of failure. Classify the sample space for this data set as being finite or infinite.

2-7. Assume that the flaw distribution shown in Table 2-2 and Fig. 2-2 fits a normal distribution. Calculate the mean $\bar{x}$ and the standard deviation σ_{n-1}, for the flaw

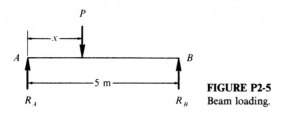

FIGURE P2-5
Beam loading.

data. If the shafts are inspected using the system described in Example 2-6 (Fig. 2-5), show that the mean flaw size after inspection $\bar{x}'$ is smaller than the previous mean $\bar{x}$? Would you expect that the sample of defects remaining after the inspection still could be assumed to be normally distributed? Explain your answer.

2-8. A collection of 1000 shafts having a flaw distribution as given in Table 2-2 and Fig. 2-2 is inspected with a system that has the flaw detection characteristics as given in Example 2-6. Assume that the defects are normally distributed and use Eq. (A-3) to develop a generalized expression for the distribution. If there are 20 shafts in the set containing one defect each, how many defects will be missed in the inspection?

2-9. A group of 700 used aircraft turbine blades are being inspected for fatigue cracks. Prior data indicate that there should be 75 blades in the set of 700 blades that are defective. If 300 blades have been inspected and set aside and 10 defective blades have been found, what is the probability that the next blade removed from the set will contain a defect?

2-10. The first launch of a 3900 series Delta rocket on 12 December 1975 was successful as was the second. Launch three failed. A series of 29 successful launches followed until the 33rd launch failed on 3 May 1986. Launch 34 was successful. Plot the probability of success $P(S)$ and the probability of failure $P(F)$ for each launch of the Delta rocket. Speculate on the effect of quality control in manufacture and design modifications on the success rate of the rocket.

Chapter 3

3-1. Failure rate data can be useful in establishing a test plan for an NDE system. List the three specific areas of a typical failure rate curve and show them on a sketch. For each area, discuss the probable material properties and loading characteristics associated with these failures. Describe how NDE could best be used to minimize failures in that region.

3-2. Give a brief but specific summary of the beneficial ways that NDE can contribute in the following areas: safety in operating equipment and in plant processes, uniform quality in raw material, conservation of raw materials, economic benefit in equipment maintenance, uniform quality in raw materials, and factor-of-safety in design.

3-3. The failure rate curves for two components are compared and it is discovered that the burn-in and constant wear regions are virtually identical. Once failure begins, however, the final failure curve for component A rises very sharply while that for component B rises very slowly. Discuss the likely differences in material and loading conditions for the two materials. Use the failure patterns shown in Fig. 3-7 in preparing your answer.

3-4. In the context of the three portions of the failure rate curve (Fig. 3-3), discuss the relationship of each of the following to final failure and nondestructive inspection:
(*a*) Critical flaw size
(*b*) Minimum detectable flaw size
(*c*) Flaw propagation rate
(*d*) Frequency of inspection

Chapter 4

4-1. A manufacturer of small parts is considering the purchase of a NDE system and the engineering staff has been asked to prepare a proposal estimating the total cost of the inspection. What information would be needed to prepare this estimate? How might this information be obtained? (See Chap. 3 also.)

4-2. Nondestructive inspection systems only deliver indications. The operator or logic network must judge the likelihood that an indication is in fact a defect or a spurious indication. Complete the following blank table using the words *correct, false call, or missed defect.*

Operator performance

Judgement	Condition	
	Defective	No defect
Defective		
Not defective		

4-3. An existing NDE system is felt to have a probability of rejection $P(R_1)$ for a particular flaw size x^*. The probability of failure for parts that are passed by the system is $P(F_1)$. At this flaw size, the total cost of inspection C_{TI} is equal to the total costs of failure C_{TF}. A proposal is made to increase the size of the allowable flaw. Discuss the likely effects of this proposed increase on the total cost per item produced, the total cost of inspection, and the total cost of failure.

REFERENCES
FOR
PART I

1. Lipschutz, S., *Theory and Problems in Probability*, Schaum's Outline Series, McGraw-Hill, New York, 1965.
2. Spiegel, M., *Theory and Problems of Statistics*, Schaum's Outline Series, McGraw-Hill, New York, 1961.
3. Ang, A. H-S., and Tang, W., *Probability Concepts in Engineering Planning and Design*, sec. 8.3.2, Wiley, New York, 1975, pp. 341–344.
4. Smith, C. O., *Introduction to Reliability in Design*, McGraw-Hill, New York, 1976.
5. *Failure Prevention and Reliability*, Design Engineering Technical Conference, Chicago, IL, 1977, H00101, American Society of Mechanical Engineers, New York, 1977.
6. Shigley, J. E., and Mitchell, L. D., *Mechanical Engineering Design*, 4th ed., McGraw-Hill, New York, 1983.
7. Packman, P. F., Pearson, H. S., Owens, J. S., and Marchese, G. B., "The Applicability of a Fracture Mechanics–Nondestructive Testing Design Criterion," Technical Report AFML-TR-68-32, Air Force Materials Laboratory, Wright-Patterson Air Force Base, OH, May 1968.
8. Williams, S., and Mudge, P. J., "Statistical Aspects of Defect Evaluation using Ultrasonics," *NDT International*, vol. 18, no. 3, pp. 123–131, June 1985.
9. Tang, W. H., "Probabilistic Updating of Flaw Information," *Journal of Testing and Evaluation*, vol. 1, no. 6, pp. 459–467, November 1973.
10. Mahon, J. B., "Delta Launch History of 177 Vehicles," private communication, National Aeronautics and Space Administration, Washington, DC, July 1986.
11. "NASA Pocket Statistics," National Aeronautics and Space Administration, Washington, DC, January 1986.
12. Duncan, A. J., *Quality Control and Industrial Statistics*, 3rd ed., Richard D. Irwin, Inc., Homewood, IL, 1965.
13. Mordfin, L., "Nondestructive Evaluation," in J. F. Young and R. S. Shane (eds.), *Materials and Processes, Part B: Processes*, Marcel Dekker, New York, 1985, chap. 30.
14. Bray, D. E., "Railroad Accidents and Nondestructive Inspection," ASME Paper 74-WA/RT-4, presented at Winter Annual Meeting, American Society of Mechanical Engineers, New York, November 1974.
15. Bray, D. E., Goff, M. M., and Zard, R. E., "Reliability Analysis of Rotary Couplers on Unit-Train Coal Cars," ASME Paper No. 83-WA/DE-13, presented at Winter Annual Meeting, American Society of Mechanical Engineers, Boston, MA, November 1983, *Reliability Engineering*, vol. 10, pp. 27–48, 1985.
16. Wulpi, D. J., *Understanding How Metals Fail*, American Society for Metals, Metals Park, OH, 1985.
17. Barer, R. D., and Peters, B. F., *Why Metals Fail*, Gordon and Breach, New York, 1970.

18. Colangelo, V. J., and Heiser, F. J., *Analysis of Metallurgical Failures*, Wiley, New York, 1974.

19. *Metals Handbook*, vol. 10: *Failure Analysis and Prevention*, 8th ed., American Society for Metals, Metals Park, OH, 1975.

20. Hutchins, F. R., and Unterweiser, P. M., *Failure Analysis. The British Engine Technical Reports*, American Society for Metals, Metals Park, OH, 1981.

21. Jacoby, G., "Fractographic Methods in Fatigue Research," *Experimental Mechanics*, vol. 5, no. 3, pp. 65–82, March 1965.

22. Hertzberg, R. W., *Deformation and Fracture Mechanics of Engineering Materials*, Wiley, New York, 1976.

23. Chin Quan, H. R., and Scott, I. G., "Operator Performance and Reliability in NDI," in R. S. Sharpe (ed.), *Research Techniques in Nondestructive Testing*, vol. III, Academic, New York, 1977, chap. 10.

24. Lewis, W. H., Sproat, W. H., and Boisvert, G. W., "A Review of Nondestructive Inspection Reliability an Aircraft Structure," in W. W. Bradshaw (ed.), *Proceeding of the Twelfth Symposium on Nondestructive Evaluation*, Southwest Research Institute, San Antonio, TX, pp. 1–5, 1979.

25. Rich, T. P., Tracy, P. G., and Cartwright, D. J., "A Survey of Fracture Mechanics Applications in the United States," AMMRC MS76-1, Army Materials and Mechanics Research Center, Watertown, MA, February 1976.

26. Johnson, D. P., "Determination of Nondestructive Inspection Reliability using Field or Production Data," *Materials Evaluation*, vol. 36, no. 1, pp. 78–84, January 1978.

27. Richardson, J. M., and Fertig, K. W., "Probabilistic Failure Prediction and Accept/Reject Criteria," in R. S. Sharpe (ed.), *Research Techniques in Nondestructive Testing*, vol. V, Academic, New York, 1982, chap. 5.

28. Balkey, K. R., Meyer, T. A., and Witt, J., "Probabilistic Structural Mechanics, Chances are...," *Mechanical Engineering*, vol. 108, no. 7, pp. 56–62, July 1986.

29. Annis, C. G., Cargill, J. S., Harris, J. A. Jr., and Van Wanderham, M. C., "Engine Component Retirement-for-Cause: A Nondestructive Evaluation (NDE) and Fracture Mechanics Based Maintenance Concept," *Journal of Metals*, vol. 33, pp. 24–28, July 1981.

30. Hagemaier, D. J., Abelkis, P. R., and Harmon, M. B., "Supplemental Inspection of Aging Aircraft," *Materials Evaluation*, vol. 44, pp. 989–997, July 1986.

31. *Metals Handbook*, vol. 11: *Nondestructive Inspection and Quality Control*, 8th ed., American Society for Metals, Metals Park, OH, 1976.

PART II

ULTRASONIC TECHNIQUES IN NONDESTRUCTIVE EVALUATION

CHAPTER 5

ELASTIC
WAVE
PROPAGATION

5-1 INTRODUCTION

Acoustic nondestructive inspection techniques utilize as the indicating parameter some particular characteristics of the propagating stress waves. For the present concern, the stress waves are generally assumed to be in the ultrasonic range, i.e., having frequencies greater than 20 kHz. Moreover, the stress waves of greatest interest here usually exist in the form of pulses of energy.

In order to further examine these propagating energy pulses and their general uses in nondestructive evaluation, the one-dimensional plane wave equation will be developed for longitudinal waves. This will be followed by discussion relative to the propagation of three-dimensional, or bulk, waves. Following that, some specific characteristics of the pulses and the accompanying wave front will be discussed. Portions of the material presented here were derived from a number of sources, notably Refs. 1–7. Where appropriate, specific references to pertinent material will be given.

5-2 ONE-DIMENSIONAL (PLANE) WAVES

In the basic sense, we assume a long, undisturbed bar of arbitrary cross section suspended by frictionless supports, as shown in Fig. 5-1. The bar is of length L and cross-sectional area A. Approaching toward the left end of the bar is an impulsive disturbance. This disturbance could be in one of several forms, e.g., a rifle bullet, hammer, electrical spark, laser pulse, piezoelectric transducer, or even

47

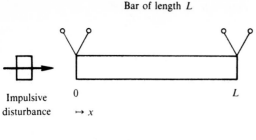

Bar of length L

Impulsive disturbance 0 $\mapsto x$ L

FIGURE 5-1
Long bar of arbitrary cross section.

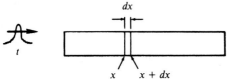

dx

t

x $x + dx$

FIGURE 5-2
Bar with approaching impulse.

a shove with the hand. The nature of the impulse governs to a large extent the response of the bar. In the one extreme, if the bar is merely shoved by a hand and released, it will oscillate on the supports and there will be no relative motion between ends $x(0)$ and $x(L)$. On the other hand, if the bar is tapped lightly by a small hammer on the left face, there will be a period in which the left face will have moved, while the right face remains undisturbed. Obviously, then, there must be an elastic deformation present in the bar in order for this relative motion to occur.

In a more general sense, assume that we have an impulsive source approaching the bar, as indicated by the wave form shown in Fig. 5-2. The task will be to investigate the elastic response of the bar to the passage of this impulse. We also have defined an arbitrary localized element of length dx somewhere along the bar.

Figure 5-3(a) and (b), respectively, show the undisturbed element as the perturbation approaches and the disturbed element during the passage of the perturbation. At some instant in time, the left face of the element will have moved a distance ξ and the right face a distance $\xi + d\xi$. For this condition, the

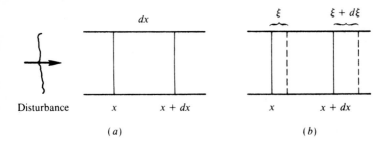

dx ξ $\xi + d\xi$

Disturbance x $x + dx$ x $x + dx$

(a)   (b)

FIGURE 5-3
(a) Undisturbed element. (b) Disturbed element.

displacement of a face of the element is a function of x and t, i.e.,

$$\xi = \xi(x, t) \qquad (5\text{-}1)$$

Since most nondestructive inspection applications involve very low displacement magnitudes, it therefore is proper to assume small displacements and to use a Taylor series expansion for the displacement. Thus, for the right face,

$$\xi + d\xi = \xi + \frac{\partial \xi}{\partial x} dx + \text{H.O.T.} \qquad (5\text{-}2)$$

For our purposes, the higher-order terms (H.O.T.) can be neglected. With this, the change in length of the element Δl is then

$$\Delta l = (\xi + d\xi) - \xi = d\xi = \frac{\partial \xi}{\partial x} dx \qquad (5\text{-}3)$$

The resulting incremental strain $\Delta \varepsilon$ for the element is

$$\Delta \varepsilon = \frac{\Delta l}{dx} = \frac{(\partial \xi / \partial x) \, dx}{dx} = \frac{\partial \xi}{\partial x} \qquad (5\text{-}4)$$

The preceding material has described the relative motion of the faces of the element during the passage of the disturbance. For completeness, now, we must add the forces in the material, i.e., in the element, that are reacting to the deformation. This will introduce into the discussion the elasticity of the material, which, therefore, specifically defines the material.

Considering the same element as previously used, the forces that would be present during the passage of the disturbance are shown in Fig. 5-4. On the left face, the instantaneous force F_x is acting. Due to the elastic nature of the material, the instantaneous reactive force on the opposite face is $(F_x + (\partial F / \partial x) \, dx)$. The difference in these forces is given by

$$dF_x = \left(F_x + \frac{\partial F_x}{\partial x} dx \right) - F_x = \frac{\partial F_x}{\partial x} dx \qquad (5\text{-}5)$$

To relate these forces, and the previously described strain, Hooke's Law is used. Hooke's Law for uniaxial (i.e., one-dimensional) deformation can be written as

$$E = \frac{\Delta \sigma}{\Delta \varepsilon} \qquad (5\text{-}6)$$

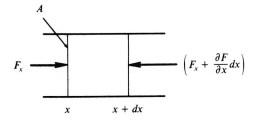

FIGURE 5-4
Forces on element during passage of the disturbance.

where $\Delta\sigma$ = incremental stress
$\Delta\varepsilon$ = incremental strain
E = Young's modulus

In a strict sense, the assumption of uniaxial deformation limits our results to propagation only in long thin bars. While this limitation is correct, it is nevertheless convenient and useful to introduce wave propagation in this manner and to later expand the results to the more common three-dimensional bulk wave case.

The incremental stress on the left face can be written as

$$\Delta\sigma = \frac{F_x}{A} \tag{5-7}$$

Using this expression and Eq.(5-4), Eq. (5-6) can be written

$$E = \frac{F_x/A}{\partial\xi/\partial x} \tag{5-6a}$$

Rearranging Eq. (5-6a) yields an expression for the force at the left face:

$$F_x = AE\frac{\partial\xi}{\partial x} \tag{5-6b}$$

Taking a partial derivative of Eq. (5-6b) gives

$$\frac{\partial F_x}{\partial x} = AE\frac{\partial^2\xi}{\partial x^2} \tag{5-6c}$$

Substituting this in Eq. (5-5) yields an expression for the net force on the element, which now includes the elastic properties

$$dF_x = AE\frac{\partial^2\xi}{\partial x^2}\,dx \tag{5-5a}$$

The dynamic nature of the propagating wave is introduced through Newton's Second Law

$$dF_x = ma \tag{5-8}$$

where m is the mass of an incremental element and a is the instantaneous acceleration of the particles in a face. With the mass and acceleration written as

$$m = \rho(A\,dx) \tag{5-9}$$

and

$$a = \frac{\partial^2\xi}{\partial t^2} \tag{5-10}$$

Eq. (5-8) can be written as

$$dF_x = \rho A\frac{\partial^2\xi}{\partial t^2}\,dx \tag{5-8a}$$

Here, ρ is the mass density of the material.

Equating Eqs. (5-5a) and (5-8a) yields

$$\frac{\partial^2 \xi}{\partial t^2} = C_l^2 \frac{\partial^2 \xi}{\partial x^2} \tag{5-11}$$

where

$$C_l = \sqrt{\frac{E}{\rho}} \tag{5-12}$$

This is the one-dimensional form of the wave equation for plane longitudinal waves. The constant C_l is the speed of travel of the propagating disturbance. The dependence of the magnitude of this speed upon the material properties E and ρ is clearly demonstrated. A similar discussion for the propagation of plane transverse (shear) waves would yield

$$C_t = \sqrt{\frac{\mu}{\rho}} \tag{5-13}$$

where μ is the shear modulus of the material.

Plane longitudinal and shear waves differ in at least two significant aspects. First, a comparison of Eqs. (5-12) and (5-13) reveals that the wave speed for the longitudinal waves C_l is greater than the speed of the shear wave C_t by a factor of $\sim 1\frac{1}{2}$ for most materials. Second, where the particle motion of the longitudinal wave is oriented parallel to the direction of travel of the pulse, the particle motion for the transverse wave is perpendicular to the corresponding direction of travel.

5-3 BULK WAVES

Ultrasonic inspection in most cases is conducted using bulk type waves or quasibulk waves. A pure bulk wave may be excited with point source, e.g., an explosion, occurring in an unbounded media. With this configuration, then, the propagation characteristics are described by a three-dimensional field as compared to the one-dimensional field of the plane waves previously described. In the plane wave case, Poisson's ratio effect was ignored, i.e., longitudinal motion was expected to produce only longitudinal deformation. For the bulk situation, a pure longitudinal disturbance produces deformation in the lateral directions as well. In this case, the expression for C_l in Eq. (5-11) is no longer valid.

Particle motion for the bulk waves may also occur along the direction of propagation as well as in the lateral directions. For bulk waves, the term *dilatational* is often used to describe the particle motion analogous to the longitudinal plane wave previously described. Similarly, torsional bulk waves are analogous to the shear or transverse waves in the plane wave case. In practice, however, the terms are often used interchangeably and care must be exercised to prevent identification errors.

For the full three-dimensional field, vector mechanics is used to produce wave equations for the dilatational and solenoidal components of the propagat-

TABLE 5-1
Typical values for elastic constants and wave speeds for engineering materials.

Material	E†† GPa	λ† GPa	μ† GPa	SI Units					
				ρ§ kg/m³	C_l†§ m/s	C_l§ m/s	C_t§ m/s	C_2§ m/s	$z = \rho C_l$§ 10^6 (kg/m² s)
Steel	207	112	81	7.7×10^3	5190	5900	3230	3230	45
Stainless steel 3XX	198	*	76	7.85×10^3	*	5760	*	3120	45
Copper	110	95	45	8.9×10^3	3670	4700	2260	2260	42
Aluminum	72	56	26	2.7×10^3	5090	6320	3130	3130	17
Plexiglas	3.4	*	*	1.18×10^3	*	2730	1430	1430	3.2
Rubber (soft)	0.2	1.0	0.0077	0.9×10^3	46	1480	*	*	1.4
Rubber (vulcan.)	*	*	*	1.2×10^3	*	2300	*	*	2.8
Fused quartz†	73	*	31.2	2.6×10^3	5300	5570	3520	3520	14.5
Polyester†	2.8	*	*	*	*	*	*	*	*
60% G, 40% PE†	42.5	*	*	*	*	*	*	*	*
Polyethylene‖	0.4–1.3	*	*	*	*	*	*	*	*
Water	*	*	*	1×10^3	*	1483	*	*	1.5

English Units

Material	E[††] lb/in²	λ[†] lb/in²	μ[†] lb/in²	ρ[§] lb/in³	C_l[†][§] in/s	C_l[§] in/s	C_t[§] in/s	C_2[§] in/s	$z = \rho C_l$[§] 10³ (lbm m/s)
Steel	29.6×10^6	16×10^6	11.6×10^6	0.28	204,000	232,000	127,000	127,000	65
Stainless steel 3XX	28.7×10^6	*	11.0×10^6	0.28	*	228,000	*	123,000	65
Copper	15.8×10^6	13.6×10^6	6.4×10^6	0.32	144,000	185,000	89,000	89,000	59
Aluminum	10.3×10^6	8.0×10^6	3.7×10^6	0.10	200,000	249,000	123,000	123,000	25
Plexiglas	0.49×10^6	*	*	0.04	*	107,000	56,000	56,000	4.3
Rubber (soft)	28.6×10^3	0.14×10^6	1.1×10^3	0.03	1,811	58,000	*	*	1.7
Rubber (vulcan.)	*	*	*	0.04	*	91,000	*	*	3.6
Fused quartz‡	10.4×10^6	*	4.5×10^6	0.10	209,000	219,000	138,000	138,000	22
Polyester‡	0.4×10^6	*	*	*	*	*	*	*	*
60% G, 40% PE (9)‡	6.1×10^6	*	*	*	*	*	*	*	*
Polyethylene (10)¶	$(0.06$–$0.19) \times 10^6$	*	*	*	*	*	*	*	*
Water	*	*	*	0.04	*	58,000	*	*	2.3

*Data omitted.

†Reference 2.

‡Reference 8.

§Reference 10.

¶Reference 9.

Note: z is the boldface italic font symbol for z.

ing wave. The procedure is similar to that presented for the one-dimensional case. For the dilatational wave and the torsional wave, the resulting expressions are

$$\ddot{\xi} = C_1^2 \nabla \theta \tag{5-14}$$

and
$$\ddot{\xi} = C_2^2 \nabla^2 \bar{\xi} \tag{5-15}$$

where
$$C_1 = \sqrt{\frac{\lambda + 2\mu}{\rho}} \tag{5-16}$$

and
$$C_2 = \sqrt{\frac{\mu}{\rho}} \tag{5-17}$$

where ξ = particle displacement at the wave front
$\ddot{\xi}$ = particle acceleration at the wave front
$\nabla \theta$ = vector notation for the dilatational motion at the wave front
$\nabla^2 \bar{\xi}$ = vector notation for the torsional motion at the wave front
λ, μ = elastic (Lamé) constants
ρ = density
C_1 = speed of propagation of the dilatational component of the wave
C_2 = speed of propagation of the torsional component of the wave.

The dilatational wave equation [Eq. (5-14)] applies where particle motion is in the direction of wave propagation and the torsional (shear) wave equation [Eq. (5-15)] describes particle motion transverse to the direction of propagation. This will be discussed in more detail in the material to follow.

Equations (5-16) and (5-17) compare to Eqs. (5-12) and (5-13). It is obvious in both cases that the longitudinal-type waves have a higher speed than the shear-type waves. In fact, for most materials, $C_1 \simeq 2C_2$ and $C_l \simeq 1.5C_t$. Typical values for E, λ, μ, and ρ are given for several common materials in Table 5-1. Some published values for the wave speed are also given. The specific acoustic impedance ρC_1 is also given, but will be discussed later.

It is useful to emphasize at this point that the elastic moduli can also be calculated from the wave speeds. Expressions for Young's modulus E, Poisson's ratio ν, and the bulk modulus k can also be written in terms of the Lamé constants as

$$E = \frac{\mu(3\lambda + 2\mu)}{\lambda + \mu} \tag{5-18}$$

$$\nu = \frac{\lambda}{2(\mu + \lambda)} \tag{5-19}$$

$$k = \frac{3\lambda + 2\mu}{3} \tag{5-20}$$

Algebraic manipulation can yield the results

$$E = \frac{\rho C_2^2 \left[3(C_1/C_2)^2 - 4 \right]}{(C_1/C_2)^2 - 1} \tag{5-21}$$

$$\nu = \frac{(C_1/C_2)^2 - 2}{2\left[(C_1/C_2)^2 - 1\right]} \tag{5-22}$$

$$k = \rho C_2^2 \left[\left(\frac{C_1}{C_2} \right)^2 - \frac{4}{3} \right] \tag{5-23}$$

The use of Eqs. (5-21)–(5-23) offers a powerful investigative technique for materials studies. With reasonable care, one can obtain values for C_1, C_2, and ρ that are sufficiently accurate to yield useful experimental measurements for the elastic properties. This is particularly true for orthotropic materials where there are important differences in these constants for different directions in the materials.

5-4 PARTICLE MOTION, WAVE FRONTS, WAVELENGTH, FREQUENCY, AND WAVE SPEED

Thus far, we have considered the movement of a single wave front through the material. Ultrasonic inspection generally deals with oscillatory motion, which requires the definition of some additional parameters.

Consider now that the one-dimensional bar of Fig. 5-1 consists of the spring-mass system shown in Fig. 5-5. Also, assume that the excitation is in the form of the rotary crank shown attached to a rigid plate. The crank can rotate with a circular (angular) frequency of ω radians per second, resulting in a harmonic excitation of f cycles per second (Hertz). If the crank rotates at a constant frequency, a series of compressions and rarefactions will progress along the length of the bar. This type of excitation is known as simple harmonic motion and the particle displacements are described by

$$\xi = A e^{j(\omega t - kx)} \tag{5-24}$$

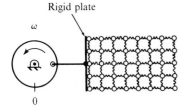

Rigid plate

ω

0

FIGURE 5-5
Excitation of harmonic waves in continuous systems.

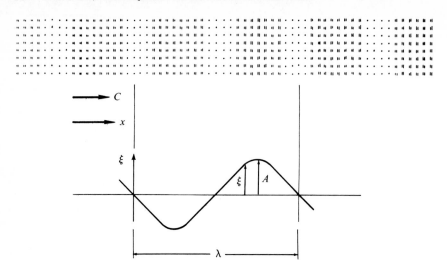

FIGURE 5-6
Particle motion during the passage of a plane longitudinal wave.

where ω is the circular frequency and k is the wave number. The maximum particle displacement is the amplitude A.

The displaced positions of particles at some time t during the passage of a longitudinal wave are shown in Fig. 5-6 overlaid on the original, unperturbed positions. Some particles have been displaced to the right, others to the left, and some not at all. Wave fronts drawn through adjacent positions of comparable motion will be separated by a distance of one wavelength λ. Wavelength, frequency, wave speed, and the wave number are related through the equations

$$f = \omega/(2\pi)$$

$$\lambda = C/f = (2\pi C)/\omega \qquad (5\text{-}25)$$

$$k = \omega/C$$

FIGURE 5-7
Particle motion during the passage of a plane transverse (shear) wave.

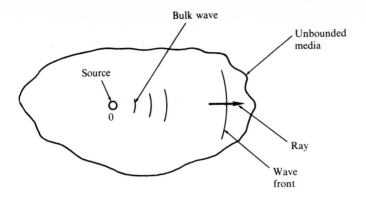

FIGURE 5-8
Bulk wave excitation and propagation.

Particle displacements for a plane shear wave are shown in Fig. 5-7 where the particle displacements are seen to be in the vertical direction rather than horizontal. Wavelengths are defined as before, between adjacent locations of identical motion. The relationships defined by Eq. (5-25) apply also to the shear wave. For the same excitation frequency f, it is clear from Eq. (5-25) that λ for the slower wave (C_t, C_2) will be shorter than for the faster one (C_l, C_1).

An acoustic ray is defined as a line drawn parallel to the direction of travel of the wave energy. The resulting line, then, would be perpendicular to the wave front.

A schematic of a bulk wave is illustrated in Fig. 5-8. Here, an excitation source, e.g., an explosion, occurs at point 0 and the energy then radiates spherically without reaching a boundary. The spherical line is the wave front and the line from the source perpendicular to the front is the ray. At some location a far distance from the source, it is easy to see that the spherical wave front would be virtually plane.

Particle motion in the wave fronts of bulk waves is somewhat different from that which occurs in the plane wave and deserves some brief discussion. Figure 5-9(a) shows the particle motion that occurs in the wave front of a spherical wave front. While we recognize that the particle motion is not truly parallel, we also realize that at significant distances from the source, the assumption of parallel particle motion does not venture far from actuality. Longitudinal particle motion is, therefore, very similar for both the plane and the bulk wave. This is not necessarily the case, however, for the transverse waves. For the bulk wave case, the particle motion in the wave front for transverse waves may exist in a rotational or torsional form as shown in Fig. 5-9(b). The special plane wave case where the particle motion is either lateral of vertical still meets the required conditions.

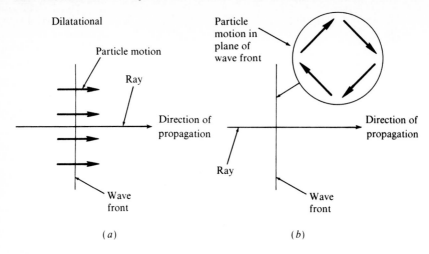

FIGURE 5-9
(a) Wave front and particle motion for bulk longitudinal wave. (b) Wave front and particle motion for bulk shear wave.

Example 5-1. Compare the wavelengths for 1.0 and 5.0 MHz longitudinal waves excited in aluminum and in Plexiglas. From Eq. (5-25), $\lambda = C/f$ and from Table 5-1, $C = 6320$ m/s.
Aluminum

$$\lambda_{1.0} = \frac{6320 \text{ m/s}}{1 \times 10^6 \text{ Hz}} = 6.32 \text{ mm}$$

$$\lambda_{5.0} = \frac{6320 \text{ m/s}}{5 \times 10^6 \text{ Hz}} = 1.26 \text{ mm}$$

Plexiglas

$$\lambda_{1.0} = \frac{2730 \text{ m/s}}{1 \times 10^6 \text{ Hz}} = 2.73 \text{ mm}$$

$$\lambda_{5.0} = \frac{2730 \text{ m/s}}{5 \times 10^6 \text{ Hz}} = 0.55 \text{ mm}$$

5-5 REFLECTION AND REFRACTION AT INTERFACES

The reflection and refraction behavior of the wave front at an interface plays an important role in ultrasonic investigations. Specifically, this is the effect that is used to determine the presence of a flaw or other anomaly. The impedance ratio at the interface is the parameter most frequently used to describe this behavior. Impedance is frequently used in both electrical and mechanical applications to describe the energy transfer characteristics at various boundaries.

For the present case, the parameter used is called the specific acoustic impedance z as expressed by

$$z = p/u \qquad (5\text{-}26)$$

where p is the instantaneous pressure at a point and $u = \dot{\xi}$ is the instantaneous particle velocity at the same point. For elastic material, the pressure term may be viewed as a stress and rewritten

$$p = E\frac{\partial \xi}{\partial x} \qquad (5\text{-}27)$$

where E is Young's modulus of the material and $\partial \xi / \partial x$ is the strain. Similarly, the particle velocity can be written

$$\dot{\xi} = \partial \xi / \partial t \qquad (5\text{-}28)$$

In both instances, the particle displacement ξ is given by Eq. (5-1).

The impedance term can be expressed as a frequency-dependent term by first taking partial derivatives of Eq. (5-24) with respect to distance x and time t and substituting these results into Eqs. (5-27), (5-28), and (5-26), yielding

$$z = -E(k/\omega) \qquad (5\text{-}26a)$$

Further, where $\omega = Ck$ [Eq. (5-25)] and, for plane waves, $E = \rho C_l^2$ [Eq. (5-12)], Eq. (5-26a) can be written

$$z = -\rho C_l \qquad (5\text{-}26b)$$

Two further simplifications result in the more common expression for specific acoustical impedance, i.e.,

$$z = \rho C_1 \qquad (5\text{-}26c)$$

First, the negative sign is dropped since the absolute magnitude of the impedance is the item of interest in the present discussion. Second, the bulk longitudinal wave speed C_1 has been substituted for the plane wave speed C_l. In most situations where ultrasonics is used for nondestructive evaluation, the bulk wave front is almost planar. With this, the conditions governing C_1 approach those governing C_l and the substitution in Eq. (5-25c) causes no serious discrepancy. It should be noted that C_1 never equals C_l.

5-5.1 Normal Incidence

To study the behavior at a boundary, first consider the plane wave situation shown in Fig. 5-10. Here, the plane wave traveling in material I approaches the boundary from the left. This wave front is parallel to the boundary and is

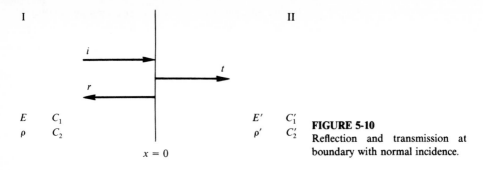

FIGURE 5-10
Reflection and transmission at boundary with normal incidence.

designated as the incident wave, indicated by the symbol i. The ray is seen to be normally incident to the boundary. Upon striking the boundary, part of the energy is reflected back into material I and part of it is transmitted into material II. The portion of energy transmitted and reflected is a function of the properties of materials I and II.

The particle velocity resulting from the incident wave can be written as

$$u_i = U_i e^{j(\omega t - kx)} \tag{5-29a}$$

where U_i = velocity amplitude of the incident wave
$\quad t$ = time
$\quad x$ = horizontal location.

The terms ω and k have been previously defined [Eqs. (5-1) and (5-25)] and the exponential term is used to designate a harmonic, traveling wave. Similarly, equations for the reflected and transmitted wave are

$$u_r = U_r e^{j(\omega t + kx)} \tag{5.29b}$$

and
$$u_t = U_t e^{j(\omega' t - k'x)} \tag{5.29c}$$

where the primed terms indicate waves traveling in material II. The choice of negative or positive sign for the kx term is associated with the direction of travel of the wave.

At the boundary we assume continuity of particle velocity, stress, and excitation frequency. Hence, for velocity continuity

$$u_i + u_r = u_t \tag{5-30}$$

since, at the boundary, the velocity of a particle must be the same in materials I and II.

Substituting from Eq. (5-29) into Eq. (5-30) yields

$$U_i e^{j(\omega t - kx)} + U_r e^{j(\omega t + kx)} = U_t e^{j(\omega' t - k'x)} \tag{5-30a}$$

The assumption of frequency continuity requires that $\omega = \omega'$. Therefore, at the boundary where $x = 0$, the exponential terms cancel leaving

$$U_i + U_r = U_t \tag{5-31}$$

The stress (pressure) at the boundary can be similarly written as

$$p_i + p_r = p_t \tag{5-32}$$

where the frequency-dependent terms are written

$$p_i = P_i e^{j(\omega t - kx)} \tag{5-33a}$$

$$p_r = P_r e^{j(\omega t + kx)} \tag{5-33b}$$

and

$$p_t = P_t e^{j(\omega' t - k'x)} \tag{5-33c}$$

The substitution of these terms into Eq. (5-32) yields

$$P_i + P_r = P_t \tag{5-34}$$

For these expressions, the sign of the kx term designates the direction of travel of the pressure gradient and the P terms designate the respective pressure amplitudes.

The specific acoustic impedance for materials I and II can be defined as

$$z_1 = P_i/U_i \tag{5-35a}$$

$$z_1 = P_r/U_r \tag{5-35b}$$

and

$$z_2 = P_t/U_t \tag{5-35c}$$

Further, dividing Eq. (5-34) by Eq. (5-31) and using Eq. (5-35c) yields

$$z_2 = \frac{P_i + P_r}{U_i + U_r} \tag{5-36}$$

Additional manipulation, using Eqs. (5-35a) and (5-35b), results in the reflection coefficient

$$R = \frac{P_r}{P_i} = \frac{z_2 - z_1}{z_2 + z_1} \tag{5-37a}$$

and the transmission coefficient

$$T = \frac{P_t}{P_i} = \frac{2z_2}{z_1 + z_2} \tag{5-37b}$$

These terms are written as ratios since the absolute amplitudes of P_t and P_r are of little interest in nondestructive evaluation. The value of the terms relative to that of the incident pressure is of considerable importance, however.

The assumption of some extreme conditions can demonstrate the implications of Eqs. (5-37). Table 5-2 lists solutions to these ratios for $z_1 = z_2$, $z_1 \gg z_2$, and $z_1 \ll z_2$. These situations, respectively, compare to no boundary, a metal and a fluid, and a fluid and a metal.

For the no-boundary case, the results are obvious, i.e., no reflection. In the second case, a high impedance material radiating into a low impedance material, the reflection coefficient of -1 may appear to violate physical laws. This is not so. It simply indicates a change in phase for the reflected wave. For the opposite

TABLE 5-2
Pressure ratios at boundaries.

	$z_1 = z_2$	$z_1 \gg z_2$	$z_1 \ll z_2$
$\dfrac{P_r}{P_i}$	0	-1	1
$\dfrac{P_t}{P_i}$	1	0	2

case (i.e., $z_1 \ll z_2$), the results indicate a pressure amplification for a wave transmitted from a low impedance material to one of a higher impedance.

A more complete explanation of the behavior of stress pulses at interfaces has been presented by Graff [7]. First, he considers the events at a fixed boundary, as would occur for a stress wave excited in a string that was tied to a fixed object. This would be analogous to a stress pulse traveling in a low impedance material and striking a boundary with a higher impedance material. Recognizing that there will be no particle displacement at the fixed boundary, he assumes that two displacement pulses, as shown to the left of Fig. 5-11a, are approaching the boundary. With one being the inverse image of the other, there will be no displacement as they pass the fixed boundary.

Since stress may be expressed as a function of the derivative of displacement with respect to x, as shown in Eq. (5-27), the sequence of events shown for the stress pulse shapes to the right of Fig. 5-11(a) are seen to demonstrate the behavior of the reflected pulse as predicted in the right column in Table 5-2. For simplicity in this demonstration, the materials on each side of the boundary are assumed to have the same Young's modulus. In this illustration, the incident pulse on the right moves to the left and approaches the boundary. After being reflected, it moves with a reversed shape to the right. In this case, the reflected stress pulse is a mirror image of the incident pulse. The additive nature of the stresses at the boundary should also be noted.

Figure 5-11(b), for a free ended string, is analogous to a stress wave traveling in a high impedance material striking a low impedance material. The sign reversal for the reflected pulse is shown here as is predicted in the middle column of Table 5-2. Also, the canceling effect of the stresses at the interface is demonstrated in these events. A similar case could be made for the transmitted pulse, showing the behavior of P_t/P_i.

Power ratios rather than amplitude ratios are often of greater importance in ultrasonic inspection. The full derivation of the power ratios will not be given here, however. Reference 1, among others, presents the development of these expressions.

The power expressions yield the rate at which energy is being transmitted per unit area of the wave front. For our purposes, power is defined as

$$\text{Pwr} = \text{stress} \times \text{displacement velocity at a point}$$

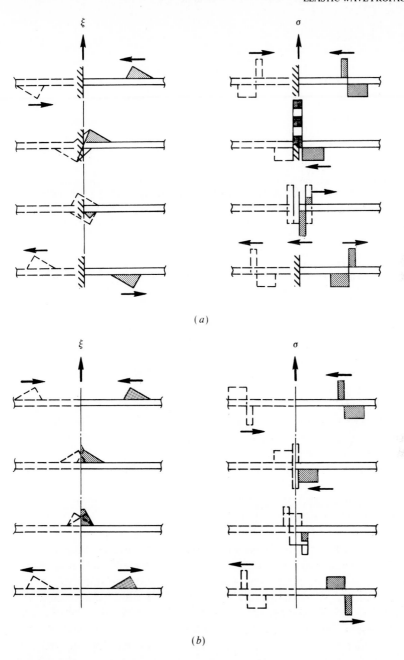

(a)

(b)

FIGURE 5-11

(a) Image displacement and stress pulse behavior for reflection at a fixed boundary. (*From Graff* [7]. *Courtesy Ohio State University Press.*) (b) Image displacement and stress pulse behavior for reflection at a free end. (*From Graff* [7]. *Courtesy Ohio State University Press.*)

TABLE 5-3
Power ratios at boundaries.

	$z_1 = z_2$	$z_1 \gg z_2$	$z_1 \ll z_2$
$\dfrac{\text{Pwr}_r}{\text{Pwr}_i}$	0	1	1
$\dfrac{\text{Pwr}_t}{\text{Pwr}_i}$	1	0	0

In a manner similar to that previously used for the pressure ratios, the respective power expressions for the incident, transmitted, and reflected waves are

$$\text{Pwr}_i = \frac{1}{2} \frac{EA_i^2\omega^2}{C_1} \tag{5-38a}$$

$$\text{Pwr}_t = \frac{1}{2}\rho'C_1'A_t^2\omega^2 \tag{5-38b}$$

$$\text{Pwr}_r = \frac{1}{2}\rho C_1 A_r^2\omega^2 \tag{5-38c}$$

Algebraic manipulation, using the particle displacement amplitudes and Eq. (5-26c), yields

$$\frac{\text{Pwr}_r}{\text{Pwr}_i} = \left[\frac{1 - z_2/z_1}{1 + z_2/z_1}\right]^2 \tag{5-39}$$

and

$$\frac{\text{Pwr}_t}{\text{Pwr}_i} = \frac{4[z_2/z_1]}{[1 + z_2/z_1]^2} \tag{5-40}$$

Power transmission and reflection ratios for the three extreme conditions are shown in Table 5-3.

The behavior of both the pressure and the power at the boundary are important in nondestructive inspection. As will be discussed later, maximizing the power transfer is important for designing efficient transducers. This is accomplished through proper impedance matching of the transducer ceramic and the receiving material. Where defects are indicated by reflection of ultrasonic energy at a boundary, the degree of impedance mismatch is quite important in generating the reflected pulse. The sign change predicted for the reflected pressure amplitude at a high to low impedance boundary can be observed in inspection practice.

Example 5-2. Determine the pressure ratios and power ratios for a Plexiglas/aluminum interface and an aluminum/Plexiglas interface, assuming normal incidence. The pressure ratios are given by Eqs. (5-37a) and (5-37b) and the power ratios by Eqs. (5-39) and (5-40).

Plexiglas/aluminum

$$R = \frac{(17 - 3.2) \times 10^6 \text{ kg/m}^2 \text{ s}}{(17 + 3.2) \times 10^6 \text{ kg/m}^2 \text{ s}} = 0.68$$

$$T = \frac{2(17) \times 10^6 \text{ kg/m}^2 \text{ s}}{(3.2 + 17) \times 10^6 \text{ kg/m}^2 \text{ s}} = 1.7$$

$$\frac{\text{Pwr}_r}{\text{Pwr}_i} = \left[\frac{1 - 17/3.2}{1 + 17/3.2}\right]^2 = 0.47$$

$$\frac{\text{Pwr}_t}{\text{Pwr}_i} = \frac{4(17/3.2)}{(1 + 17/3.2)^2} = 0.53$$

Aluminum/Plexiglas

$$R = \frac{(3.2 - 17) \times 10^6 \text{ kg/m}^2 \text{ s}}{(3.2 + 17) \times 10^6 \text{ kg/m}^2 \text{ s}} = -0.68$$

$$T = \frac{2(3.2) \times 10^6 \text{ kg/m}^2 \text{ s}}{(17 + 3.2) \times 10^6 \text{ kg/m}^2 \text{ s}} = 0.32$$

$$\frac{\text{Pwr}_r}{\text{Pwr}_i} = \left[\frac{1 - 3.2/17}{1 + 3.2/17}\right]^2 = 0.47$$

$$\frac{\text{Pwr}_t}{\text{Pwr}_i} = \frac{4(3.2/17)}{(1 + 3.2/17)^2} = 0.53$$

5-5.2 Oblique Incidence

Wave transmission and reflection at oblique interfaces is of considerable importance in ultrasonic inspection. Solid/solid, fluid/solid, and fluid/fluid interfaces can each be encountered. Additionally, the response at the boundary is different for an incident longitudinal and an incident shear wave. Moreover, the polarization of the shear wave affects the response. Rather than attempt to cover all of these possibilities, a summary, largely from Refs. 10 and 11, for a longitudinal wave incident on a solid/solid interface will be presented first.

A typical solid/solid interface is shown in Fig. 5-12. Here, the incident longitudinal wave, traveling at speed C_1 in material I, strikes the boundary at incident angle θ_1. Obeying Snell's Law for a homogeneous, isotropic solid, a reflected longitudinal wave and reflected shear wave are emitted as well as a transmitted longitudinal wave and shear wave. The appropriate angles are indicated in the figure. Snell's Law for this situation can be written

$$\frac{\sin \theta_1}{C_1} = \frac{\sin \theta_2''}{C_2} = \frac{\sin \theta_1'}{C_1'} = \frac{\sin \theta_2'}{C_2'} \tag{5-41}$$

where the single primed symbols indicate travel in material II.

At normal incidence, a normal longitudinal wave would be excited in material II traveling at a wave speed of C_1'. Reflected back into material I would be a longitudinal wave traveling at the same speed as the incident wave. For these

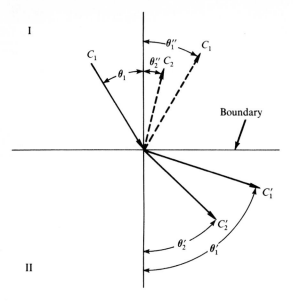

FIGURE 5-12

Reflection and refraction at a solid/solid interface at an oblique angle.

conditions, no shear wave is predicted for either the reflected or refracted components. As the incidence angle is increased from 0, the refracted longitudinal wave at speed C_1' leads the refracted shear wave traveling at the slower speed C_2', according to Snell's Law. The relationship of θ_1' and θ_2' is determined, respectively, by the longitudinal and shear wave speeds in material II. If θ_1 is continuously increased, θ_1' eventually reaches 90°. The incident angle θ_1 which results in $\theta_1' = 90°$ is known as the first critical angle. In this situation, the longitudinal wave is traveling just below the surface of the material interface. As the angle is further increased, the longitudinal component essentially disappears, leaving only the shear component C_2'. A further increase in θ_1 leads eventually to $\theta_2' = 90°$, the second critical angle. Beyond the second critical angle, the ultrasonic energy is concentrated on the surface in the form of Rayleigh waves, which are described in more detail later.

Applications of Snell's Law are most frequently encountered in angle-beam inspection where the refracted longitudinal wave, at C_1' and θ_1', and/or the shear wave, at C_2' and θ_2', are used as the inspecting wave in material II. These waves are typically excited with the incident longitudinal wave at C_1 and θ_1 in material I. The reflected longitudinal and shear waves (shown by dotted lines) usually may be neglected since they do not directly contribute to the flaw detection. They may, however, serve as an indicator of test difficulties in angle-beam inspection. This will be discussed in more detail in a section to follow.

The pressure transmitted at oblique angles across the interface is a function of the incident angle and Snell's Law. The distribution of energy between the various reflected and transmitted waves is of considerable importance in angle-beam inspection since the strength of the inspecting wave is affected. The variation of pressure ratios with the incident angle is treated extensively in several sources, e.g., Ref. 10, and only will be summarized here.

Citing the work of Kuhn and Lutsch [12], Krautkramer and Krautkramer [10] define the pressure transmission ratio D_{ll} for the longitudinal transmitted wave C_1' as

$$D_{ll} = \frac{2\rho'C_1'(C_2')^2 C_1 \cos^2(2\theta_2')\cos(2\theta_2'')}{\rho(C_2)^4 \cos(2\theta_1)\sin(\theta_1)N} \tag{5-42}$$

where

$$N = 2\cot(\theta_2) + \frac{C_1\cos^2(2\theta_2'')}{2C_2\cos(\theta_1)}$$

$$+ \frac{2\rho'(C_2')^4\cot(\theta_2')}{\rho(C_2)^4}$$

$$+ \frac{\rho'C_1(C_2')^3\cos^2(2\theta_2')}{2\rho(C_2)^4\cos(\theta_1)} \tag{5-43}$$

In all cases, the incident wave is the longitudinal wave C_1 at the incident angle θ_1. The material properties are as shown and the incident, reflected, and refracted angles are obtained from Eq. (5-41) as required. Similarly, the pressure transmis-

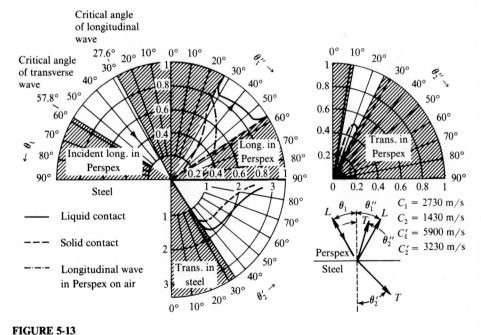

FIGURE 5-13
Pressure ratios for refracted transverse wave for Plexiglas and steel interface. (*From Krautkramer and Krautkramer [10]. Courtesy Springer Verlag.*)

sion ratio for a refracted transverse wave is

$$D_{tl} = \frac{2\rho'(C_2')^3 C_1 \cos(2\theta_2'')}{\rho(C_2)^4 \sin(2\theta_1) N} \tag{5-44}$$

Solution of Eqs. (5-42) and (5-44) yields the pressure ratios for the respective waves.

Figure 5-13 shows the pressure ratios for a transverse wave excited in steel with a longitudinal wave in a Plexiglas wedge. Only the region between the first and second critical angles is shown since this is the usual circumstance for angle-beam inspection.

The pressure transmission ratio for the shear wave rises sharply after the first critical angle is passed. Following that, the value increases smoothly until after a refracted angle of 70°, where the increase becomes more rapid.

5-6 PULSE CHARACTERISTICS— DISPERSION

The energy bursts, or pulses, that are typically used in nondestructive inspection require some special consideration. Up to this point, there has been little mention of frequency except to assume that it was constant with time and distance. Even though ultrasonic transducers will typically have a nominal resonant frequency, the practical fact is that a large number of frequencies will be excited by a typical transducer. This is discussed in the following material. References 7, 10, and 11 contain particularly useful discussion on pulse dispersion.

If the arrangement shown in Fig. 5-5 is initially at rest, i.e., $\omega = 0$, it is reasonable to assume that when it is started, it does not instantly jump to some single frequency of oscillation. Inertial effects and other practical considerations demand that the oscillation frequency gradually increase from 0 to some value ω_r, a resonant frequency. Similarly, if the driving source is cut off, ω will then decrease to 0. Thus, an observer at some distance to the right would see an earliest arrival of the lower frequencies, followed by increasingly higher frequencies until continuous excitation was reached at ω_r. Then there would be a similar decrease in frequency as the full pulse continued to pass.

A time domain display of a typical ultrasonic pulse is shown in Fig. 5-14. While the previous example using the crank indicates that the spectrum of the pulse includes only frequencies from 0 to ω_r, the actual fact is that pulses may contain a larger number of frequencies somewhat evenly spread about ω_r.

If the material is nondispersive, the pulse will travel undistorted. Few materials, however, are truly nondispersive, and pulse distortion to some extent is inevitable. Some situations where dispersion exists are described in the following.

A typical pulse as shown in Fig. 5-14 can be created by the superposition of several continuously excited wave forms. For example, assume that two waves A and B, having different frequencies as shown in Fig. 5-15, simultaneously exist in a test sample. Further, assume that the phase relationship between the two wave

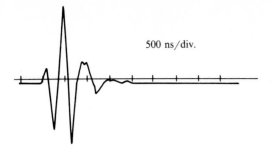

500 ns/div.

FIGURE 5-14
Pulse form excited by typical ultrasonic system.

forms at some instant is also as shown. Superposition of the two waves results in the pulse shape shown at the bottom in Fig. 5-15. If the speeds of each of the waves are identical, and remain so, the pulse will travel undistorted. If, on the other hand, there are differences in the travel speeds, distortion will occur.

Two typical situations often found in ultrasonic inspections are significant generators of pulse dispersion. These are attenuation and wave guide effects. These will be discussed in more detail later. For the present, it is sufficient to note that if pulse A were to lose amplitude relative to that of pulse B, the result of the superposition of the pulses would be different from that shown, even though the frequency and phase relationship remained constant. It is well known that for certain conditions the higher frequencies suffer greater attenuation than lower frequencies, thus creating a situation where pulse dispersion would be expected. Further, waves traveling at different frequencies in bars, rods, and plates can

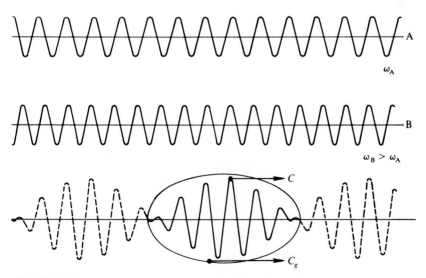

FIGURE 5-15
Formation of pulses from superposition of continuous wave forms.

propagate at speeds different from those of other frequencies, also creating conditions for pulse dispersion. Pulse dispersion may not be a serious problem for an item having large dimensions or small grain structure. It can exist, however, for other conditions as just described.

The phenomena of pulse dispersion creates the need for a differentiation between phase speed C and group speed C_g. The phase speed is the rate of travel of a point of constant phase in the pulse and the group speed is the travel speed of the entire pulse. In Fig. 5-15, each of waves A and B may travel at a different speed. For the nondispersive case, $C_A = C_B$ and the group speed of the pulse will be equal to the phase speed. For dispersive pulses, the superposed pulse may move at a rate different from the phase speed.

For the pulse shown at the bottom in Fig. 5-15, the phase speed is the speed of travel of a point of constant phase, e.g., the peak. This is the speed experienced by a surfer at the beach riding the peak of an ocean wave. As the surfer advances toward the beach, an increase or decrease in the heights of the peaks may be observed. As this occurs, the group speed may be noted to be traveling at a rate different from the phase speed experienced by the surfer. The phenomenon is a result of pulse dispersion similar to that which occurs in ultrasonic wave propagation.

More detailed analysis could show the following relationship between the phase and group speeds

$$C_g = C + k(dC/dk) \tag{5-45}$$

Here, C_g, C, and the wave number k are as previously defined. The last term on the right inputs the varying nature of the group speed. If the phase speed is unvarying with frequency, there is no dispersion and $C = C_g$. A varying phase speed, then, would have the expected effect on the group speed.

It is important to recognize that the frequency components of a typical ultrasonic transducer having a resonant frequency f_0 will be closely grouped around f_0, i.e., the amplitudes of components away from f_0 may be quite small. Thus, the group speed is in fact determined by frequencies very near to the central frequency of the probe, as will be discussed further in relation to Fig. 6-9(a).

For many pulse-echo inspection problems where the dimensions of the side of the test piece are approximately 10 times greater than the nominal wave length of the excited pulse, the distinction between phase speed and group speed is of little concern. In many other situations, as previously described, where material dimensions are small or anisotropic conditions exist, pulse dispersion can significantly affect test results [11].

5-7   ATTENUATION AND SCATTERING

An acoustic wave traveling through engineering materials will lose energy for a variety of reasons. This behavior can account for a loss in amplitude as well as for a change in its appearance. Pulse attenuation is treated very thoroughly in several of the references (e.g., [3], [7], [10], and [11]) and will only be briefly reviewed here.

There are three basic processes that account for loss of pulse energy, namely, beam spreading, absorption, and scattering. Beam spreading is primarily a geometric function where the intensity is decreasing with the square of the distance traveled. This may be observed by noting that the initial pulse energy is being distributed over a larger spherical area as the wave front advances. Absorption accounts for the mechanical energy converted to heat energy as the wave front passes. Essentially, this energy is permanently lost and this type of attenuation is not of much further use in materials inspection.

Scattering results from reflections at grain boundaries, small cracks, and other material nonhomogeneities. This phenomena may, on the one hand, account for serious energy losses that can render an item uninspectable. On the other hand, it may be useful in materials studies where, for example, there is a need to nondestructively measure grain size in metals.

Attenuation is generally expressed in the form

$$P = P_o e^{-aL} \tag{5-46}$$

where P_o = original pressure level at a source or other reference location
P = Pressure level at second reference location
a = Attenuation coefficient (Nepers per centimeter)
L = Distance of pulse travel from original source to second reference location

Ultrasonic pulse attenuation is typically expressed in units of decibels (dB). Decibels are based on a logarithmic scale and are a convenient unit to use when the magnitude of the parameter being measured varies over a very large range. Ultrasonic pulses decay rather rapidly, due to the processes just described.

The relative sound pressure level (SPL) of a propagating wave is

$$\text{SPL} = 20 \log \frac{P}{P_o} \text{ dB} \tag{5-47}$$

where P is the effective pressure of the ultrasonic wave at some observation point and P_o is the previous pressure at an earlier reference point. Considering two points in the path of an ultrasonic wave, the sound pressure level loss for a wave passing between points 1 and 2 is given by

$$\text{SPL}_1 - \text{SPL}_2 = 20 \log \frac{P_1}{P_2} \text{ dB} \tag{5-48}$$

If the stations are separated by a distance L and the material has an attenuation coefficient α, Eq. (5-48) can be written

$$\alpha L = 20 \log \frac{P_1}{P_2} \text{ dB} \tag{5-49}$$

Typically, L would be expressed in units of length (e.g., meters) and α in decibels per meter.

Example 5-3. If the pressure level observed at location 2 is one-half that previously observed at location 1, what would be the sound pressure loss in decibels from location 1 to 2?

$$\text{SPL}_1 - \text{SPL}_2 = 20 \log \frac{P_1}{P_1/2} = 20 \log 2 = 6.0 \text{ dB}$$

Example 5-4. Attenuation tests performed on a sample of cold-rolled aluminum 50 mm in length showed that a pulse that traveled 200 mm in the material had an amplitude 70% as great as one that which had traveled 100 mm. Calculate the attenuation coefficient for this material.

$$\alpha = \frac{20}{L} \log \frac{P_1}{P_2} = \frac{20}{100 \times 10^{-3}} \log \frac{1}{0.7} = 31.0 \text{ dB/m}$$

The value of the coefficient α varies considerably with the material and state of cold-work and/or heat treatment. As summarized by Krautkramer and Krautkramer, values for α range from less than 10 dB/m for worked steel and aluminum to over 100 dB/m for some plastics, cast steels, irons, coppers, bronzes, and brasses [10]. Intermediate values may be found for other plastics, some cast steels, and irons as well as worked copper and copper alloys.

Experimental data obtained at frequencies typically used for ultrasonic inspection are shown in Table 5-4 for some engineering materials. The rail material was a pearlitic (0.8% C), hot rolled steel while the hypoeutectoid steel showed a significant amount of ferrite around the pearlite. The microstructure was not established for the stainless steel or aluminum. The acrylic plastic was a type typically used for making wedge transducers.

TABLE 5-4
Typical attenuation coefficients for engineering materials.

Material	Freq. (MHz)	Mode	α (dB / m)	σ
Rail, pearlitic steel*	1	Long.	5.3	2.0
Rail, pearlitic steel*	2.25	Long.	5.6	2.6
Rail, pearlitic steel*	5	Long.	6.1	2.4
Rail, pearlitic steel*	2.25	Shear	8.8	
Hypoeutectoid steel, normalized	2.25	Long.	70	
Stainless steel, 3XX	2.25	Long.	110	
Aluminum, 6061-T6511	2.25	Long.	90	
Plastic (clear acrylic)	2.25	Long.	380	

*Reference 14.

Over 20 samples of railroad rail were used for the longitudinal wave data. Each was 1 m in length. The shear data for the rail was obtained with one rail sample with the ends inclined so that the distance would vary as the fixed angle probe was moved. The other material samples were < 50 mm in length. The rather large standard deviation for the longitudinal wave data for the rail indicates the large variation in the attenuation coefficient that was experienced for this material.

These data show the frequency dependence of attenuation, as well as the higher value obtained for shear waves as compared to longitudinal waves for the same frequency and material. The significantly higher value for the normalized, hypoeutectoid steel may be explained by the ferrite surrounding the pearlite. Worked steel, as predicted, shows a lower attenuation due to a breakdown of the grain boundries. Values obtained for the aluminum and plastic material are also within the range given by Ref. 10.

5-8 TEMPERATURE EFFECT ON WAVE SPEEDS IN SOLIDS

Wave speed variations with temperature may play a significant role in ultrasonic NDE since they affect both travel times and the angle of refraction in angle-beam inspection. Experimental results reported by Egle [14] on wave speeds in pearlitic steel and in Plexiglas show the magnitude of these speed changes. For longitudinal waves in steel traveling parallel to the rolling direction of the forged steel, the wave speed variation was found to be

$$C_1 = C_1^\circ - (dC/dT)\,\Delta T \tag{5-50}$$

where C_1° = longitudinal wave speed at a reference temperature
dC/dT = speed change constant $(m/s^\circ C)$
ΔT = temperature change in $^\circ C$

For a nominal longitudinal wave speed of 5900 m/s at 25°C (77°F) the results are

$$C_1 = 5900 \text{ m/s} - 0.55(T - 25)^\circ C \tag{5-51}$$

Shear wave speed changes in the same material are given by

$$C_2 = 3228 \text{ m/s} - 0.38(T - 25)^\circ C \tag{5-52}$$

Longitudinal waves in Plexiglas at a nominal speed of 2690 m/s at 25°C were found to vary according to

$$C_1 = 2690 \text{ m/s} - 2.3(T - 25)^\circ C \tag{5-53}$$

Comparing the expected wave speed changes over a typical temperature range of 2.8–47.2°C (37–117°F) shows that both the longitudinal wave speed and the shear wave speed in the steel vary by just less than one-half of 1%. Longitudinal wave speeds in Plexiglas, on the other hand, vary by just less than 4% over the same temperature range.

5-9 SUMMARY

The physical principles of excitation and propagation of ultrasonic waves have been developed with an intent that they be used in materials studies and for flaw detection. While the basic techniques of flaw detection have been long established, there will be rapidly developing opportunities to apply ultrasonics to broad material investigations as well as to the more precise definition of defects. The emphasis of general principles in this section is meant to lead into the more specialized examples of ultrasonic nondestructive evaluation that follow.

CHAPTER 6

WAVE
PROPAGATION IN
GUIDED
WAVE
MODES

6-1 INTRODUCTION

Most of the preceding material has described waves propagating in a bulk material where the travel path was unaffected except for an encounter with a reflection boundary. In this section, special wave forms will be described. The path of travel and, in some cases, the speed, are affected by the surfaces encountered in plates, bars, and rods.

6-2 SURFACE (RAYLEIGH) WAVES

Surface waves present some unique characteristics that justify particular attention to their usage in ultrasonic inspection. The unique feature of these waves is that they will follow complex curvatures that can often provide a path to defect areas that are virtually inaccessible to other waveforms.

Perhaps the most vivid illustration of surface waves is that seen on the surface of a still pond once a pebble is dropped into it. The cross section shown in Fig. 6-1 shows that waves radiate from the source at the surface, independent of other wave forms generated as the pebble falls through different depths in the material. Waves excited on solid surfaces behave in a similar manner and have

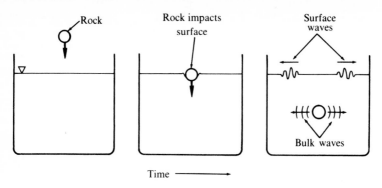

Time ⟶

FIGURE 6-1
Excitation of surface (Rayleigh) wave on a fluid surface.

the additional advantage of following the curvature of the surface, whether straight, convex, or concave.

The fact that the energy of the wave is restricted to the surface provides both an advantage of travel along a curved path as well as a disadvantage of being severely affected by ordinary anomalies that appear on machine and structural parts, e.g., pits on castings, tool marks, etc. Nonetheless, there are particular applications where surface waves are the answer to an inspection problem.

References 1–7, 10, 11, and 16–18 contain extensive discussions and bibliographies on surface waves that will not be repeated here. Much of the material to follow is extracted from these sources, however.

A cross-sectional view of a surface wave is shown in Fig. 6-2. As the harmonic wave front passes, an individual particle will travel in a retrograde elliptical pattern, moving backward and up, further backward and down, further down and forward, and then up and forward to begin the cycle again. The arrows in Fig. 6-2 show the displaced position of a surface particle during the passage of the surface wave.

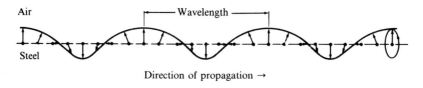

Direction of propagation →

FIGURE 6-2
Particle motion for surface (Rayleigh) waves. (*From Krautkramer and Krautkramer [10]. Courtesy Springer-Verlag.*)

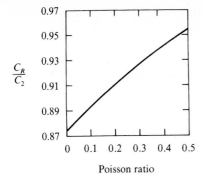

FIGURE 6-3
Surface (Rayleigh) wave speed and Poisson's ratio calculated from Eq. (6-1).

The velocity equation for the fundamental Rayleigh wave has a form more complex than those previously used for uniaxial and bulk waves, namely,

$$R^6 - 8R^4 + (24 - 16T^2)R^2 + 16(T^2 - 1) = 0 \qquad (6\text{-}1a)$$

where

$$R = \frac{C_R}{C_2} \qquad (6\text{-}1b)$$

$$T = \frac{C_2}{C_1} = \frac{1 - 2\nu}{2(1 - \nu)} \qquad (6\text{-}1c)$$

and C_R is the speed of the fundamental Rayleigh wave. Figure 6-3 and Table 6-1 show solutions to C_R/C_2 for various values of Poisson's ratio. It should be noted here that the specification of Poisson's ratio alone will not yield a unique solution to Eq. (6-1a) since the shear wave velocity C_2 must also be specified.

The material properties also affect the depth of penetration of the Rayleigh wave. Figure 6-4 shows the amplitude of the particle displacements at depths relative to the wavelength and the material. For example, simple calculations will show that a 2.25 MHz surface wave pulse in aluminum will have a wavelength of approximately 1.3 mm in both steel and aluminum. The wave will penetrate deeper into the aluminum, however, because of its higher Poisson ratio.

TABLE 6-1
Rayleigh wave velocity for various materials [18].

Material	ν	C_R / C_2	C_2 / C_1
Steel	0.29	0.9258	0.544
Copper	0.34	0.9335	0.492
Aluminum	0.34	0.9335	0.492

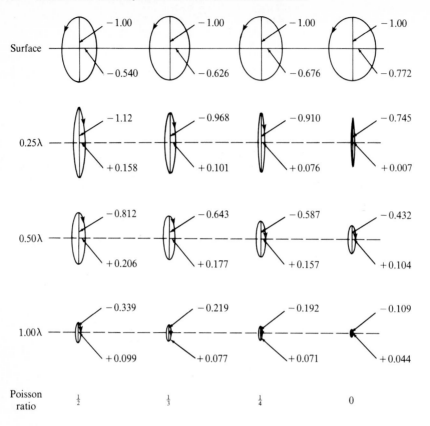

FIGURE 6-4
Particle displacements for a surface wave at various depths below the surface. Note reversal in oscillation direction below the surface. (*From Cook and Van Valkenberg [16]. Courtesy American Society of Testing and Materials.*)

Rayleigh waves propagating around curved surfaces and corners may encounter speed changes as well as some reflection. Where the ratio of radius of curvature and wavelength r/λ is large, Sezawa [19] has shown that the speed is the same as that experienced on a flat surface. As the ratio decreases, i.e., for lower frequencies or sharper corners, the speed increases. At $r/\lambda = 2.5$, the Rayleigh wave speed is approximately equal to the shear wave speed C_2. At a further reduction, where $r/\lambda = 0.7$, the speed has increased almost linearly to $\sim 1.28C_2$. Of greater importance in NDE are the reflection characteristics. Figure 6-5, from Viktorov [17], shows that, for large values of r/λ, the reflection coefficient R_r approaches, 0, while that of the transmission coefficient T_r approaches 1. As the r/λ ratio decreases, the reflection ratio increases with an accompanying decrease in the transmission coefficient. The erratic behavior near $r/\lambda = 0.5$ is attributed by Viktorov to phase differences as the half-waves fit into the rounded arc.

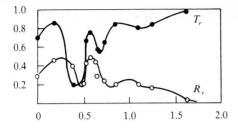
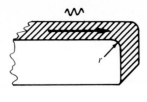

FIGURE 6-5
Rayleigh wave reflection and transmission coefficients at a cylindrical surface. (*From Viktorov [17]. Courtesy Plenum Press.*)

6-3 WAVES IN PLATES, BARS, AND RODS

Where the characteristics of the surface wave were a result of the presence of just one surface, the Lamb wave is propagated in situations where two parallel surfaces are found, as in plates. Similar behavior is observed for waves in rectangular bars.

A plate can be described as a structure having two parallel surfaces separated by a thickness t. The remaining parallel surfaces are assumed to be separated by much larger distances l and w that do not affect the wave propagation. This is shown in Fig. 6-6. The waves to be described in this section are those excited at the origin O, midway between the boundaries, and propagated in direction x. The manner of propagation is a function of the excitation frequency and the plate thickness t. The parameter used to describe this behavior is often either the ratio of plate thickness to wavelength (t/λ) or the product of frequency and thickness ($f \times t$).

Two extreme conditions for plate wave propagation may be described using material previously discussed. At very low frequencies, the wavelength may be many times longer than either of the lateral dimensions t or w. In this case, the conditions for plane wave propagation exists and the corresponding solution for Eq. (5-12) applies. For very short wavelengths, on the other hand, separate waveforms are excited. In the interior, bulk waves propagate, while on the surface, Rayleigh waves are excited. The intervening conditions are the subject of the present material.

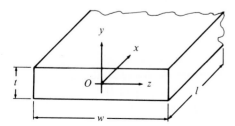

FIGURE 6-6
Plate for Lamb wave excitation.

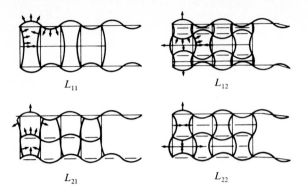

L_{11}

L_{12}

L_{21}

L_{22}

FIGURE 6-7
Symmetrical (L_{11}, L_{12}) and asymmetrical (L_{21}, L_{22}) Lamb waves in a sheet, the amplitude being represented on an exaggerated scale. At some points the vectors of oscillation are marked. (*From Lehfeldt* [20]. *Courtesy Materialprüf.*)

At the intervening frequencies, numerous plate modes having both symmetric and asymmetric configuration are excited by the ultrasonic signal. Figure 6-7 shows the first two modes of each type [20]. Here, the symbol L designates a Lamb wave mode. The first subscript designates a symmetric (1) or an asymmetric (2) mode, and the rank order of each mode is designated by the second subscript.

Lamb waves possess some characteristics that make them particularly useful for some specialized applications in NDE. As guided waves, they will travel within flat or mildly curved plates, giving a possible path of inspection to an otherwise inaccessible area. Equally useful is the dispersive nature of the waves where the propagation speeds of the waves are a function of the excitation frequency and the plate thickness, as shown in Fig. 6-8(a) for the phase speeds [21].

The previously described speeds for the extreme conditions can be observed in this figure where the upper and lower dotted lines indicate the bulk longitudinal and the Rayleigh wave speeds, respectively. For example, at very low frequencies, or for very thin plates, it is seen that the speed of the fundamental symmetric mode (L_{11}) approaches that of the plane wave speed in a bar. At very high frequencies, where $\lambda < t/10$, the Lamb wave speed approaches the speed of surface waves while the bulk wave travels unaffected by the presence of the surfaces [11]. The group speeds for Lamb waves in steel are given in Fig. 6-8(b).

Lamb wave behavior provides a practical example of boundary induced pulse dispersion and the phase speed and group speed distinction given in Eq. (5-45). A cutoff frequency is observed where some of the phase speeds of the higher-order modes approach infinity, e.g., L_{12} and L_{22} at $f \times t = 2$. This indicates nonexistence of the mode at frequencies below that value.

Since ultrasonic pulses are composed of a number of frequencies, each frequency component in the excited pulses will travel at a varying phase speed, as seen in Fig. 6-8(a). However, due to transducer characteristics, the amplitudes of the frequency components decrease away from the probe central frequency f_0. Also, components having different phases may tend to cancel. The overall result is that contributions to the pulse amplitude are diminished for components away

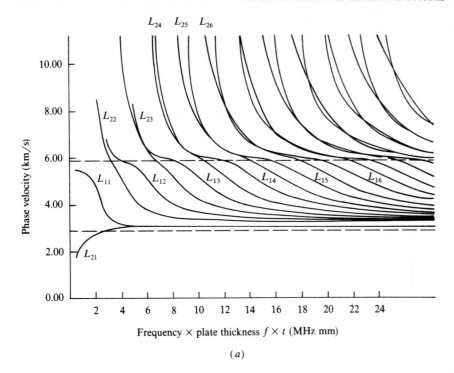

(*a*)

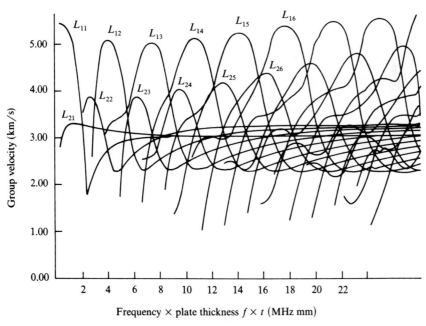

(*b*)

FIGURE 6-8
(*a*) Phase speeds for Lamb waves in steel with excitation frequency f (MHz) and plate thickness t (mm). (*Adapted from Egle* [21].) (*b*) Group speeds for Lamb waves in steel with excitation frequency f (MHz) and thickness t (mm). (*Adapted from Egle* [21].)

from f_0. An additional effect is due to the fact that frequency components different from the central frequency will travel at different phase speeds. The result is a group speed determined by the frequency components containing the dominant energy. For example, a 1 MHz single frequency wave excited in an 8 mm (0.315 in) thick plate would produce a wave containing components traveling at varying phase speeds as seen where $f \times t = 8$ in Fig. 6-8(a).

The dominant components of a typical ultrasonic pulse having a central frequency of 1 MHz would be in the form of the L_{13} mode and would travel at a group speed of just less than 5000 m/s, as seen in Fig. 6-8(b). Other components, traveling at slower or faster speeds, and therefore not remaining with the pulse, would not contribute significantly to the group speed of the pulse. The components having frequencies nearer to the 1 MHz central frequency would contribute the most to the shape and group speed of the pulse. The effect observed would be that of a dispersive pulse, which is often observed as a cluttered arrival pattern that creates difficulty in interpreting ultrasonic test results in bars and plates.

Despite these complications, however, Lamb waves are useful in ultrasonic NDE, particularly in the inspection of metal and composite plates for horizontal separations. Further discussions on applications of Lamb waves in ultrasonic NDE will follow in Chap. 8.

When the lateral dimension w in Fig. 6-6 is near to the thickness, the configuration now is that of a rectangular bar. Additional dispersion effects occur because of this additional surface. The results are similar to those described for the Lamb wave, however, and will not be repeated here. Reference 16 contains more detail on the propagation characteristics in rectangular and round bars.

6-4 SUMMARY

Guided wave modes have seen limited but rather specialized applications in nondestructive evaluation. For the surface waves, the particular advantage that is exploited is the ability of the wave to follow the curvature of a part to an expected defect location. Somewhat the same advantage exists for Lamb waves, although the dispersive nature of these waves adds some difficulty in their usage.

CHAPTER
7

ULTRASONIC CIRCUITRY AND TRANSDUCERS

7-1 INTRODUCTION

The modern, pulsed ultrasonic circuit has evolved largely from the electronics advances achieved in World War II. As reported in Refs. 1 and 2, the primary successes before that were analytical with few reports of experimental work. In 1945, Erwin [22] reported pioneering work on ultrasonic thickness measurement and discussed the limitations of contemporary equipment on ultrasonic research. Shortly thereafter, Firestone [23, 25] and Simmons [24] reported results describing the use of pulsed circuits in materials inspection. It was from these early researchers that the modern ultrasonic pulse–echo nondestructive evaluation system evolved.

The solid-state electronics used in present equipment offers considerable improvement over the earlier instrumentation. With the reduced power and size requirements, present equipment offers more capabilities in a more compact unit. These developments notwithstanding, the fundamental principles of operation of the key components of a typical ultrasonic circuit remain as described in 1955 by Hueter and Bolt [26].

Advances in numerical analysis have also benefited ultrasonic NDE. For example, prior to the development of the modern computer, calculations of the speeds for the various types of dispersive wave forms required considerable human effort. The availability of calculated data such as furnished for Lamb waves was quite limited. Additionally, the ability of computers to retrieve, store, analyze, and report the vast amount of data generated in most ultrasonic tests has increased both the breadth of applications as well as the confidence in the results.

7-2 PIEZOELECTRIC TRANSDUCERS

For most systems in use today, piezoelectric transducers provide the mechanism for converting electrical pulses into mechanical, or stress, waves. The material presented here is largely summarized from Refs. 10, 18, and 26–36 with other specified sources. Most modern piezoelectric materials are ceramics such as barium titanate (BaTi), lead zirconate titanate (PZT), and lead metaniobate (PMN). The thin disk is the most common shape of piezoelectric ceramic that is used for nondestructive testing applications. In this situation, an electric field applied to the disk through leads soldered to its faces (Fig. 7-1) will cause a proportional change in the thickness t of the disk to some new value t'. Depending on the polarity of the field, the disk will either contract or expand. An increase in field strength may cause further contraction or expansion of the disk and, conversely, a reduction in strength will relax it. These piezoelectric materials also have the ability to perform the opposite function in converting a mechanical pulse into an electrical signal.

Construction techniques for ultrasonic probes have been given by Egerton [29] and by Washington [30]. A typical transducer is shown in Fig. 7-2 and the material conditions are shown in Fig. 7-3. Here the load is the test piece, while the backing material is part of the probe assembly. The wear face material is thin and, therefore, its effect on the wave transmission can be ignored. The mechanical, or acoustic, impedance of the load, the transducer, and the backing material are shown as z_1, z_0, and z_2, respectively. Each impedance is determined by the density ρ and the dilatational wave velocity C_1 of the material, i.e.,

$$z = \rho C_1 \tag{7-1}$$

When a transducer is excited by an electrical spike, impulses of opposite sign will be excited at each interface, as shown by the spikes on each side of the interfaces in Fig. 7-3. The resulting stress waves will travel back and forth within the ceramic as the oscillations occur. Each time a stress wave strikes a boundary,

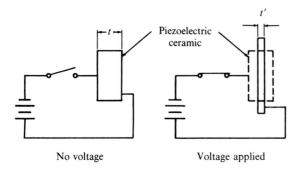

No voltage   Voltage applied

FIGURE 7-1
Piezoelectric effect.

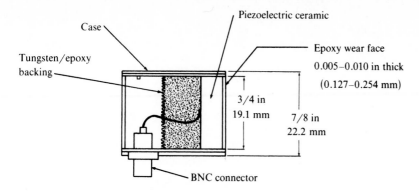

FIGURE 7-2
Typical ultrasonic probe, 19 mm ($\frac{3}{4}$ in) diameter.

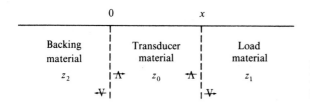

FIGURE 7-3
Material impedances for ultrasonic probes.

the amplitude and phase of the reflected and transmitted waves will be governed by the relative impedance of the two materials making up the boundary. Transducer modeling techniques, as described by Redwood [28] and Silk [27], use these reflection characteristics to create ultrasonic pulse shapes for further study.

The reflection coefficients at the boundaries, from Eq. (5-37a), are

$$R_x = \frac{z_0 - z_1}{z_0 + z_1} \qquad (7\text{-}2a)$$

and

$$R_0 = \frac{z_0 - z_2}{z_0 + z_2} \qquad (7\text{-}2b)$$

where R_x = reflection coefficient at transducer–load interface
R_0 = reflection coefficient at transducer–backing material interface

The transmission coefficients from Eq. (5-37b), would be

$$T_x = \frac{2z_1}{z_0 + z_1} \qquad (7\text{-}3a)$$

$$T_0 = \frac{2z_2}{z_0 + z_2} \qquad (7\text{-}3b)$$

Maximum energy transfer happens when the boundary impedances are equal, i.e., zero reflection will occur. For load or backing material impedances

TABLE 7-1
Typical acoustic impedances ($z = \rho C_1$) for transducer materials.

Material	$z \times 10^6$ kg $/$ (m^2 s)
Piezoelectric materials	
Quartz [27]	15.2
BaTi (barium titanate) [31]	31.2
PZT (lead zirconate titanate) [27]	33.0
PMN (lead metaniobate) [27]	20.5
LSH (hydrated lithium sulfate) [27]	11.2
PVDF (polyvinyl film) [27]	4.1
Mounting and backing materials	
Araldit casting resin [10]	2.8–3.7
Casting resin [18]	5.2
Tungsten/epoxy (200 : 100) [18]	9.4

greatly different from the transducer impedance, strong reflections will occur. Acoustic impedances of several materials are listed in Tables 5-1 and 7-1.

Silk [27] reports that an initial decrease occurs in the impedance when tungsten is added to the araldite (epoxy) mixture. This is shown to be a function of an initial decrease in speed for the araldite as amounts up to 50% by volume of tungsten powder are added. This speed drop is countered by an increase in density due to the tungsten. Finally, impedances equal to the values seen for the piezoelectric ceramics may be achieved with larger amounts of tungsten added to the araldite, thereby achieving a no-reflection condition at the back boundary.

Example 7-1. An ultrasonic probe with a PZT ceramic is to be used for inspecting both steel and aluminum parts. For this probe, calculate the transmission and reflection coefficients at the front interface for each of the inspected materials.
Steel

$$R_x = \frac{33 - 45}{33 + 45} = -0.15$$

$$T_x = \frac{2(45)}{45 + 33} = 1.15$$

Aluminum

$$R_x = \frac{33 - 17}{33 + 17} = 0.31$$

$$T_x = \frac{2(17)}{17 + 33} = 0.68$$

Example 7-2. Compare the reflection and transmission coefficients at the rear interface for an air-backed probe and one backed with the 200 : 100 tungsten/epoxy mixture from Table 7-1.
T/E-backed

$$R_0 = \frac{33 - 9.4}{33 + 9.4} = 0.56$$

$$T_0 = \frac{2(9.4)}{33 + 9.4} = 0.43$$

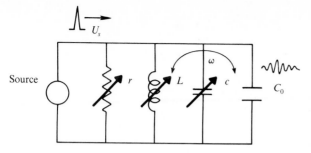

FIGURE 7-4
Basic ultrasonic pulser–receiver circuit.

Air-backed

$$R_0 = \frac{33 - 0}{33 + 0} = 1$$

$$T_0 = \frac{2(0)}{0 + 33} = 0$$

The tungsten/epoxy mixture may be used as a damping material for the probe in order to reduce the ring-down time since the reflected pressure is reduced by 43% each time the pulse strikes the rear interface.

7-3 ULTRASONIC CIRCUITRY FOR PIEZOELECTRIC TRANSDUCERS

Knowledge of the operation of the pulsed ultrasonic circuit is essential for understanding the characteristics of pulsed ultrasonic inspection. In the basic circuit, as shown in Fig. 7-4, the piezoelectric ceramic is seen as an electrical capacitance C_0. A high-voltage electrical spike U_s strikes the piezoelectric ceramic, causing it to oscillate. When excited by the electrical spike, the transducer will, in the ideal case, oscillate at the resonant frequency determined by

$$f_{re} = \frac{1}{2\pi\sqrt{LC_0}} \tag{7-4a}$$

where L is the inductance of the coil.

A piezoelectric transducer also has a mechanical resonant frequency determined by the equation

$$f_{rm} = \frac{1}{2\pi}\sqrt{\frac{s}{M}} \tag{7-4b}$$

where s is the mechanical stiffness of the transducer material and M is the mass. Optimum performance occurs at a frequency f_r where the electrical resonant frequency is matched to the mechanical resonant frequency

$$f_r = f_{re} = f_{rm} \tag{7-5}$$

Since for a particular transducer s, M and C_0 are fixed, tuning may be achieved by varying the inductance L. The electrical damping and, hence, the pulse length can be adjusted through the variable resistance r. Slight adjustments in the circuit capacitance can be made through variable capacitor c.

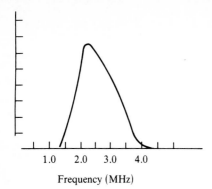

Frequency (MHz)

FIGURE 7-5
Frequency spectrum of pulse shown in Fig. 5-14 from a typical ultrasonic probe. $f_r = 2.25$ MHz, $d = 19$ mm, aluminum test material.

A typical output pulse of an ultrasonic transducer is shown in Fig. 5-14 and on the right in Fig. 7-4. The magnitude and shape of the initial pulse obviously affects the characteristic of the emitted wave form. Typical voltages for commercial pulsers are on the order of 300–500 V. Triangular, square-wave, and more specialized initial pulse shapes are available. The significance of the effect of the pulser on the performance of ultrasonic transducers has been reported by Posakony [32].

7-4 TRANSDUCER CHARACTERISTICS

Commercial ultrasonic transducers are available with a variety of nominal frequencies, diameters, and damping characteristics. The actual frequency spectrum and the damping quality of a transducer, however, can vary considerably from the specified value, largely due to manufacturing irregularities. The report of Kwun, Burkhardt, and Teller [33] documents some of these inconsistencies. Extensive reviews of ultrasonic transducer characterization have been furnished by Silk [27], Bredael [34], and Papadakis [35].

The spectrum envelope of the output pulse of a nominal 2.25 MHz normal beam 19 mm diameter transducer is shown in Fig. 7-5. For this pulse, the transducer was placed in direct contact with an aluminum sample. It is clear from the figure that the nominal frequency is quite near to the specified 2.25 MHz. The envelope is reasonably symmetric, with a slight skew toward the right. It is significant to observe from this figure that this nominal 2.25 MHz transducer certainly contains a considerable amount of energy at frequencies both below and above the nominal resonance, e.g., in the 2–3 MHz range.

The bandwidth of an ultrasonic transducer is a measure of its inherent damping characteristics. Bandwidth B may be represented by

$$B = f_b - f_a \tag{7-6}$$

as illustrated in Fig. 7-6 where f_a and f_b are the frequency locations corresponding to the half-power points of the power spectrum curve. The internal damping of the probe is a function of the damping characteristics of the piezoelectric material as well as the reflections occurring at the front and rear interfaces of the

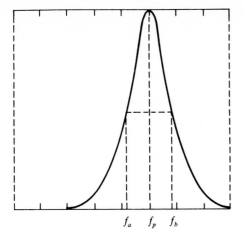

FIGURE 7-6
Spectrum power curve for ultrasonic transducers.

$f_a \quad f_p \quad f_b$

piezoelectric material. The value of the bandwidth increases with the amount of damping. Thus, a lightly damped transducer would have a rather narrow spectrum curve and a small bandwidth.

Bandwidth is often expressed as the percentage bandwidth as given by

$$B^* = \frac{f_b - f_a}{f_c} \times 100 \tag{7-7}$$

where the central frequency is

$$f_c = \tfrac{1}{2}(f_b + f_a) \tag{7-8}$$

Noting that for most ultrasonic probes the spectrum curve shows a skewed rather than a Gaussian or normal shape, a measure of the amount of skew in the spectrum offers another parameter that may be used to evaluate a transducer performance. Skewness may be represented by

$$f_{sk} = \frac{f_p - f_a}{f_b - f_p} \tag{7-9}$$

where f_p is the peak-power (resonant) frequency.

Figure 7-7 shows a typical skewed spectrum for a nominal 5 MHz probe placed in contact with an aluminum block. The parameters in the figure are defined as

$$f_a = \text{Flo}$$
$$f_b = \text{Fhi}$$
$$f_p = \text{Fcur}$$
$$f_c = \text{Fcen}$$
$$B^* = \text{Bndw}$$
$$f_{sk} = \text{Fskw}$$

Ideal probe performance would be where the spectrum curve was normal, i.e., $f_{sk} = 1$ and $f_p = f_c$ from Eq. (7-5).

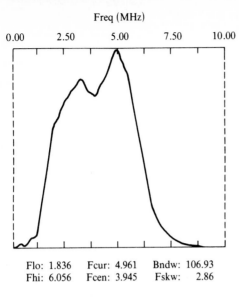

Freq (MHz)

| Flo: 1.836 | Fcur: 4.961 | Bndw: 106.93 |
| Fhi: 6.056 | Fcen: 3.945 | Fskw: 2.86 |

FIGURE 7-7
Skewed power spectrum for damped ultrasonic probe.

Narrow bandwidth units having a rather sharp peak in the spectrum curve require very close tuning between the inductance coils and the capacitance of the ceramic in order to obtain peak performance. Broad-band transducers, on the other hand, are flatter across the peak and, therefore, will excite and respond to a wider range of frequencies. Close tuning of the inductance and capacitance does not significantly affect their performance. Both transducer types are used for particular applications in ultrasonic NDE. These will be discussed in more detail in later sections.

7-5   BEAM CHARACTERISTICS

Ultrasonic transducers are frequently designated as *search units*, a label that is certainly correct for the typical flaw detection application. With the search unit designation in mind, it is proper to discuss in some detail the characteristics of the probe beam.

For the present discussion, the transducer will be assumed to consist of a thin, circular piezoelectric element in direct contact with an infinite medium, as described by Kinsler et al. [3]. While there are a myriad of other shapes used in ultrasonic NDE, the circular shape is most common and will suffice for a general discussion of the beam characteristics. Figure 7-8 shows the typical main lobe of the beam of a circular piston radiator. The lobe is a constant pressure (isobaric) plot of the emitted energy. Parameters of interest are the acoustic pressure P at some location p in the field of the energy emitted by a circular element of diameter d. The element is assumed to be excited at a single frequency ω. The coordinates of p are r, the distance from the center of the element, and $\Phi/2$, the

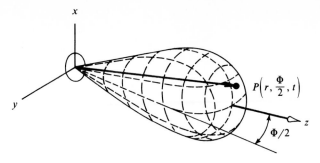

FIGURE 7-8
Pressure pattern of the major lobe for a piston radiator. (*From Kinsler et al.* [3]. *Courtesy John Wiley.*)

angle from the center axis of the element. Since the emitted pressure is assumed to exist in the form of a traveling wave, the time t also is included as a parameter.

First consider the pressure pattern along the z axis, i.e., at $\Phi/2 = 0$. In general, the pressure fluctuations in the near field immediately adjacent to the probe are erratic. Beyond the near field is a region where the pressure decays smoothly as a function of the distance from the probe. This smooth decaying zone is the far field of the transducer.

The strong pressure fluctuations in the near field are described by Huygen's principle, as discussed in more detail in a number of sources, e.g., Refs. 3 and 10. As r is moved along the axis away from the probe surface, pressure levels will be observed to vary as shown in Fig. 7-9 in a series of minima and maxima from 0 to $2\rho_0 C U_o$. The final maximum pressure will occur at a point defined by

$$\frac{r'}{d} = \frac{d}{4\lambda} - \frac{\lambda}{4d} \tag{7-10}$$

Defining the distance r' to be the near field length N, Eq. (7-10) may be rearranged to the form

$$N = \frac{d^2 - \lambda^2}{4\lambda} \tag{7-10a}$$

In most situations, NDE investigators are not concerned with the pressure behavior within the near field, but are rather more concerned with knowing the length of the zone so that inspection within the region may be avoided. As shown in Eq. (7-10a) and recalling Eq. (5-25), the length of the near field is found to be a function of the probe diameter d, the frequency of excitation f, and the wave speed C of the material to be inspected.

Following the presentation of Kinsler et al., the far field is defined by the conditions $r/d \gg 1$ and $r/d \gg kd$ which limit it to the region where r is much greater than the probe diameter and the wavelength λ. In the far field, the pressure amplitude P along the z axis may be defined to be

$$\frac{P_z(r)}{R} = \frac{1}{8}\rho_0 C U_0 \frac{d}{r} kd \tag{7-11}$$

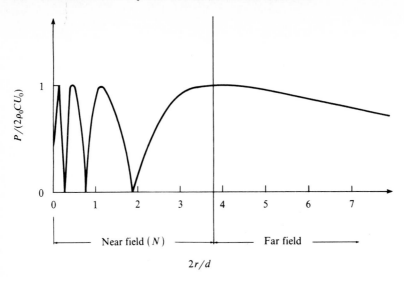

FIGURE 7-9
Pressure fluctuations along the axis of a plane, circular radiator $d/\lambda = 8$, N-near field. (*From Kinsler et al.* [3]. *Courtesy John Wiley.*)

where ρ_0 = volume density
C = wave speed
U_0 = speed amplitude at surface of piston radiator
r = distance along the z axis
k = wave number
d = probe diameter

Rearranging the terms leads to

$$\frac{P_z(r)}{R} = \frac{d/\lambda}{r/d} \qquad (7\text{-}11a)$$

where

$$R = \frac{\pi}{4}\rho_0 C U_0$$

Several parametric effects on acoustic pressure may be observed from Eq. (7-11a). First, the asymptotic decay of pressure for increasing r in the far field is indicated by the location of the distance r in the denominator. Further, larger probe diameters and higher frequencies are seen to result in higher pressures.

Pressure fluctuation patterns along the z axis in both the near field and far field are shown in Fig. 7-9 for a d/λ value of 4.5, which is representative of typical ultrasonic NDE values. The pressure values given by Eq. (7-11a) at the r' given by Eq. (7-10) are not in exact agreement. The inconsistency is small, however, and the curve given in Fig. 7-9 is a good general representation of the pressure fluctuations in a typical probe field.

The acoustic pressure of the probe is dependent also on the angle away from the beam axis. Figure 7-10 shows the pressure change with the angle $\Phi/2$ in the far field for the conditions $\lambda = \pi d/5$. The major lobe is shown centered along

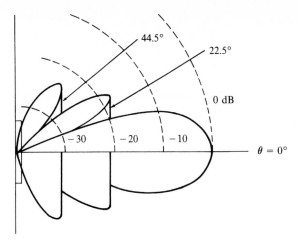

FIGURE 7-10
Beam pattern for a circular, plane piston for $\lambda = \pi d/5$. (*From Kinsler et al.* [*3*]. *Courtesy John Wiley.*)

along the z axis. As the angle $\phi/2$ is increased, the constant pressure boundary moves toward the probe until a node is reached. For the given conditions, this occurs at 22.5°. The minor lobes that appear at larger angles are of little concern in most NDE applications. Decreases in beam pressure are given in decibels with the maximum at the center as the reference. The half-power points at 6 dB down from the maximum at the center are used in NDE for flaw size estimation, amongst other applications.

Using Schlieren visualization techniques, studies of the total beam divergence angle Φ have been conducted by several investigators ([10], [34], [36], and [37]). Figure 7-11, from the work of Hall, shows the divergence of an ultrasonic

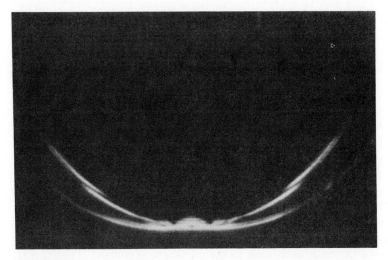

FIGURE 7-11
Diverging ultrasonic pulse in water excited by piezoelectric film, showing waves from transducer edges trailing the main wave front. (*From Hall* [*37*]. *Reprinted with permission from Materials Evaluation, vol. 42, no. 7, 1984, The American Society for Nondestructive Testing, Columbus, OH.*)

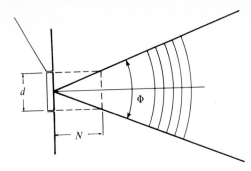

FIGURE 7-12
Beam divergence for 1 MHz 19 mm probe with steel as the load material.

pulse in water. For circular probes, the divergence angle is given to be

$$\Phi = 2 \sin^{-1}\left[1.2 \times 10^{-3}\frac{C_1}{fd}\right] \tag{7-12}$$

where C_1 = speed of a dilatational wave in the load material (m/s)
$\quad\quad f$ = probe frequency (MHz)
$\quad\quad d$ = probe diameter (mm)

For the wavelength and probe diameter restrictions just described, the near field, diameter, and divergence angle are shown in Fig. 7-12.

Example 7-3. Calculate and sketch the divergence angle Φ and the near field N for a 1 MHz longitudinal wave 19 mm diameter probe on a piece of steel.

$$f = 1.0 \times 10^6 \text{ Hz} \quad\quad d = 19 \text{ mm} \quad\quad C_1 = 5900 \text{ m/s}$$

$$\lambda = \frac{C_1}{f} = \frac{5900}{1 \times 10^6} = 5.9 \times 10^{-3} \text{ m}$$

$$\Phi = 2 \sin^{-1}\left[1.2 \times 10^{-3}\frac{5900}{(1)(19)}\right] = 43.8°$$

$$N = \frac{d^2 - \lambda^2}{4\lambda} = \frac{\left[(19)^2 - (5.9)^2\right] \times 10^{-3}}{4(5.9)} = 13.8 \text{ mm}$$

These results are shown in Fig. 7-12.

7-6 FOCUSED PROBES

Optical principles using curved lenses of materials having wave speeds greatly different from the transducer material may be used to reduce the length of the near field by concentrating ultrasonic energy at localized regions. This effect may be created in a number of ways, e.g., a curved probe face immersed in water or lenses inserted in the beam path near to the piezoelectric element. Phased array techniques also may be used for focusing. Additional details on using focused probes are available from a number of sources, e.g., Ref. 10.

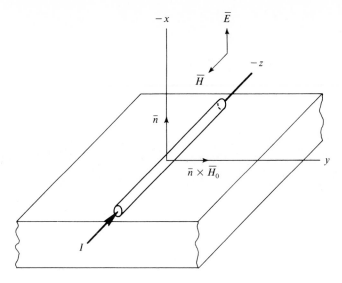

FIGURE 7-13
Excitation of Lorentz forces in a magnetic field.

7-7 ELECTROMAGNETIC–ACOUSTIC (EMAT) PROBES

The electromagnetic–acoustic transducers offer the potential advantage of performing ultrasonic inspection without the need for a physical couplant or other contact media between the probe and the test piece. This would facilitate automated inspection of linear items such as pipe, pipelines, railroad rail, etc., where there is relative motion between the probe and the test item. Further, the need for increased speed in inspection creates an additional advantage for these probes since the dynamic effects from higher speeds are minimal. EMAT probes may also be used to inspect metals covered with a protective coating. Information on principles and applications of EMAT probes is available from a number of sources, notably Refs. 27, 38, and 39.

The EMAT system uses the principle that an electromagnetic wave incident on the surface of an electrical conductor induces eddy currents within the skin of the conductor. In the presence of a static magnetic field and in the region where the eddy current density is nonvanishing, ions are subjected to oscillatory forces. It is these oscillatory forces (the Lorentz mechanism) generated within the skin depth that are the sources of electromagnetically induced acoustic waves.

The Lorentz forces that are the source of the acoustic waves are excited with a current carrying coil located near to the surface of a test sample. The geometry for the excitation of the Lorentz force is shown in Fig. 7-13 where $\bar{n}$ represents the normal to the surface of the sample to be inspected and $\bar{H}_0$ is the direction of the current flow in the excitation coil. As illustrated, the direction of

Lorentz stress ($\bar{n} \times \bar{H}_0$) at the surface is mutually perpendicular to the directions of the coil current and the surface normal.

Shear or longitudinal waves may be generated by applying the static magnetic field ($\bar{B}_0$) either perpendicular or parallel to the material surface, as shown in Fig. 7-14(a) and (b). Since the electron eddy current density is oriented parallel to the Lorentz forces, the ion displacement $\bar{u}$, which excites the acoustic

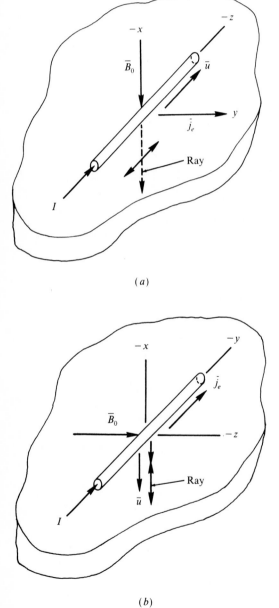

(a)

(b)

FIGURE 7-14
(a) Excitation of shear waves in electromagnetic field. (b) Excitation of longitudinal wave in electromagnetic field.

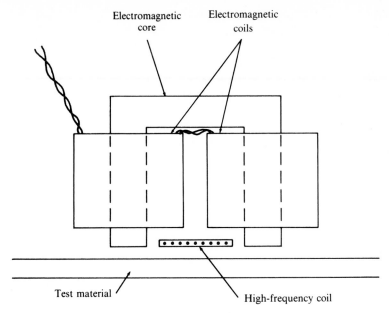

FIGURE 7-15
Electromagnetic transducer (EMAT) probe location for exciting longitudinal waves.

waves, would need to be either parallel to the surface for shear wave generation or perpendicular to the surface for longitudinal wave generation.

The coil location in the magnetic field will determine the type of ultrasonic wave that is generated. The coil location for exciting longitudinal waves is shown in Fig. 7-15. Shear waves would be excited if the coil were placed under one of the legs of the magnet. A variety of coil shapes may be used to generate special wave forms and beam shapes.

It is well known that the signal-to-noise ratio for EMAT probes is low and very large magnets are generally required to excite ultrasonic signals with a satisfactory amplitude. This limitation has hampered their broad adoption for industrial inspection. In installations such as pipe mills, where the magnets and the sensors may be fixed, however, the size and weight of the system appears to not be a serious handicap and the advantages of noncontact ultrasonic inspection may be utilized.

7-8   LASER GENERATION AND RECEPTION OF ULTRASOUND

Ultrasonic waves may be excited in materials when the surface is struck by laser pulses. The heating of the impulse creates expansions at the localized region where the laser strikes the surface. Ultrasonic waves are generated with these expansions and subsequent contractions. Discussions of laser excitation and reception of ultrasound are given by Refs. 27, 40, and 41, amongst others.

There are several characteristics of laser excitation and reception of ultrasound that should be discussed. First, as has been described, large amplitude ultrasonic pulses may be excited in a true noncontact mode. Since the excitation is accompanied by a large energy input at a surface, however, there may be some concern about material damage at the excitation point. Further, laser excited ultrasound tends to be impulsive with the result that a wide frequency spectrum is present. Depending on the application, this may be an advantage or a disadvantage. The reception of ultrasound with laser technology is not done with the same system that is used to excite the signal. Generally, reception is accomplished with an optical beam which is capable of measuring small surface displacements.

The ability to perform high-power, noncontact ultrasonic inspection is a very important advantage for laser ultrasonics and the topic is a very active one in NDE research.

7-9 ARRAY TRANSDUCERS

An array of single probes each coupled through multiplexer circuits may be used to create very specialized inspection processes. A typical multiplexer and array circuit is shown in Fig. 7-16. In this case, each circuit A through J consists of a separate pulser and transducer. Each probe element is assumed to be a point, or very small, source and the firing sequence of the pulsers is controlled through

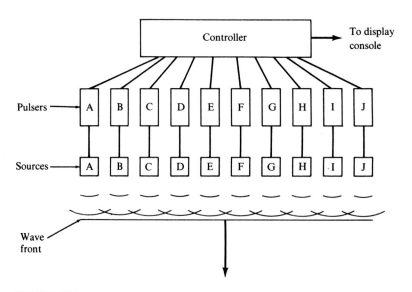

FIGURE 7-16
Multiplexer arrangement for array transducers. Normally refracted, longitudinal wave shown.

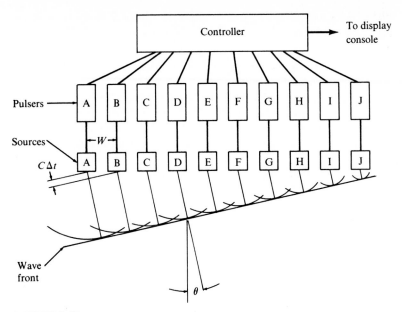

FIGURE 7-17
Angle-beam multiple probe array transducer.

other circuitry. Where each pulser fires simultaneously, a normal wave front as shown in Fig. 7-16 will be generated.

Pulsing the probes in phased sequences can create a variety of beam patterns, including angle-beam and focus configurations. For example, the angle-beam pattern shown in Fig. 7-17 results when the firing sequence is evenly phased from A through J. The inclination angle θ is given by

$$\theta = \sin^{-1} \frac{C\Delta t}{w} \tag{7-13}$$

where C = phase speed in the material
Δt = phase delay in pulsing each probe
w = probe center-line spacing

Varying the delay Δt between pulses would change the angle θ. Firing the pulsers in the reverse order, J through A, would result in an angle beam with the opposite orientation. Further, firing from the outside to the inside, i.e., A and J simultaneously, then B and I, etc., will create a focused beam.

A wide variety of sources may be used to construct an array transducer, including piezoelectric elements as well as EMAT and laser sources. The expense of the system is significantly increased since each circuit is essentially a separate ultrasonic system controlled by the multiplexer. The increased speed and versatility of inspection that the array transducers permit, however, may offer advantages that counter the added expense of the system. Additional discussion on the construction and use of phased arrays and multiplexer circuits is given in Refs. 6, 10, and 27.

7-10   DISCUSSION AND SUMMARY

This section has presented a cursory view of the fundamental electronic circuit and components used in ultrasonic nondestructive inspection and the relationship of the electrical and mechanical properties in the piezoelectric transducer. More in-depth discussions on these topics are available from a number of the previously referenced sources.

The transducer, of course, is a critical component both in the electronic circuit as well as in its direct application to flaw detection and materials evaluation. Transducers come in a variety of configurations other than the circular, plane item that has been described here. For example, both piezoelectric and EMAT probes are available in square and cylindrical shapes for specialized uses. Several ceramic materials may also be purchased that have different electromechanical properties that may be used for particular applications.

Typical damping can range from air-backed, i.e., no damping, to a very heavily damped probe. The air-backed units will excite long pulse trains, which are useful when narrow frequency ranges are desired [42]. The disadvantage is that the initial oscillations (i.e., the ring-down) may continue for too long a time, masking early signals that could be important. Heavily damped probes, on the other hand, can be made where essentially one short oscillation is all that occurs. These probes are most useful for investigations of very thin samples where a minimum ring-down is required.

A send–receive probe combination may be useful where a narrow-band pulse is required but either the ring-down interferes with the test or the attenuation is too severe to permit a pulse–echo probe arrangement. For this circumstance, a pulse is excited with a narrow-band transducer and received with a broad-band unit having a center frequency somewhat above that of the narrow bandwidth unit. This prevents performance problems that could occur due to slight mismatches in the peak frequencies for narrow-band probes. This combination is often used for travel-time measurements in nondispersive material. Also, two broad-band probes used where one is the sender and the other is the receiver will be more likely to operate together satisfactorily than will two narrow-band probes used in the same fashion. This is due to the flatter region near to the peak frequency for the broad-band units. The justifications for these comments will be more obvious after covering the material in later sections.

A variety of transduction devices other than piezoelectric ceramics offer current and future advantages in ultrasonic NDE. Notable amongst these evices are the piezoelectric film (PVDF), magnetostrictive, laser, and electromagnetic–acoustic (EMAT) devices. Piezoelectric films are discussed by Silk [27] and Hall [37], amongst others. Silk [27] and Hueter and Bolt [26] describe magnetostrictive techniques. EMAT and laser methods have been discussed in some detail in previous sections.

CHAPTER

8

INSPECTION PRINCIPLES AND TECHNIQUES

8-1 INTRODUCTION

Ultrasonic inspection is typically performed using one of the following basic procedures, namely: (1) normal beam pulse–echo; (2) normal beam through-transmission; (3) angle-beam pulse–echo; (4) angle-beam through-transmission. The material presented in this section will describe the fundamentals of these methods. Some more specialized applications of the procedures will also be discussed in order to demonstrate more clearly the unique capabilities of each.

8-2 NORMAL INCIDENCE PULSE–ECHO INSPECTION

As an ultrasonic energy packet (i.e., pulse) travels through a test sample and strikes a crack or some other anomaly, it will be reflected much as light, sonar, or radar pulses are reflected when they strike an object. As previously shown in Fig. 5-10, the incident beam strikes the interface, resulting in a reflected beam returning in the opposite direction in the same material and a refracted (transmitted) beam passing through the interface into the other material. Ultrasonic energy that is reflected and returned to the probe is the source of the indications shown on the instrument screen in Fig. 8-1. All sound energy that travels completely through the test piece will be reflected at the end, giving the large back-echo indication located at 5 on the horizontal scale. Once excited in the material, the ultrasonic pulse will continue to reflect from the parallel surfaces

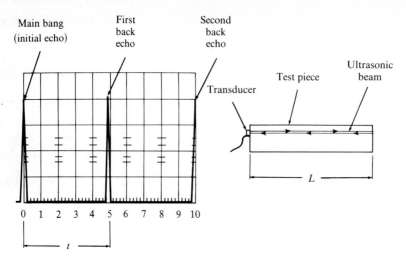

FIGURE 8-1
Pulse–echo inspection in bar of length L.

shown, creating a multiple echo display when the time base is sufficiently extended. The second multiple echo appears at 10 in this figure.

It is convenient at this point to describe in more detail the operation of a typical ultrasonic inspection instrument. While the basic circuitry is as previously described in Chap. 7, the arrangement of the components for a laboratory setup is such that the function of each may be more clearly understood. Figure 8-2 shows a typical arrangement of laboratory equipment used in ultrasonic inspection. In laboratory instruments, the pulser and receiver are normally contained in the same unit, which is connected to an oscilloscope. Commercial flaw detectors include all of the components shown in Fig. 8-2 fitted within one compact unit. Modifications to provide specialized inspection capabilities can be made in both the laboratory and commercial instruments.

The pulser is the starter for the system, firing the high-voltage spike that sets the probe into oscillation and the trigger pulse for starting the oscilloscope display. The main bang, seen at the left of the screen in Figs. 8-2 and 8-1, results from the high-voltage spike striking the probe. The main bang is sometimes labeled as the initial echo. In sequence, once the spike is released, the pulser becomes an open nonconducting electrical circuit, while the receiving circuit is waiting for the return signal to strike the probe. Where no defects are present, the back echo would result from the initial pulse traveling through the test piece, reflecting off of the end, and returning to the probe. Oscillations of the probe are governed by the tuned LC circuit previously discussed, while the firing rate of the pulser, i.e., the repetition rate, is adjustable to the circumstances of the test. Since the test material would normally have a constant velocity, the time base of the oscilloscope is directly correlated to the distance traveled. The first back echo will

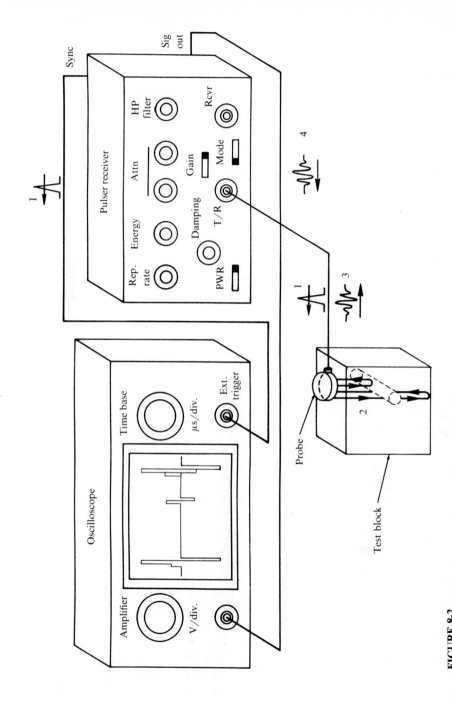

FIGURE 8-2

Typical laboratory setup for ultrasonic pulse–echo inspection.

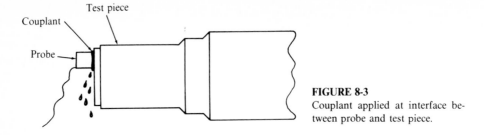

FIGURE 8-3
Couplant applied at interface between probe and test piece.

therefore occur at a time $t = 2L/C$, where L is the length and C is the wave speed in the material.

Energy is lost from the ultrasonic pulse each time that it strikes an interface. For the example just cited, energy loss would occur at both the probe–test piece interface as well as at the back end of the test piece. This behavior is a function of the impedances of the materials, as described previously in Chapter 5.

A coupling medium as shown in Fig. 8-3 is normally required in order to introduce the ultrasonic energy from the probe into the test material. For contact inspection, the couplant may be one of several materials, e.g., motor oil, glycerin, honey, water, or special gel couplants designed for particular applications. Usually, the couplant layer is very thin and does not significantly contribute to any energy losses.

8-3 NORMAL INCIDENCE THROUGH-TRANSMISSION INSPECTION

Many incidents arise where the pulse–echo probe arrangement may not provide the required test information. This may occur where a flaw or other anomaly does not provide a suitable reflection surface or where the orientation or location of the flaw affects difficult access. Additionally, high attenuation material such as polymers are often inspected with the through-transmission method. Where the pulse–echo method uses one probe as both the sender and the receiver, the through-transmission technique uses two probes, one as the sender and another as the receiver. Another name for the through-transmission technique, then, is the pitch–catch method. This arrangement can be visualized in Fig. 8-1 if a separate receiving probe is placed on the opposite end of the sample, in the line of travel of the pulse. With the ultrasonic instrumentation reset for through-transmission inspection, the sending and receiving probes are electronically separated. It is easy to see, then, that the first pulse to appear on the screen after the initial pulse would be at a time $t/2$ as compared to time t for the pulse–echo probe arrangement. Still, only two echos would appear on the screen, both displaced to the left by a time $t/2$.

8-4 ANGLE-BEAM PULSE–ECHO INSPECTION

Angle-beam transducers provide access to areas that are inaccessible to normal beam probes. A general arrangement for an angle-beam probe and a cylindrical hole reflector is shown in Fig. 8-4. Angle-beam inspection is accomplished mostly with SV (shear vertical) waves, i.e., those having particle motions perpendicular to the ray and parallel to the plane of the paper. The refraction angle θ_2' is the key parameter specified in angle-beam inspection. It is a function of the material properties and the incident angle, as given by Snell's Law Eq. (5-41). While Fig. 8-4 shows a shear wave as the only one present, a full description of the behavior at the interface shows that a longitudinal wave will be present in the test material at lower incident angles; reflected shear and longitudinal waves will always occur in the wedge material.

A special consideration in the use of angle-beam probes is the fact that the energy transmitted across the interface is a function of the angle of incidence as well as the material. For pulse–echo angle-beam inspection, the greatest interest is in the total energy transmission ratio of the returned pulse relative to the source pulse. Where the previous material in Chap. 5 considered only longitudinal waves, the various transmission ratios of interest in a shear wave inspection have been given in Ref. 10. The transmission ratio for a refracted transverse wave arising from an incident longitudinal wave is given by Eq. (5-44). For the return wave, i.e., a refracted longitudinal wave in the probe, caused by the incident shear wave in the test material, the transmission ratio is given by

$$D_{lt} = \frac{4\rho'(C_2')^2 C_1' \cos 2\theta_2' \cos \theta_2''}{\rho(C_2)^4 (\sin 2\theta_1) N} \tag{8-1}$$

where N is given by Eq. (5-43) and the parameters are as listed in Chap. 5. The item of principle interest here is the echo transmittance E, which is the product of the transmission ratio of the source pulse times that of the received pulse. Thus

$$E = D_{tl} D_{lt} \tag{8-2}$$

Figure 8-5 shows a plot of the echo transmittance for a Plexiglas wedge probe on

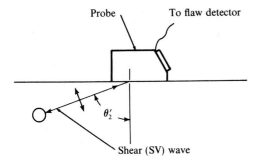

Probe   To flaw detector

θ_2'

Shear (SV) wave

FIGURE 8-4
Angle-beam inspection with shear waves.

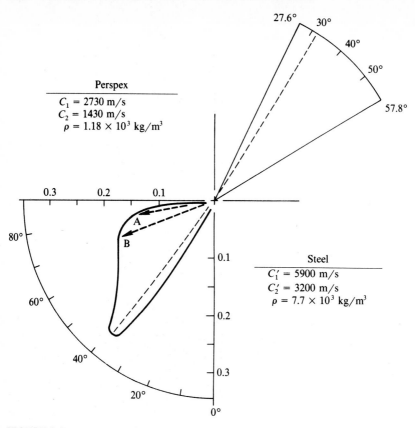

FIGURE 8-5
Echo transmittance at the interface of Perspex and steel. (*From Krautkramer and Krautkramer [10]. Courtesy Springer Verlag.*)

steel. Refraction angles are again governed by Snell's Law [Eq. (5-41)]. The maximum echo transmittance occurs at a refracted beam angle of 39°, excited at 32.4° in the wedge, as shown by the light dotted line. At the higher angles, shown by lines A and B, the magnitude of the echo transmittance falls off quite sharply.

Probe temperature may affect the refracted angle of a shear wave probe due to the temperature dependence of the longitudinal wave speed in the Plexiglas, as described in Chap. 5, and the refraction characteristics as governed by Snell's Law. For example, assume that a nominal 70° probe has been selected for the inspection of a weld in steel plate. This is the nominal angle in the steel θ_2'. The specification of θ_2' in the steel, however, fixes the angle in the probe θ_1. Snell's Law shows that if θ_1 is fixed, a change in either C_1 in the probe or C_2' in the steel rail can cause a change in the refracted angle in the steel θ_2'.

The earlier discussion showed that there is a significant change of wave speed in the Plexiglas with temperature change. For example, for three typical

TABLE 8-1
Effect of temperature variations on the entry angle in steel for nominal 70° probe.

Probe temp. (°F)	C_1 in probe		θ_2' (deg)
	in / s	m / s	
32	108,110	2746	67.4
68	106,220	2698	70
120	103,780	2636	74.1

temperatures, the wave speeds in Plexiglas and the resulting entry angle in the steel are as given in Table 8-1. For the temperature range given, the wave speed change in the Plexiglas in the probe will result in a refracted angle variation of 3.6–4.1°.

8-5 ANGLE-BEAM THROUGH-TRANSMISSION INSPECTION

Just as was the case for the normal beam situation, there are also conditions demanding the use of separate sending and receiving probes in angle-beam inspection. As will be seen more clearly in later discussion on diffraction techniques, a good knowledge of the geometry of the test piece is required to perform this type of inspection.

8-6 CRITERIA FOR PROBE SELECTION

Before embarking on more complete descriptions of ultrasonic inspection procedures, it is useful to discuss several general characteristics of transducers that affect their selection.

8-6.1 Flaw Size Sensitivity

The frequency of the probe is one of the most important factors to be considered when the minimum detectable flaw size is of concern. While there are a number of other parameters that affect the flaw sensitivity, the detectability of a flaw is a direct function of the wavelength, which varies inversely with the frequency. Most favorable detection conditions exist when the flaw is somewhat larger than the wavelength. As the wavelength becomes larger than the flaw, the likelihood of detection decreases considerably.

8-6.2 Beam Divergence

Beam divergence was described in Chap. 7 as being a function of both the probe specifications and the material being inspected. Figure 8-6 illustrates how beam

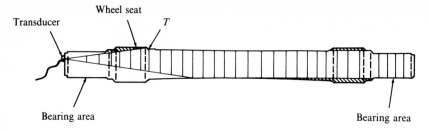

FIGURE 8-6
Beam divergence in a typical railroad freight car axle.

divergence may affect an inspection. While the axle shown is similar to one used in railroad cars and locomotives, the geometric characteristics of changing cross sections, fillets, etc., are often seen in other rotating machinery applications, such as turbine rotors.

For these shafts, or rotors, if an inspection is to be made using a normal beam longitudinal wave probe from either end of the long axle, the geometry is such that there will be areas that are not reached by the ultrasonic beam. These are identified as shadow zones, indicted by the cross-hatched areas in the figure. Moreover, significant spurious echos may occur because of reflections and mode conversions occurring within the beam divergence angle. Typical effects caused by these side reflections will be discussed in material to follow.

Since divergence is a function of the probe diameter, frequency, and material wave speed, the selection of the proper probe is a significant factor in assuring a satisfactory inspection.

8-6.3   Penetration and Resolution

Attenuation plays an important role in ultrasonic NDE in a variety of ways. Not only does it account for the loss of signal height for equal reflectors at increasing distances from the probe, it is also a useful diagnostic tool for several types of inspection. Of the sources of attenuation discussed in Chap. 5, the ones most affecting ultrasonic NDE are beam spreading and scattering. In general, the use of lower frequency probes will minimize the attenuation and maximize the penetration due to the fact that the longer wavelength pulse is less affected by scattering off of grain boundaries, etc. Lower frequency probes, then, are said to have a greater penetration ability. As is often the case, however, what is an advantage on one hand is a disadvantage on the other.

Typical examples of the effect of probe frequency on penetration are shown in Fig. 8-7(a) and (b). Using a cast steel block 40 mm thick and with visible porosity as the test piece, Fig. 8-7(a) shows that with a 1 MHz probe a back echo is obtained even though it is accompanied by a rather large amount of "grass."

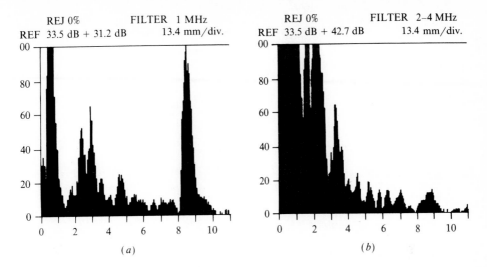

| REJ 0% | FILTER 1 MHz | | REJ 0% | FILTER 2–4 MHz |
| REF 33.5 dB + 31.2 dB | 13.4 mm/div. | | REF 33.5 dB + 42.7 dB | 13.4 mm/div. |

(a) (b)

FIGURE 8-7
(a) Reflection from porosity and large grain structure with 1.0 MHz probe. Back echo is at 8.
(b) Internal scattering with 2.25 MHz probe.

Locating a small defect echo in the reflection field would be difficult. With a 2.25 MHz probe, as shown in Fig. 8-7(b), there is no back echo, no matter how much the amplification is increased. In this case, the scattering off of the grain boundaries and porosity completely obliterate the back echo.

What is an advantage on the one hand is often a disadvantage on the other. Lower frequency probes have a decreased ability to resolve closely spaced reflectors. The ability to resolve closely spaced reflectors is termed resolution and the example illustrated in Figs. 8-8 and 8-9(a) and (b) shows the importance of resolution in ultrasonic inspection. Where the distinction between the flaw and the shoulder is clear in Fig. 8-9(a), an operator might easily assume the response in Fig. 8-9(b) to be only off of the visible shoulder and thereby completely miss the nearby crack.

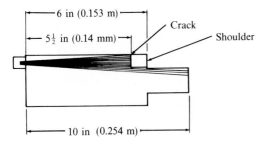

FIGURE 8-8
Closely spaced reflectors.

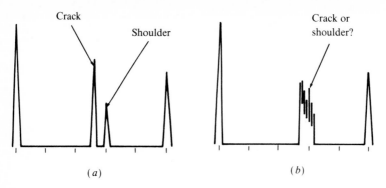

FIGURE 8-9

(*a*) Good resolution of closely spaced reflectors. (*b*) Poor resolution of closely spaced reflectors.

8-7 OPERATION OF ULTRASONIC INSPECTION INSTRUMENTS

The control functions and the operating procedures for each of the commercially available ultrasonic flaw detector are to some extent identical. Differences that occur between manufacturers will generally rest in functional capabilities resulting from individual variations in the internal electronic circuitry. The several manufacturers will emphasize different aspects of the instrument, banishing some controls to rear or hidden panels while maintaining prominence of other controls on the front panel. In some situations, the function of the knob or control will be indicated with a word or a number or an identifiable symbol.

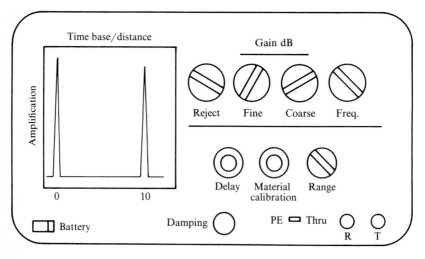

FIGURE 8-10

Generic ultrasonic flaw detector.

TABLE 8-2
Functional control description for ultrasonic instruments.

Display mode

Ultrasonic response information generally is displayed in either a full wave or a rectified format. For the full wave, or RF, display, the zero voltage line is at the center of the screen and wave excursions extend above and below the zero line. This type display is normally used where the signal shape needs to be observed. The rectified display is where the negative voltage amplitudes are folded above the zero amplitude line and the zero amplitude line is then moved to the bottom of the screen. Although some signal information is lost, the amplitude scale is increased to full screen height. Most commercial flaw detectors give a rectified display, although some allow a selection of either the RF or rectified shape.

Test method selection

A selection of either the pulse–echo or through-transmission inspection method must occur at the start of the test and the probe cables must be connected to the corresponding terminals.

Battery charge indicator

Portable units contain battery packs that can provide very long-lived, reliable service if they are properly maintained. An indicator light or meter is normally used to indicate current battery status.

Frequency

Many instruments may be switched for either broad-band or fixed frequency operation. In fixed frequency, or narrow band, operation, a selector switch on the instrument is set at the nominal frequency for the probe to be used. While this provides optimal performance when frequencies are correctly matched, it also can cause poor performance for mismatched conditions. In broad-band operation, the instrument will operate with much the same response for all probes within the stated frequency range.

Pulse length or damping

This control varies the value of the resistor r in series with the oscillating LC circuit shown in Fig. 7-4. High damping will shorten the pulse length without significantly affecting the amplitude.

Pulse tuning

The pulse tuning control varies the value of the internal inductance L in Fig. 7-4 in order to obtain peak amplitude. This aids in matching the transducer to the internal circuit.

Horizontal adjustment—delay

Since the horizontal scale represents time, the main bang must be correctly aligned in order to provide correct travel time or distance information. For normal beam, contact probe inspection, the main bang should be adjusted to the left as shown in Fig. 8-1. Other situations may demand that the main bang be moved considerably to the left of the screen so that different areas of the test piece may be viewed with more detail. This adjustment is used in conjunction with the material calibration, or range, control.

Material calibration—range

The sweep rate of the display is varied with this control. Typically, both coarse and fine adjustments are allowed in order to increase the accuracy of the travel time or distance information obtained. On most instruments, the units for these controls are in length, i.e., millimeters or inches. Used in conjunction with the delay control, very good accuracy may be obtained for distance measurement with most ultrasonic instruments.

Gain—amplification

The signal returned from the probe is amplified before it is displayed on the screen. Coarse and fine adjustments are usually available. Typically, these units of measure are in decibels. As this control is adjusted, all signals to be displayed will be affected.

Reject

Some materials generate a large amount of "grass" or spurious echoes along the base line of the screen. These may be removed with the reject knob, which essentially chops off the bottom portion of the display. This control must be used with caution, howaver, since important information often occurs at small amplitudes and may be lost in the grass if care is not used.

Distance amplitude correction (DAC)

With the natural decrease in pulse energy that occurs due to ordinary attenuation, it is clear that returned signal amplitudes may not always be a reliable indicator of the size of the reflector. Compensation for this decrease in signal amplitude for equal reflectors at greater distances from the probe is obtained through the use of a circuit that increases the amplification for signals arriving at later times.

Automatic signal recognition—gates

Occasions arise where information is needed only within a certain time, or distance, window within the piece being inspected. A gated area, as shown in Fig. 8-11, is set to respond only to signals that occur within the chosen window. The start, stop, and sensitivity within the gate may be adjusted. Recognition of a signal within the region may be obtained with an audible or visual alarm or through logic information for data storage systems.

A typical portable ultrasonic flaw detector is shown Fig. 8-10 and the functions of the various controls are discussed in Table 8-2. This generic instrument is meant to acquaint the reader with the functions of the controls; full familiarity with a particular instrument can be acquired only with diligent studying of the manual furnished with the equipment.

Calibration is required in a test procedure in order to establish that the instrument is working properly and that the control functions are correctly set.

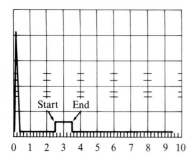

0 1 2 3 4 5 6 7 8 9 10

FIGURE 8-11
Gated area in typical ultrasonic display.

Among other things, this requires that the time base and amplification controls are set so that the test can be properly performed. This may be accomplished in one of several ways. A standard of the same material and with a known length may be used. Also, one similar in size and shape to the item to be inspected may be required. Each time that the unit is restarted or when the test conditions are changed, it is usually a good practice to recalibrate the instrument.

8-8 TECHNIQUES FOR NORMAL BEAM ULTRASONIC INSPECTION

At a risk of being trivial, it must be emphasized that the basic role of an ultrasonic instrument is to obtain and display information. Interpretation of the data must come from other sources. Many times, the ultrasonic operator or the responsible engineer will supply the interpretation. Automatic interpretation offers great potential, provided the correct interpretive data have been furnished to the instrument. In order to demonstrate the basics of ultrasonic inspection, a few examples of typical defects will now be presented and some appropriate normal beam inspection schemes will be discussed.

8-8.1 Fatigue Cracks

In the simplest sense, fatigue cracks as described in Chap. 3 are assumed to be planar in shape, with boundaries rather well defined, providing sharp, distinct display echoes. For example, Fig. 8-12 shows the expected screen appearance

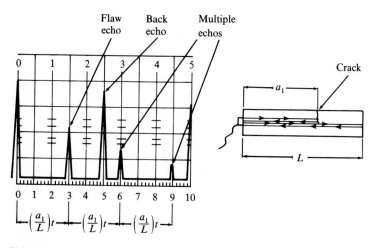

FIGURE 8-12
Typical pulse–echo response from a planar reflector.

from such a defect. With length L and total time t for a pulse reflecting off of the bar end, the defect echo would appear at time $(a_1/L)t$ on the screen. Multiple (secondary) echoes of the flaw echo would also appear. If the probe is moved down in this example, the defect echo should disappear, while the back echo remains. For a circular shaft with a planar flaw shaped as shown in Fig. 3-6, moving the probe in a circular fashion around the outer extremity of the end of the bar should generate a pattern of rising and falling flaw echo that will indicate the approximate size and shape of the flaw.

Where access to a suitable inspection location is available, it is useful to confirm the presence and shape of a flaw by approaching it from a different direction. Figure 8-13(a) and (b) shows a probe positioned at opposite ends of the same cracked bar. Figure 8-13(c) and (d) illustrates the expected screen appearance from each location. The change in travel length has caused the same size flaw to yield different amplitude echoes at the different distance from the probe. The distance amplitude correction (DAC) feature previously discussed may be adjusted to yield echo amplitudes at different distances from the probe end that are proportional to the size of the flaw. While this offers potential advantages in flaw size interpretation, it too must be used with some caution.

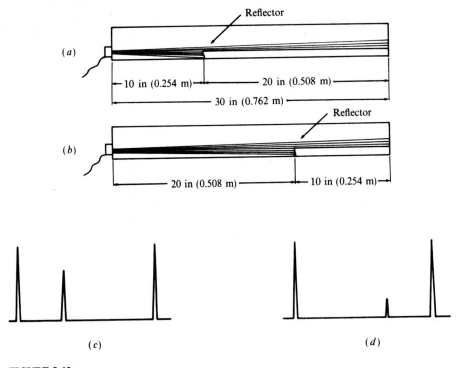

FIGURE 8-13
(a, b) Defect confirmation by inspecting from opposite ends of the same bar. (c) Screen appearance for (a). (d) Screen appearance for (b).

8-8.2   Inclusions, Slag, Porosity, and Large Grain Structure

Many manufactured products contain internal defects that give responses quite differently than that seen for the fatigue flaw. Such defects are inclusions, slag, porosity, and large grain structure, to name a few. Figure 8-14(a) and (b) indicates the screen appearance and the general reflection behavior for inclusions, slag, and porosity. In these situations, where there is no flat plane reflector, the scatter may be sufficient to destroy the back-echo signal. In this event, the loss of the back echo may be adequate justification for the removal of the part from service. Often, attenuation measurements along with ultrasonic spectroscopy may improve the likelihood of correct identification of these classes of defects. This will be discussed in more detail in a section to follow.

8-8.3   Press Fits

Press fitted items such as gears, turbine discs, roller bearings, wheels, pulleys, etc., can generate ultrasonic responses due to the acoustic energy traversing the interface. Often this can generate a false indication in an area particularly susceptible to fatigue cracks. Some investigative effort may aid in determining that the source of reflection may be from a press fit. For example, for the gear pressed onto the axle, as shown in Fig. 8-15, if the probe is moved in a circular fashion along the outer extremity of the shaft end, an indication arising from a press fit should give a reasonably steady appearance, while one originating from a

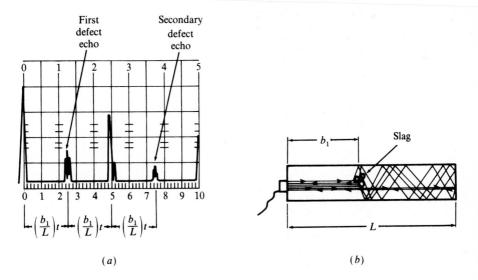

(a)

(b)

FIGURE 8-14
(a) Screen appearance from slag, inclusions or porosity. (b) Reflection pattern for slag, inclusions or porosity.

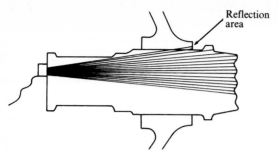

FIGURE 8-15
Ultrasonic energy traversing a press fit.

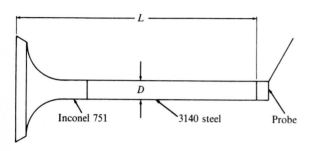

FIGURE 8-16
Inspection arrangement for ultrasonic inspection of diesel engine valve. (Inconel is a registered trademark of the INCO Family of Companies.)

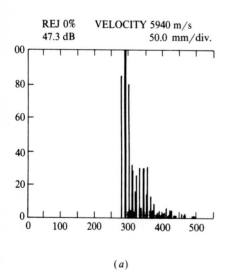

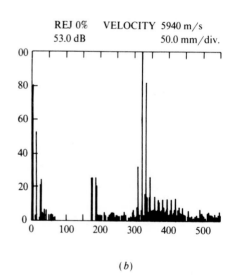

(a)

(b)

FIGURE 8-17
(a) Ultrasonic display from a good engine valve. 5.0 MHz, $D = 14.2$ mm, $L = 278$ mm.
(b) Ultrasonic display from a defective engine valve. 5.0 MHz, $D = 15.5$ mm, $L = 292$ mm.

fatigue crack should give a regular pattern, rising and falling as previously described for Fig. 8-12. This ability of a press fit to transmit ultrasonic energy suggests that ultrasonics may be useful in evaluating the quality of the fit [10].

8-8.4 Bonding between Dissimilar Materials

Mechanical bonds between dissimilar materials may be inspected ultrasonically. One example, as reported by Vezina [43], is in the inspection of diesel engine exhaust valves. Typically, the head portion of the valve may be made of Inconel 751, a heat resistant material, while the stem portion may be made of a more wear-resistant material, such as 3140 steel. The two materials are joined by an inertial weld.

Ultrasonic indications from the defective bond area have proved to be effective in eliminating this type of failure in diesel engines. The method of inspecting the valve is shown in Fig. 8-16. Figure 8-17(a) shows the back reflection obtained from a good valve and Fig. 8-17(b) shows the ultrasonic reflections obtained between the initial pulse and the back echo. This is from a defective bond. It should be noted that the valves used in this example differ slightly in length and diameter.

8-8.5 Spurious Reflections at Fillets on Machined Shafts

Reflections at oblique angles in machined parts can result in clear yet false ultrasonic indications in areas suspected of containing cracks. An example of such a situation is shown in Fig. 8-18(a) and (b). With the probe located in the center of the end of a 407 mm long shaft machined as shown in Fig. 8-18(a), the A-scan display in Fig. 8-18(b) was obtained.

The spurious defect indication at 180 mm is a result of the reflection pattern shown in Fig. 8-18(a). Following the work of Hall [37], who has studied this situation with photoelastic models, the conditions are such that a ray may be traced within the beam spread from the probe center to point T on the fillet, diagonally across to the opposite surface and then back to the probe. The emitted beam is a longitudinal wave, but the reflected diagonal beam has been converted to a transverse wave. At the opposite fillet, conversion back to a longitudinal wave occurs. The result of this scenario is the appearance on the screen of a strong reflector at a distance along the axle from point T equal to the diagonal distance there.

8-8.6 Effects of Material Structure on Ultrasonic Inspection

Material structure may affect both the attenuation and the speed of ultrasonic waves. Generally, ultrasonic waves suffer greater attenuation in cast metals than in worked metals. In worked metals, the grain boundaries are broken down and

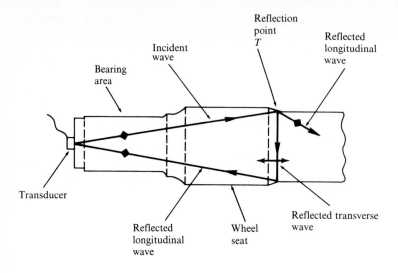

(a)

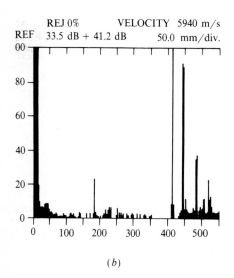

(b)

FIGURE 8-18
(a) Ultrasonic ray path causing spurious echo in machined axle. (b) A-scan indication of spurious echo from fillet as shown in (a). Shaft length is 409 mm, probe to point $T = 130$ mm, wheel seat diameter is 52 mm.

less scattering results. Anisotropy in the form of residual stress and material texture may exist in metals due to working at low temperatures and uneven temperature during the heating and cooling processes. Both of these effects may alter the speed and direction of travel of ultrasonic waves with the effect most severe for shear waves. Materials such as polymers and composite structures may be quite attenuative and severe anisotropy also is frequently encountered.

Materials such as cast irons, cast steels, polymers, etc., may have quite large attenuation rates and an item that actually is suitable for service may give reflected pulses of such magnitude and scatter that judgement of the quality of the part is impossible. Often times, through-transmission methods can be used to satisfactorily inspect this type of item, since there would be much less attenuation for one trip through the material, rather than the round trip required for pulse–echo inspection.

8-8.7 Thickness Measurements—Corrosion Detection

Ultrasonic travel-time measurements are conveniently used for thickness determination in piping, tubings, and pressure vessels. Thickness measurement, of course, is crucial in the prevention of failures caused by corrosion. Since the longitudinal wave speed is essentially constant for most engineering materials, changes in material thickness may be determined quite accurately using the back echo of a conventional normal beam ultrasonic system.

Thickness measurement is based on a travel-time comparison, i.e., travel times first are established for known thicknesses of similar material and the comparable travel times are then obtained for the item being inspected. Step blocks, furnished with machined, incremental thicknesses for the range of interest, are most often used in the calibration of the ultrasonic instrument for thickness measurement. Adjustment of both the delay and the material calibration controls is required in the initial setting-up process.

A highly damped, normal beam transducer is first chosen for thickness measurement, minimizing the ring-down effect. The choice of a higher frequency probe is sometimes beneficial since this, too, will minimize the decay time of the transducer. Using the pulse–echo mode, the initial calibration procedures are as follows:

1. Place the probe on the thickest section of the calibration block and adjust the screen to obtain a display similar to that shown in Fig. 8-1. It is convenient to set the first back echo at a point on the horizontal scale equal to some multiple of the thickness being measured, e.g., for a 12 mm thickness, set the first back echo at 6 on the horizontal scale. For a $\frac{1}{2}$ in thickness, set the first back echo at 5.

2. Alternately use the delay adjustment and the material calibration controls to align the first back echo at 6, or 5, for this example. Do not attempt to maintain the initial pulse at 0.

3. Move the probe to the next thinnest section and repeat step 1.

4. Repeat these procedures on all sections until the desired accuracy is achieved.

5. A suitable back echo is now obtained from the item being inspected and the thickness of the item may be read directly from the horizontal scale on the screen.

Ultrasonic thickness measurement is also accomplished with digital instruments that automatically determine the travel time to the first back echo. Proper calibration is also required for these instruments. In all cases, the manufacturers' instruction manuals should be followed in order to obtain reliable results.

8-8.8 Attenuation Measurement

Several NDE inspections call for the measurement of attenuation properties in materials. This may be because of suspected bad grain structure in a material or because of irregular cracking, such as occurs with hydrogen attack on metals.

Attenuation may be determined by obtaining multiple echoes within a material and recording the echo height of several echoes past the first back echo. The echo amplitude is a representation of pulse pressure, as shown in Fig. 8-19(a) and (b) for a rectified display. In this case, the signal height may be measured to determine the signal amplitude. For the RF display, the amplitude of the positive and negative amplitudes should be averaged to obtain the signal

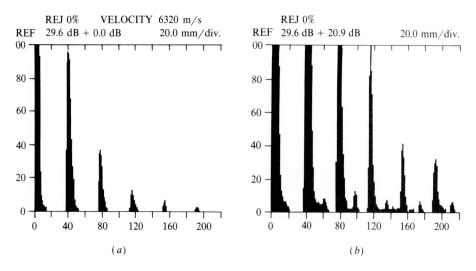

(a)   (b)

FIGURE 8-19
(a) First back echo at full screen height for attenuation measurement, aluminum, 38 mm thick, 2.25 MHz. (b) Third back echo at full screen height for attenuation measurement, aluminum, 38 mm thick, 2.25 MHz.

strength. Application of Eq. (5-49) will yield the attenuation coefficient α for the material. Due to common experimental errors, the data should be sufficiently repeated to yield confidence in the results.

Most commercial instruments are fitted with amplitude controls that are directly calibrated in decibels. For these, attenuation of one pulse arrival relative to another may be obtained by first adjusting the amplitude to obtain full screen height for the first pulse and noting the setting of the amplitude control. The increase in amplification required to obtain full screen height for a succeeding arrival will yield a direct indication of the total attenuation between the two pulses. Dividing this by the appropriate travel length will then yield the attenuation coefficient.

Attenuation data in Fig. 8-19(a) and (b) may be obtained by noting that in Fig. 8-19(a) the first back echo is at full screen height and the amplification is seen to be at 29.6 dB. In Fig. 8-19(b), the amplification has been raised by 20.9 dB to give full screen height for the third echo. With this and the thickness of the specimen, attenuation can be directly calculated.

It must be strongly emphasized that a number of variables may affect attenuation and attenuation measurement. Realizing that attenuation is a function of frequency, care must be exercised to insure that probe characteristics are exactly as expected. Typical sources of error include probes with actual frequencies away from the specification, differences in damping for probes of the same frequency, and probe misalignment in the initial construction. Moreover, there may be differences between materials that are assumed to be similar. Nonetheless, reliable attenuation data is an important contribution in ultrasonic NDE.

8-8.9 Intergranular Cracks—Hydrogen Attack, Corrosion

Attenuation measurements often are used to detect defects that have irregular growth patterns. An example of this is the intergranular defect associated with hydrogen attack and stress corrosion cracking, as shown in Fig. 8-20. The usual effect on the ultrasonic pulse of this type of defect is a scattering of the energy

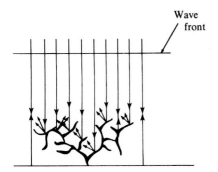

Wave
front

FIGURE 8-20
Scattering at intergranular type defects.

with a resulting loss of amplitude and little, if any, reflected energy [41, 44]. These irregular cracks are usually detected with a loss of amplitude test using a normal beam pulse–echo probe. Calibration for the test starts with a multiple echo display obtained using a defect-free sample of the item to be inspected. Time base and amplitude adjustments are fixed so that the fourth back echo is full screen height in the calibration piece. A significant defect is felt to be present in the item being inspected when the fourth back echo falls to less than 20% of the full screen.

8-9 TECHNIQUES FOR ANGLE-BEAM ULTRASONIC INSPECTION

As previously discussed, angle-beam inspection is typically accomplished using vertically polarized shear waves inclined at some specified angle θ_2', as shown in Fig. 8-4. This is the inspection angle and is the angle specified on the probes.

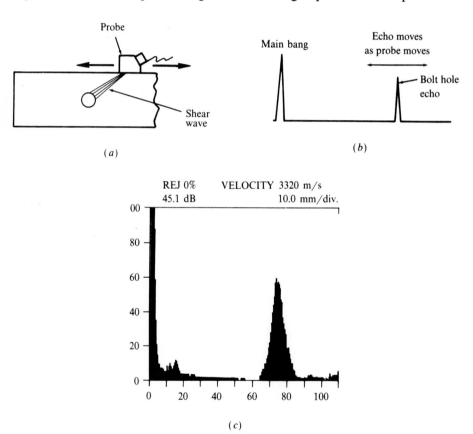

FIGURE 8-21
(a) Angle-beam inspection of bolt holes. (b) A-scan display for bolt hole inspection. (c) Echo envelope for bolt hole inspection.

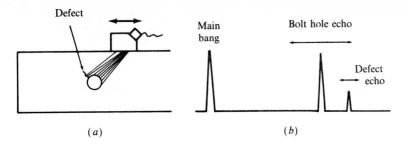

FIGURE 8-22
(*a*, *b*) Locating a bolt hole crack using angle-beam transducers.

Inspection procedures with angle-beam probes can be introduced using the typical example of bolt hole inspection as shown in Fig. 8-21(*a*), (*b*), and (*c*). There is a frequent need to inspect bolted and riveted connections for fatigue at the joint. Obviously, there is an advantage in an ultrasonic inspection that would not require the removal of the cover plates.

The inspection shown in Fig. 8-21(*a*) assumes a nominal 45° probe. Inspection is begun by moving the probe in the plane of the paper and observing that the echo pulse from the hole will rise and fall as the inspection beam is moved past the hole, as shown in Fig. 8-21(*b*). The display shown in Fig. 8-21(*b*) is significant in that there is no back echo. In the normal beam inspection, the presence of the back echo gives the inspector confidence that ultrasonic energy is being introduced into the material. This does not occur for most angle-beam inspection configurations and proper caution must be exercised to guarantee the quality of the test.

As the probe is moved, the peak of the echo will trace the "echo envelope" of the defect. The maximum indication will occur when the center line of the refracted beam is located along the diagonal of the hole. The shape of the echo envelope may be used as an indicator of flaw size and shape.

The introduction of a typical 45° fatigue crack at the bolt hole creates the display shown in Fig. 8-22(*a*) and (*b*), where the defect echo is seen trailing behind the echo from the bolt hole. In this example, the presence of the bolt hole echo indicates that ultrasonic energy is reaching the area of interest and shows that a proper test has been accomplished.

8-9.1 Inspection for Planar Cracks using Angle-Beam Probes

Ultrasonic angle-beam inspection is frequently used for locating planar cracks such as those occurring in welds and oil country tubulars, for example. In the previous case of the bolt hole, the ultrasonic beam would always be impinging on the hole at a diagonal, providing the optimum perpendicular reflector surface for

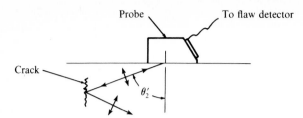

FIGURE 8-23
Reflection of angle-beam incident wave from a vertical planar flaw.

most orientations of the probe. For the flat planar crack, however, this is not the case. Typical inspection conditions might be as shown in Fig. 8-23, where Snell's Law would predict that all of the energy would be either reflected down, further into the material, or transmitted through the defect. The fact that any energy at all is reflected back to the probe is dependent on the irregular shape of the defect as it is drawn.

Photographs of reflections at an inclined flaw in a solid have been obtained by Hall using photoelastic visualizations [37]. Shown in Fig. 8-24(*a*) is a 50°

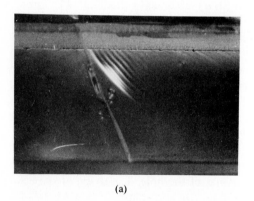

(a)

(b)

FIGURE 8-24
(*a*) Photoelastic visualization of shear wave incident on an inclined, planar reflector. (*From Hall* [*37*]. *Reprinted with permission from Materials Evaluation, vol. 42, no. 7, 1984, The American Society for Nondestructive Testing, Columbus, OH.*) (*b*) Photoelastic visualization of reflected wave at an inclined flaw. (*From Hall* [*37*]. *Reprinted with permission from Materials Evaluation, vol. 42, no. 7, 1984, The American Society for Nondestructive Testing, Columbus, OH.*)

shear wave approaching the flaw, which is inclined at 20°. Using Snell's Law and recalling the differential wave speeds for the longitudinal and shear waves, one can easily identify the reflected longitudinal wave and the reflected shear wave at 40° and 20° to the defect, respectively. They are moving down and to the right, away from the interface. The wave traveling down and to the left has passed to the side and around the defect. Obviously, a more favorable orientation of the inspection beam relative to the flaw would greatly enhance the effectiveness of the test. If the incident shear wave struck the flaw surface at normal incidence then the maximum shear energy would be reflected back to the probe. For the conditions shown here, however, there would be little likelihood that the flaw would be detected.

The photographs by Hall were obtained using a glass material having wave speed values similar to steel, thus yielding photographs representative of wave behavior that might be encountered in actual conditions. The photographs dramatically demonstrate the importance of proper orientation of the inclined beam relative to the expected flaw inclination. The problem becomes quite severe in the case of perpendicular flaws, such as those typically encountered in welds, for example. Since it is impossible to orient the inspection beam at 90° through the upper surface, it might be assumed that there is an advantage to obtaining as high an inspection angle as possible. While the higher angles appear to offer an advantage from the standpoint of reflection, problems may occur due to the increased energy losses at the interface at these high angles. This has been discussed earlier in the material relative to Fig. 8-5.

Other problems at the higher angles may be encountered when inspecting anisotropic material where the wave speeds may vary from the expected values. The effect of material anisotropy on refraction in angle-beam inspection has been reported by Silk [45] and Kupperman and Reimann [46] for austenitic welds and by Bray for railroad rail [47]. In the case of the welds, the anisotropy was induced by the welding process in the stainless steel, while cold-working in service caused the anisotropic conditions on the rail head. Generally, a "no-test" condition, which may result from the upward movement of the refracted shear wave, will be indicated by an increase in the internal reflections within the incident angle-beam probe. These internal reflections will be visible on the screen as a significant increase in the length of the initial pulse.

8-9.2 Weld Inspection

Weld inspection is quite frequently accomplished using ultrasonic angle-beam probes. Typical defects include the vertical planar-type of flaw as well as the slag, inclusions, and porosity that were discussed previously. The type of defect expected dictates the choice of the inspection angle of the probe. The obvious choice for flat planar defects would be a higher angle probe (i.e., 70°), while porosity, etc., might well be detected with a lower angle probe, e.g., 45°.

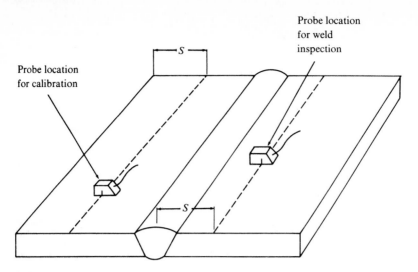

FIGURE 8-25
Typical arrangement for angle-beam weld inspection.

A typical weld inspection arrangement is shown in Fig. 8-25. Defects are assumed to be possible within the weld metal as well as within the heat affected zone. Since there is no back echo for this inspection, the correct proximity of the probe to the inspection area must be established. A method for doing this is to first obtain a corner reflection as shown in Fig. 8-26(a). In this case, the probe is manipulated along a line parallel to the edge at a distance estimated to be equal to the skip distance S, as shown in the figure. The skip distance, of course, may be estimated using the specified probe inspection angle. Once the skip distance is established, a line is drawn on the piece to be inspected parallel to the weld and at a distance away from it and equal to the skip distance. Inspection is accomplished by using the skip distance line as a reference and moving the probe near and away from the weld while rotating it slightly in the plane of the plate. Defect indications will be similar to those obtained for the nondefective bolt hole. Obviously, multiple defects, etc., could create a more complicated result.

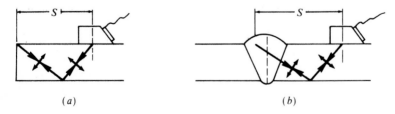

(a) (b)

FIGURE 8-26
(a) Skip distance for calibration of angle-beam probe. (b) Beam path for inspection.

8-9.3 Pipe Inspection

Pipe of virtually all sizes can be inspected ultrasonically for internal and surface defect. As with most structural steel items, both manufacturing and fatigue flaws may be present.

A typical arrangement for a pulse–echo inspection of a pipe is shown in Fig. 8-27(*a*), where the defect is simulated by the drilled hole. This scheme would detect longitudinal flaws that generally originate from the manufacturing process. As the probe is moved toward and past the hole, the echo will rise and fall as with conventional angle-beam inspection. It should be noted that the probe may be quite close to the expected defect when a strong echo appears. In some cases, this defect echo may be masked by the internal reflections in the wedge. A method for overcoming this problem is to use the through-transmission technique

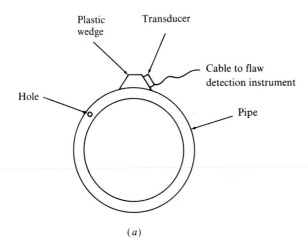

(*a*)

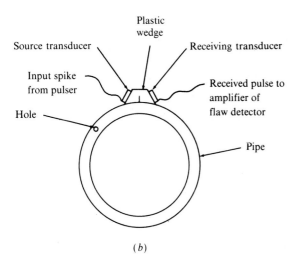

(*b*)

FIGURE 8-27
(*a*) Pulse–echo angle-beam inspection of piping. (*b*) Through-transmission angle-beam inspection of piping.

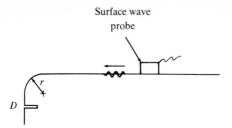

FIGURE 8-28
Surface wave excited on a test block.

shown in Fig. 8-27(*b*). In this case, the presence of an unchanging signal at the receiving probe would indicate sound material. A change in the signal characteristics, as indicated by ultrasonic spectroscopy, or a loss of signal strength could indicate the presence of a defect.

A fatigue flaw common to oil country tubulars is one that starts from a corrosion pit on the inside wall and propagates circumferentially. This type defect could be detected in a manner similar to the one just discussed, with the probe oriented longitudinally, parallel to the pipe axis.

Corrosion, of course, is a serious matter in piping. This condition may be monitored with the thickness measurement techniques previously discussed.

8-10   TECHNIQUES FOR SURFACE WAVE INSPECTIONS

Surface waves possess unique characteristics that make them useful for some very specialized inspection problems. As previously discussed, surface waves are generated when energy is excited at a surface, typically with either a longitudinal wave probe placed on the surface with no couplant or with a 90° surface wave probe, as shown in Fig. 8-28. These probes are designed so that the refracted shear wave is just past the second critical angle. With the concentration of energy at the surface, these waves are obviously quite sensitive to surface defects. Unfortunately, though, they are also quite sensitive to any sort of surface anomaly, such as pitting, oil, droplets, or other reflectors. Nonetheless, they do have advantages for certain problems.

The large reflected pulse at 6 in Fig. 8-29 shows the surface wave reflection obtained from the arrangement in Fig. 8-28. The wave excited only in one direction travels around the curvature of radius *r* until it reaches the reflector at *D*.

Normally, the crack that is represented by *D* cannot be seen by the naked eye. The precise location may be obtained, however, by tapping with a slightly oily finger in the travel path of the surface wave. As long as the damping caused by the oily finger occurs between the probe and the defect, the amplitude of the echo will be affected. Once the tapping passes the defect, however, it will no longer affect the echo and the crack, therefore, has been located.

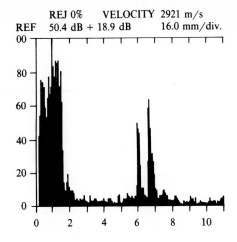

REJ 0%   VELOCITY 2921 m/s
REF   50.4 dB + 18.9 dB   16.0 mm/div.

FIGURE 8-29
Surface wave reflection obtained from crack shown in Fig. 8-28.

Previous discussion on the Rayleigh wave has indicated that for propagation on curved surfaces, the travel speed and the reflection coefficient are functions of the ratio of the radius of curvature and the wavelength. In actual inspection situations, the additional reflections occurring at sharper corners may cause some difficulties in performing inspections around curved and oblique corners.

8-11   INSPECTION WITH LAMB WAVES

Lamb waves are most frequently used to detect anomalies in plate and sheet that are associated with a change in section thickness. The thickness change will result in an excitation of different modes with an effect on the dispersion characteristics of the wave. Typical examples of these types of defects are laminations and corrosion in metal plates and delaminations in composite material. Lehfeldt and Höller [48] describe plate inspection for laminations using Lamb waves and an extensive discussion on Lamb wave inspection is given by Krautkramer and Krautkramer [10].

8-12   FLAW CHARACTERIZATION TECHNIQUES

The expanding use of fracture mechanics techniques in engineering design and maintenance has placed greater emphasis on the correct characterization of flaws. With correct information on the flaw location, shape, orientation, etc., the engineering team will be able to confidently make decisions on the appropriate action for crucial equipment that has been found to have a flaw. Discussion in this section will describe flaw characterization techniques with an emphasis on sizing, while the two sections to follow will describe imaging and frequency analysis techniques.

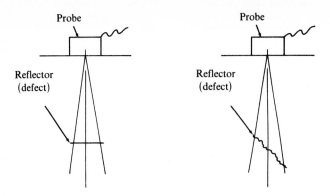

FIGURE 8-30
Flaws of different size and orientation that could give equal ultrasonic response.

Echo amplitude alone is not sufficient for correctly sizing flaws. As shown in Fig. 8-30, flaws of unequal size but with different orientations can yield equal amplitudes. It should be noted in this figure that errors in flaw sizing will generally result in an estimated flaw size smaller than the actual size. This is unfortunately true for most flaw sizing methods.

8-12.1   6 dB Down Method

One of the most widely used flaw sizing methods for sizing large flaws is the 6 dB down method, as shown in Fig. 8-31(*a*) and (*b*). With the probe situated to obtain maximum echo response, as in Fig. 8-31(*a*), the echo is adjusted to full screen height. Next, the probe is moved laterally until the echo amplitude has fallen by one-half power or 6 dB. For the large, planar flaw as shown here and a perfectly normally oriented transducer, this method will correctly locate the edge

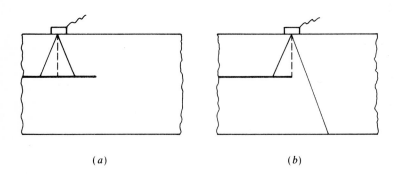

(*a*)   (*b*)

FIGURE 8-31
(*a*) Probe location for full screen response. (*b*) Probe location for 6 dB down response.

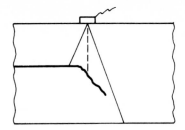

FIGURE 8-32
Flaw shape that yields erroneous response for 6 dB down sizing technique.

of the flaw. It is easy to see that the plan shape of the flaw could be traced with this method.

This method has the inherent problem of being unreliable for flaws of irregular shape and varying roughness, as shown in Fig. 8-32. While the probe location shown in the figure here may give a 6 dB down response, the estimated flaw size based on that location will be smaller than the actual size. Silk and Liddington [49] give a more thorough discussion of flaw sizing using the 6 and 20 dB down methods for flaws of varying shapes and roughnesses.

Flaws also may be sized by the 6 dB down technique with angle-beam probes, as indicated in Fig. 8-33. First, the surface location of the maximum return echo is determined (position A). For lateral sizing, the probe is moved to the side until the signal drops off by one-half amplitude, i.e., 6 dB. For vertical sizing, the locations B and C are similarly located, considering also the normal signal loss due to attenuation. By measuring the distance d on the surface and using the known entry angle of the probe (θ_2'), simple triangulation gives a reasonable estimate of the flaw size in the vertical direction. As was the case for the normal beam 6 dB down technique, flaw orientation and reflection variations will lead to some inaccuracies in flaw size estimation.

A temperature induced change in the entry angle of the probe also may lead to erroneous results in estimating the size of the flaw. For example, using the vertical flaw shown in Fig. 8-33, i.e., a transverse defect 12 mm (0.5 in) in diameter with the top 6 mm (0.25 in) below the top of the surface, the probe

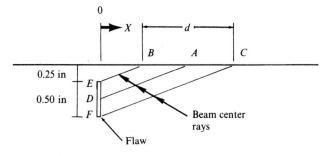

FIGURE 8-33
Angle-beam flaw sizing using the 6 dB down technique.

TABLE 8-3
Flaw size estimates at varying temperatures
by angle-beam 6 dB down technique.

Probe temp. (°F)	d (in)	d (mm)	s (in)	s (mm)
32	1.20	30.5	0.44	11.2
68	1.37	34.8	0.50	12.7
120	1.75	44.4	0.64	16.2

displacement values (d) and flaw size estimates (s) for an inspection performed at the temperatures previously used would be as indicated in Table 8-3.

If the $\frac{1}{2}$ in flaw were correctly sized at 68°F, the results obtained at 32 and 120°F, respectively, would be 0.44 and 0.64 in. At the high temperature, the temperature change in the Plexiglas probe material would cause a 25% error in the sizing of the flaw.

8-12.2 Diffraction Techniques

Ultrasonic energy diffracted from crack tips has added a new dimension in the correct sizing of flaws and anomalies. Using the photoelastic visualization results from Hall, Silk [50] has produced the sequential description of crack tip diffraction shown in Fig. 8-34, where the light regions show areas of high stress intensity. Once the longitudinal wave labeled C_1 strikes the slot, the diffracted shear S_3 and longitudinal C_3 waves are excited along with other reflected and transmitted waves. These diffracted waves are excited because the tip of the crack has become a point radiator of acoustic energy.

It is easily seen that flaw length information could be obtained from the arrival times of C_3 and S_3 on the surface to the right of the slot at position X. Considering that the first signal returning to the source probe is also a portion of C_3, the difference in the arrival time of C_3 at the source probe and at X should yield information about the flaw length.

Crack tip diffraction techniques have been used in ultrasonic NDE for some time. The delta technique, as described by Cosgrove [51], appears to be based on crack tip diffraction. Following that work, Silk and Liddington [52], Thompson [53], Gruber, Hendrix, and Schick [54], and Gruber and Jackson [55] have reported results showing the application of these techniques to the accurate sizing of flaws.

An example of crack tip diffraction to the sizing of flaws with angle-beam probes is shown in Fig. 8-35. As shown on the right of the illustration, the primary echo returned to the probe is from the corner reflection at the base of the flaw. This is reflection R. Energy originating from the tip, however, will arrive at the probe sooner than the echo R. This is labeled the satellite pulse D. The difference in arrival time Δ_1 will therefore be an indicator of the crack length. As

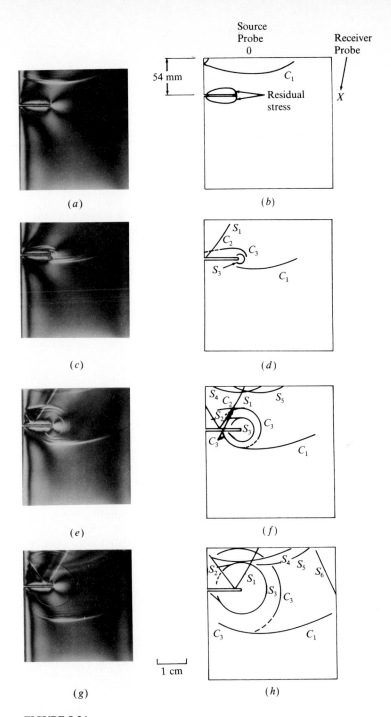

FIGURE 8-34
A series of photographs from a visualization experiment showing the diffraction of longitudinal waves. C_1, incident longitudinal wave; S_1, shear wave produced by reflection of C_1 from surface; C_2, reflection of C_1 from face of slit; C_3, diffracted longitudinal wave; S_2, reflected shear wave from face of slit; S_3, diffracted shear wave; S_4, S_5, and S_6, shear waves produced by mode conversions at block surface. (*Photographs by K. G. Hall, British Rail, Research and Development Division, Interpretation by M. G. Silk, UKAERE Harwell. From Silk [50]. Courtesy Academic Press.*)

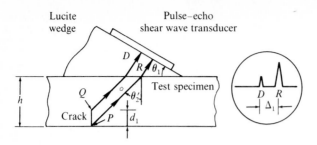

FIGURE 8-35
The interaction of a shear wave with a surface crack resulting in a large reflected pulse (R) and a small leading tip-diffracted satellite pulse (D). (Lucite is a trademark of the DuPont Corp.) (*From Gruber et al.* [54]. *Reprinted with permission from Materials Evaluation, vol. 42, no. 4, 1984, The American Society for Nondestructive Testing, Columbus, OH.*)

described by Gruber et al., a myriad of other anomalies such as circles, imbedded flaws, porosity, etc., may be characterized using the satellite pulse technique.

While these methods appear to be very powerful, it must be understood that the amplitude of the satellite pulse may be very small when compared to normal signals and noise on the screen. A person attempting to use satellite pulses for flaw sizing should be thoroughly familiar with the technique before applying it.

8-13 FLAW IMAGING

Communicating the flaw size and shape to others is a very crucial part of the NDE process. Decisions on removal of the part, repair, or continued service need full information on the flaw characteristics. The discussion to this point has assumed the common A-scan as the display mode for the flaw detector. It remains the task of the operator or engineer to convert this to information that is usable in deciding future action. The information that is available in A-scan, however, is basically one dimensional, since echo depth is all that is indicated. Interpretation, with accompanying sketches and calculations, is required to characterize the flaw. It is obvious, then, that there is considerable potential for error and variation in this part of the NDE process. Imaging systems using automated probe positioning control and computerized data analysis and storage will greatly enhance the quality of the communication at this stage.

For the present discussion, an x-y-z coordinate system that is used might be represented by a plate lying in the plane of the paper. The x-y plane is in the plane of the paper with the $(0,0)$ position at the top left corner. The x axis is the horizontal axis and the y axis is the vertical axis with the z axis representing the depth below the surface.

Ultrasonic imaging techniques are continually being enhanced and new applications are being found. A review of several methods and applications of ultrasonic imaging techniques is given in Ref. 56.

8-13.1 B-scan

The B-scan presentation gives a profile of the plate with the y axis held constant and the x position of the probe varied from left to right as shown in Fig. 8-36(a). The x position of the probe is directed into the data collection system for use in constructing the B-scan display.

It is useful to perceive the B-scan as an A-scan display rotated by 90° clockwise. Data points for each probe position are placed on the screen based on the existence of an echo above a preset threshold amplitude. Aside from the threshold setting, magnitude is not indicated in the B-scan. At position 1 as shown in Fig. 8-36(a), a point for the initial echo would appear at the $(0,0)$ position on the screen in the top left corner. The back surface would be indicated by a point directly below the initial point at a distance proportional to the pulse travel time in the test piece. If the probe is moved across a defect-free plate with parallel sides, the resulting B-scan display would be two horizontal lines. A defect as shown in the plate in Fig. 8-36(a) would give an echo at a proportional distance between the initial echo and the back echo. In the B-scan, this would be indicated by a point displayed on the screen at the same proportion of the distance between the two horizontal parallel lines. The resulting display is shown in Fig. 8-36(b). It should be noted that the B-scan gives an indication based on the first received echo at a probe position. Defects under the reflector nearest to the probe will not be indicated.

Several specialized variations in the B-scan display are used. For example, if both the vertical y dimension of the probe and the horizontal back-echo trace of the B-scan are incremented slightly, a profile map showing defect location and depth can be produced. This is called a multiple B-scan display.

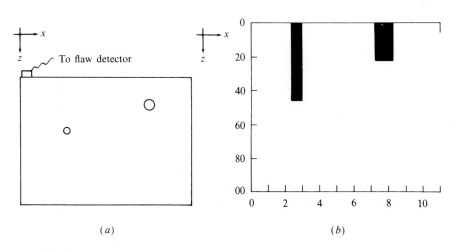

(*a*)   (*b*)

FIGURE 8-36
(*a*) Probe positioning for B-scan inspection. (*b*) Screen response for B-scan inspection.

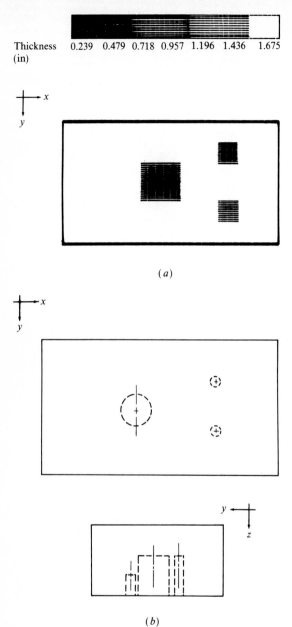

Thickness   0.239   0.479   0.718   0.957   1.196   1.436   1.675
(in)

(a)

(b)

FIGURE 8-37
(a) C-scan display for block in (b).
(b) Plan and end views of block.

B-scan also can be effectively used for displaying the data obtained with either normal beam or angle-beam inspection of long sections such as railroad rail. The results of the normal beam inspection would be the same as initially described except that the screen would require constant updating in order to accommodate new information. For angle-beam inspection, the display would produce a map of the reflector that, according to the discussion relative to bolt hole inspection (Figs. 8-21 and 8-22), would be inclined since the length from the probe to the reflector is changing as the probe approaches the bolt hole.

8-13.2   C-scan

A C-scan display is a plan view of the part with the horizontal and vertical positions of flaws indicated on the screen or a paper print-out. Additionally, flaw depth may be indicated with different shades of grey or different colors.

A grid system is used to coordinate the position of the probe and the indicated result on the screen. The x-y position of the probe, furnished to the instrument through either an electromechanical device or an air sonar locator system, is coordinated on the screen with the proper plan view position for the item being inspected. For each position, a no-flaw situation will result in a no-indication output to the screen. Typically, this will result in a blank output. A flaw echo appearing between the initial pulse and the back echo will result in a positive output, which can be indicated on the screen grid. A varying voltage input that is proportional to the depth of the flaw can be used to create a grid where a choice of several colors can be used to visually indicate the depth of the flaw. For black and white systems, this can be accomplished with varying shades of grey. Figure 8-37(a) shows a typical grey scale C-scan display for the block shown in Fig. 8-37(b).

A C-scan display is a very effective way to report data on flaw depths since the presence of the flaw as well as its severity can be indicated directly on a drawing of the part being inspected. This has been particularly useful in corrosion inspection in pipes and pressure vessels and in the detection of delaminations in composite materials.

8-14   IMMERSION TESTING

Inspecting either large parts or large quantities of the same parts can create rather laborious tasks simply in the manipulation of the probe and the recording of data. As mentioned in the previous section, data input and output can be readily handled with available storage and analysis systems. A convenient method for automating the probe manipulation is to utilize an immersion system as shown in Fig. 8-38. With this system water typically serves as the coupling medium and the probe can be readily indexed in the x and y plane for direct input into the data analysis system.

The A-scan display for an immersion system is shown in Fig. 8-39(a). It will contain an initial portion that indicates the pulse travel time in the water

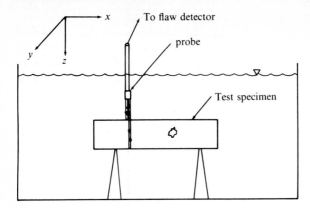

FIGURE 8-38
Immersion inspection.

followed by the echoes obtained from the pulse travel in the test piece. Since the portion showing the water travel is of no interest, the delay and time base (i.e., material calibration) can be used to develop the typical A-scan display as shown in Fig. 8-39(*b*). The delay adjustment is used to move the display in Fig. 8-39(*a*) to the left until the front echo is at the zero time position at the left of the screen. The material calibration is then used to expand the display to yield the presentation shown. Once this is accomplished, either B- or C-scan presentations can be obtained using automated scanning techniques in much the same manner as previously described.

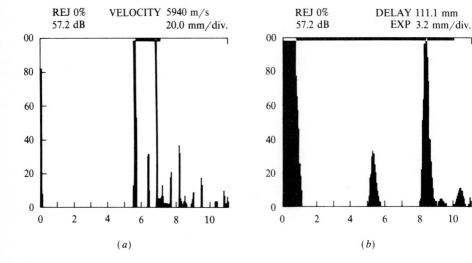

FIGURE 8-39
(*a*) A-scan display for immersion inspection. (*b*) Expanded A-scan display.

8-15 THROUGH-TRANSMISSION WITH AIR PROPAGATION TRANSDUCERS

Air propagation transducers have existed for some years, generally being used only as proximity sensors because of the serious interface losses occurring due to the high impedance mismatch. More recently, Brunk [57] has reported experiments showing that these probes can be used for ultrasonic inspection of materials. Using composite materials, where echo patterns may not be reliable indicators of delaminations, the air propagation probes have shown some ability to detect delaminations when used in the through transmission mode.

8-16 FREQUENCY ANALYSIS TECHNIQUES

The excitation of a pulse by a typical compressional wave transducer is shown in Fig. 8-40(a), using the work of Hall [37]. Probe frequency is 2.25 MHz. The leading compressional wave is seen in Fig. 8-40(a) approaching the drilled hole

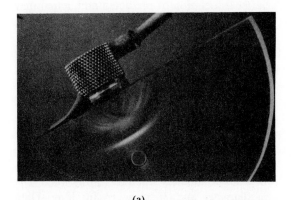

(a)

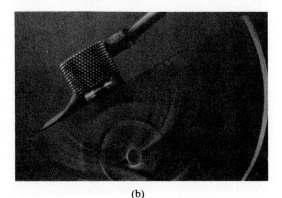

(b)

FIGURE 8-40
(a) Approach of a compressional wave toward a circular reflector. (*From Hall* [37]. *Reprinted with permission from Materials Evaluation, vol. 42, no. 7, 1984, The American Society for Nondestructive Testing, Columbus, OH.*) (b) Wave diffraction at the drilled hole after being struck by a compressional wave. (*From Hall* [37]. *Reprinted with permission from Materials Evaluation, vol. 42, no. 7, 1984, The American Society for Nondestructive Testing, Columbus, OH.*)

and in Fig. 8-40(b), the resulting effect of the hole on the diffracted pulse is shown. Where the wave front of the incident pulse is relatively flat, the wave front of the diffracted wave is circular, clearly showing that the character of the reflected wave has been reshaped by the circular reflector. It is easy to perceive that the effect on the diffracted wave of a flat reflector would result in different changes in the character of the reflected wave. Recognizing that the size and the shape of the flaw may affect the frequency content of the diffracted wave, it follows that a spectrum analysis of pulses reflected from various flaw sizes and shapes could develop additional parameters that could be further used in flaw characterization.

The effect of various flaw characteristics on pulse spectrum has been studied by a number of investigators. Adler, Cook, and Simpson [58] used several artificial defects of various shapes and sizes to show that significant correlations between frequency spectrum and flaw characteristics existed only when the flaw was radiated at an oblique angle. At high angles for the same flaw, the frequency spectrum shifted toward the lower values and multiple peaks occurred. They further emphasized that broad-band transducers should be used for frequency analysis studies since a wider range of frequencies would thus be available. In comparisons, they showed a much better association of predicted and actual flaw sizes for the frequency analysis techniques as compared to simple amplitude methods.

Later work reported by Brown [41] showed applications of ultrasonic spectroscopy (frequency analysis) to detecting machined notches, grain size, and bonding defects in various materials. Observed changes in the frequency were explainable from flaw and material conditions. Considerable attention is directed toward the design of transducers and test systems most suitable for spectroscopy.

A useful illustration of the application of ultrasonic spectroscopy is in comparing the response of a lack of fusion butt weld defect with a reference defect such as a side-drilled hole. For this example, the weld plate is 12 mm ($\frac{1}{2}$ in) thick and the lack of fusion is approximately 19 mm ($\frac{3}{4}$ in) long. The side-drilled hole is 3 mm ($\frac{1}{8}$ in) in diameter and 38 mm ($1\frac{1}{2}$ in) long. Data parameters obtained were the central frequency f_c, the peak power frequency f_p, the bandwidth B^*, and the skewness f_{sk}. An average of 14 data points for each reflector is shown in Table 8-4.

An analysis of the results, using Eq. (A-5) from Appendix A, shows that there is a significant difference in both the center frequency and the peak frequency for the two data sets. In fact, Table A-1 indicates that there is $> 99\%$ probability that these frequencies come from different reflectors. Comparatively, the bandwidth shows $> 99\%$ probability of being different and the skew factor shows basically no difference.

While these results indicate that differences may exist between reflectors, considerable evaluation of similar results for other reflectors must be undertaken in order to build a confident data base. While only four parameters are used in the example here, greater confidence in the results might be obtained with the inclusion of either more or different parameters for further comparison. There is

TABLE 8-4
Frequency analysis parameters for weld defect and side-drilled hole in 12 mm ($\frac{1}{2}$ in) plate angle beam inspection with 70° 2.25 MHz probe.

	f_c	f_p	B^*	f_{sk}
		Hole		
$\overline{X}$	1.697	1.674	14.60	1.57
σ_{n-1}	0.099	0.111	2.55	0.51
n	14	14	14	14
		Defect		
$\overline{X}$	2.115	2.09	12.29	1.55
σ_{n-1}	0.145	0.140	4.33	0.58
n	14	14	14	14
	Comparison of hole and fusion defect			
z	8.91	8.74	1.72	0.096

some indication that inspecting through a couplant might introduce more uncertainty into the frequency analysis data and that more reliability and repeatability might be achieved with immersion inspection. An example of automated data analysis for plastic pipe welds using a PC-based system is given by Whalen et al. [59].

8-17 SUMMARY

The attempt in this chapter has been to introduce a rather broad range of examples of ultrasonic NDE and thereby demonstrate some typical applications as well as problems. By no means are these examples complete. Each individual must seek solutions to the particular task at hand. Where the needed information is not furnished here, the various references should be consulted.

It is a particular challenge to the NDE engineer to match the proper technique to the inspection need. Where ultrasonics is concerned, the proper method obviously could range from the simple pulse–echo test with one probe to the multiprobe arrays, each directed into data analysis systems. In the decision process, it should be remembered, however, that the behavior of the sound pulse in the material is solely a function of the pulse and the material with its anomalies. The successful NDE engineer will extract from the reflected or refracted pulse the information necessary to reach the proper conclusion.

CHAPTER
9

ULTRASONIC TECHNIQUES FOR STRESS MEASUREMENT AND MATERIAL STUDIES

9-1 INTRODUCTION

Efficient design and satisfactory service life demand intimate knowledge of the properties of the material being used. Of notable importance are the nature of the internal stresses as well as the isotropy and homogeneity of the material structure. To be able to nondestructively investigate for abnormal material conditions would enable the engineer not only to optimize the design but also to optimize the maintenance cycle. Topics to be reviewed in this chapter represent some novel ultrasonic methods available for stress measurement, texture studies, grain size determination, and inspection of layered material for interface defects. Other ultrasonic techniques proposed for similar investigations are described in the several references that are listed.

9-2 ULTRASONIC STRESS MEASUREMENT TECHNIQUES

The basis of the ultrasonic stress measurement technique is the stress-induced anisotropic behavior of solids. It is caused by nonlinearities in the strain displacement and constitutive relations of the material. The acoustoelastic effect refers to the changes in the speed of elastic wave propagation in a body that is simultaneously undergoing static elastic deformation. In this section, which is based on the work of Egle as described in Ref. 60, techniques and results of the measurement of the acoustoelastic and third-order elastic constants are presented for pearlitic steel.

Much of the previous work on the use of the acoustoelastic effect for stress measurement [61–66] has concentrated on using the difference in speed of shear waves polarized parallel and perpendicular to the uniaxial stress and traveling perpendicular to the stress axis. The present approach is oriented primarily toward uniaxial stress fields and speed changes for longitudinal waves propagating parallel to the direction of stress application are used as the inspection criteria. Previous experimental studies have shown that these longitudinal waves are more sensitive to stress change than polarized shear waves traveling across the stress field, perpendicular to the direction of stress application.

9-2.1   Theoretical Background

Hughes and Kelly [67] derived expressions for the speeds of elastic waves in a stressed solid using Murnaghan's theory of finite deformations [68] and third-order terms in the strain energy expression. They showed that the speeds of plane waves propagating as shown in Fig. 9-1 in the 1 direction and having particle displacements in the 1, 2, or 3 directions in an initially isotropic body subjected to a homogeneous triaxial strain field are given by

$$\rho_0 V_{11}^2 = \lambda + 2\mu + (2l + \lambda)\theta + (4m + 4\lambda + 10\mu)\alpha_1 \qquad (9\text{-}1a)$$

$$\rho_0 V_{12}^2 = \mu + (\lambda + m)\theta + 4\mu\alpha_1 + 2\mu\alpha_2 - \tfrac{1}{2}n\alpha_3 \qquad (9\text{-}1b)$$

$$\rho_0 V_{13}^2 = \mu + (\lambda + m)\theta + 4\mu\alpha_1 + 2\mu\alpha_3 - \tfrac{1}{2}n\alpha_2 \qquad (9\text{-}1c)$$

where ρ_0 = initial density

V_{11}, V_{12}, V_{13} = speeds of waves propagating in the 1 direction with particle displacements in the 1, 2, and 3 directions, respectively

λ, μ = Lamé or second-order elastic constants

l, m, n = Murnaghan's third-order elastic constants

$\alpha_1, \alpha_2, \alpha_3$ = components of the homogeneous triaxial principal strains in the 1, 2, and 3 directions

$\theta = \alpha_1 + \alpha_2 + \alpha_3$

The 1 and 2 directions are to the right and up, respectively, and direction 3 is out of the plane of the paper.

For a state of uniaxial stress, there are five unique wave speeds that may be determined from Eqs. (9-1). First consider the stress acting in the 1 direction. The strains are then

$$\alpha_1 = \varepsilon, \qquad \alpha_2 = \alpha_3 = -\nu\varepsilon$$

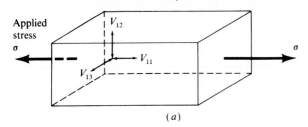

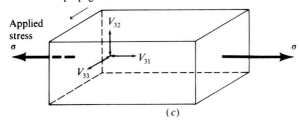

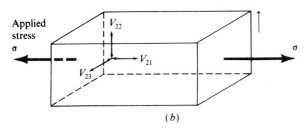

FIGURE 9-1
Speed of plane waves and stress field on orthogonal coordinate system.

where ν is Poisson's ratio. Equations (9-1) reduce to

$$\rho_0 V_{11}^2 = \lambda + 2\mu + \left[4(\lambda + 2\mu) + 2(\mu + 2m) + \nu\mu(1 + 2l/\lambda)\right]\varepsilon \quad (9\text{-}2a)$$

$$\rho_0 V_{12}^2 = \rho_0 V_{13}^2 = \mu + \left[4\mu + \nu(n/2) + m(1 - 2\nu)\right]\varepsilon \quad (9\text{-}2b)$$

The speeds of plane waves traveling perpendicular to the uniaxial stress may also be determined from Eqs. (9-1) and are

$$\rho_0 V_{22}^2 = \lambda + 2\mu + \left[2l(1 - 2\nu) - 4\nu(m + \lambda + 2\mu)\right]\varepsilon \quad (9\text{-}3a)$$

$$\rho_0 V_{21}^2 = \rho_0 V_{31}^2 = \mu + \left[(\lambda + 2\mu + m)(1 - 2\nu) + n\nu/2\right]\varepsilon \quad (9\text{-}3b)$$

$$\rho_0 V_{23}^2 = \rho_0 V_{32}^2 = \mu + \left[(\lambda + m)(1 - 2\nu) - 6\nu\mu - n/2\right]\varepsilon \quad (9\text{-}3c)$$

The relative changes in wave speed with axial strain may be calculated from Eqs.

(9-2) and (9-3) if it is assumed that the relative changes are small. The resulting equations are

$$\frac{dV_{11}/V_{11}^0}{d\varepsilon} = 2 + \frac{\mu + 2m + \nu\mu(1 + 2l/\lambda)}{\lambda + 2\mu} \tag{9-4a}$$

$$\frac{dV_{12}/V_{12}^0}{d\varepsilon} = 2 + \frac{\nu n}{4\mu} + \frac{m}{2(\lambda + \mu)} \tag{9-4b}$$

$$\frac{dV_{22}/V_{22}^0}{d\varepsilon} = -2\nu\left[1 + \frac{m - \mu l/\lambda}{\lambda + 2\mu}\right] \tag{9-4c}$$

$$\frac{dV_{21}/V_{21}^0}{d\varepsilon} = \frac{\lambda + 2\mu + m}{2(\lambda + \mu)} + \frac{\nu n}{4\mu} \tag{9-4d}$$

$$\frac{dV_{23}/V_{23}^0}{d\varepsilon} = \frac{m - 2\lambda}{2(\lambda + \mu)} - \frac{n}{4\mu} \tag{9-4e}$$

In Eqs. (9-4), the superscript 0 indicates the wave speed at zero axial strain.

The third-order constants l, m, n may be evaluated in terms of the relative changes in wave speeds by inverting the set of three equations consisting of either Eqs. (9-4a) or (9-4c) and two of Eqs. (9-4b, d, e). Using the last three of Eqs. (9-4), one may express the constants as

$$l = \frac{\lambda}{1 - 2\nu}\left[\frac{1 - \nu}{\nu}\frac{dV_{22}/V_{22}^0}{d\varepsilon} + \frac{2}{1 + \nu}\left(\frac{dV_{21}/V_{21}^0}{d\varepsilon} + \nu\frac{dV_{23}/V_{23}^0}{d\varepsilon}\right) + 2\nu\right] \tag{9-5a}$$

$$m = 2(\lambda + \mu)\left[\frac{\nu}{1 + \nu}\frac{dV_{23}V_{23}^0}{d\varepsilon} + \frac{1}{1 + \nu}\frac{dV_{21}/V_{21}^0}{d\varepsilon} + 2\nu - 1\right] \tag{9-5b}$$

$$n = \frac{4\mu}{1 + \nu}\left[\frac{dV_{21}/V_{21}^0}{d\varepsilon} - \frac{dV_{23}/V_{23}^0}{d\varepsilon} - 1 - \nu\right] \tag{9-5c}$$

Combining Eqs. (9-4b, d) and Eqs. (9-4a, c, d, e) results in the following relations among relative changes:

$$\frac{dV_{12}/V_{12}^0}{d\varepsilon} = 1 + \nu + \frac{dV_{21}/V_{21}^0}{d\varepsilon} \tag{9-6a}$$

$$\frac{dV_{11}/V_{11}^0}{d\varepsilon} = \frac{dV_{22}/V_{22}^0}{d\varepsilon} + \frac{2}{1 - \nu}\left[\frac{dV_{21}/V_{21}^0}{d\varepsilon} + \nu\frac{dV_{23}/V_{23}^0}{d\varepsilon}\right] + \frac{(1 + \nu)(1 + 2\nu)}{2(1 - \nu)} \tag{9-6b}$$

9-2.2 Experimental Measurement of Acoustoelastic Constants

Stress-induced changes in wave speed were measured in three test specimens machined from pearlitic steel used for railroad rail. The specimens were loaded uniaxially in tension and compression with a conventional testing machine.

The changes in speed for waves propagating perpendicular to the load axis were measured with commercially available wide band ultrasonic transducers having a maximum center frequencies of 2.5 or 5 MHz. The waves propagating parallel to the load axis were generated and detected with longitudinal transducers mounted on Plexiglas wedges and inclined at 28° for longitudinal and 55° for shear waves. The source transducer was a 25.4 mm square PZT-5 plate having a nominal resonant frequency of 1.6 MHz. A 5 MHz wide band longitudinal transducer was used as a receiver.

Travel-time data were obtained using a modification of the pulse–echo–overlap method as used by Hsu [65]. Changes in path length due to strain were considered in the calculation of the loaded speeds. The maximum error was estimated to be four parts in 10^5.

The relative changes in wave speed as a function of axial strain for the five types of waves mentioned previously are shown in Fig. 9-2. As predicted by the theory, the wave speed changes are linear functions of the strain and the largest relative change in wave speed is associated with longitudinal waves propagating parallel to the applied load. The smallest relative changes are associated with V_{23}, shear waves propagating perpendicular to the load and polarized perpendicular to the load axis. The acoustoelastic constants defined by Eqs. (9-4) are the slopes of the lines in Fig. 9-2 and are given in Table 9-1.

9-2.3 Force Measurement with L_{CR} Waves

Field experiments using the critically refracted longitudinal (L_{CR}) waves for stress measurement were conducted in 1976 on continuously welded rail (CWR) on the mainline of the Atchison, Topeka, and Santa Fe Railway and on similar CWR track at the U.S. Department of Transportation's Transportation Test Center at Pueblo, Colorado [15, 69]. Rail temperatures during the tests ranged from -12 to $+15.5°C$ (11–60°F). The rail at the Pueblo test track was strain gauged at four stations approximately 91.4 m (100 yd) apart. The probe used in these studies was clamped to the web, oriented along the longitudinal neutral axis of the rail. Experiments on the repeatability of probe measurements showed the measurements to repeat to within ± 6.9 MN/m^2 (1 ksi).

Ultrasonic travel-time data were well behaved at two of the stations, but showed considerable scatter at the other two. Since the stations showing the scatter were situated near the ends of the original rail sections and the two well behaved stations were situated near the middle portion of the rail lengths, it was suspected that stresses induced in the adjustment of the ends for welding might be the source of the scatter.

The results from the two well behaved stations are shown in Fig. 9-3. Using the standard error bands of force (stress) based on travel-time measurement, as shown by the dotted lines, it may be stated that for this data, a travel-time measurement of 36.20 μs would indicate with a 68% probability that the actual stress change was in the range from 19–36 MPa (2.75–5.2 ksi). The correspond-

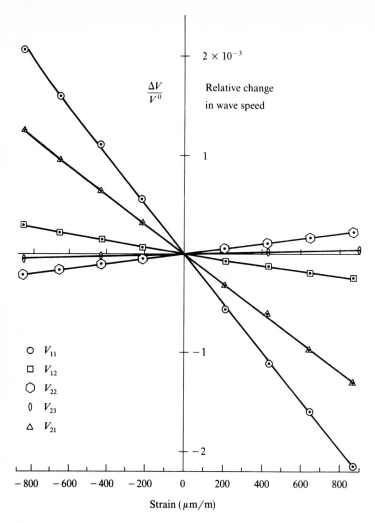

FIGURE 9-2
Relative changes in wave speeds with strain [60].

TABLE 9-1
Acoustoelastic constants for rail steel [60]

Load	$\dfrac{dV_{21}/V_{21}^0}{d\varepsilon}$	$\dfrac{dV_{23}/V_{23}^0}{d\varepsilon}$	$\dfrac{dV_{22}/V_{22}^0}{d\varepsilon}$	$\dfrac{dV_{11}/V_{11}^0}{d\varepsilon}$	$\dfrac{dV_{12}/V_{12}^0}{d\varepsilon}$
Tension	-1.5	-0.9	$+0.27$	-2.38	-0.15
Compression				-2.45	

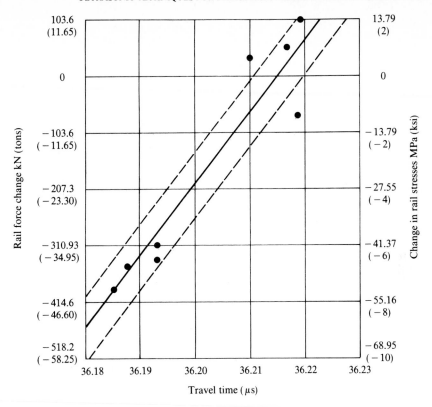

FIGURE 9-3
Travel times for critically refracted longitudinal (L_{CR}) waves in railroad rail with changing stress [69].

ing force change would be in the range of -0.14 to -0.26 MN (-15.7–29.7 ton).

The slope for the experimental data shown in Fig. 9-3 is approximately 0.57 ns/MPa (3.86 ns/ksi) which is greater than the 2.45 ns/ksi value obtained from the laboratory data as reported in Table 9-1. This lack of close agreement could come from several sources. First, it is likely that the acoustoelastic constant for railroad rail in place in the track may be affected by the track structure as well. Further, there are likely to be directional variations in the acoustoelastic constant due to mild texture in the rail web. Nonetheless, the force changes predicted by the ultrasonic measurement were in general agreement with those shown by the strain gauges. It was clear that the data would not permit an absolute measurement of the force or stress state in a previously untested rail, based on ultrasonic travel-time data. As will be discussed later, factors such as residual stress and material texture are known to cause travel-time variations greater than those induced by the applied stress change.

9-2.4   Changes In Wave Travel Time Associated With Varying Elastic Properties

One of the problems associated with using the acoustoelastic technique for stress measurement lies in the fact that nominally identical materials may exhibit slight differences in wave speeds and in acoustoelastic constants. The wave speed variations may originate from texture as well as from residual stresses. Differences in acoustoelastic constants most likely are caused by texture effects, as described by Thompson, Smith, and Lee [70] and discussed in a later section. Where there may be consistency in the texture for items being considered for stress measurement, however, measuring changes in stress appears to be achievable. In addition, data collection schemes designed for specific applications may yield reasonable estimates of absolute stress levels.

Where the variations of wave speed due only to elastic property changes can be measured, an absolute measure of stress becomes feasible. If we assume that the variation in elastic properties is caused by variations in the rolling process by which rail, for example, is formed, then, following Bradfield [71], it is reasonable to assume that the bulk modulus of the rail is constant. Experimental data reported in Ref. [60] show that this assumption is correct.

The expected effects on the wave speed of variations in residual stress and texture in the material to be investigated must be reckoned with before any measurement of applied force can be considered [72]. Again using the rail as the subject of discussion, a rail under active load would yield a travel-time measurement that would contain the combined effects of residual stress, texture, and temperature, as well as the active load. This can be expressed by

$$t = t^* + \Delta t_F + \Delta t_{RS} + \Delta t_{TX} + \Delta t_T \qquad (9-7)$$

where t = measured travel time

t^* = travel time for a homogeneous, isotropic, stress-free rail at a standard temperature

Δt_F = travel-time effect of the applied (active) force

Δt_{RS} = travel-time effect of the residual stresses

Δt_{TX} = travel-time effect of the material texture

Δt_T = travel-time effect of temperature difference at time of measurement from a standard temperature

For ultrasonic stress measurement, the parameter of interest is Δt_F, while the parameter measured is t. The task remaining, then, is to devise a scheme to either measure the effects of the remaining parameters or to collectively deal with them as a group. The effects of temperature are known, as described in Refs. 15 and 69. The possibility of separately identifying t^*, Δt_{RS}, and Δt_{TX} is one that is appealing, yet obviously very difficult. Some particular characteristics of the manufacture and service of railroad rail suggest a scheme where actual stress measurement may be possible without separate knowledge of the texture and residual stress effects.

The separation of residual stress from applied stress is a difficult problem since they are both affecting the wave speeds in the material. The question to be asked is whether or not the separation of the two is necessary. Where stress fields causing fatigue crack propagation are of interest, there may be some justification for this separation. For stress measurement in structural steel where buckling or high tensile stresses are of interest, or for measurement of weld stresses over a large region, it may not be necessary to have knowledge of the individual values for these two effects.

An approach to obtaining a reasonable measurement of applied stress in structural steel, similar to that described here, has been discussed by Williams [73]. He proposed obtaining travel-time data at several locations with the intent that textural and residual stress variations will be averaged to a low level to where the constant applied stress value will emerge as the dominant measured parameter. The experimental results reported in Ref. 72 show that there is a reasonable expectation that this averaging technique may be useful for the measurement of applied stress in railroad rail.

9-2.5 Using the Empirical, Zero Force Time Base for Applied Force Measurement

The results showing that travel-time averaging may yield force-free travel-time data suggest that a procedure may be developed that will yield a zero force travel time that includes the combined effects of residual stress and texture. The result of this development could be the ability to measure applied force with a relatively simple system.

Consider Eq. (9-8) where t_0 is a force-free time

$$t_0 = t^* + \Delta t_{RS} + \Delta t_{TX} \tag{9-8}$$

Rearranging Eq. (9-8) and substituting into Eq. (9-7) yields

$$t = \Delta t_F + t_0 + \Delta t_T \tag{9-7a}$$

which represents the travel-time data that would be collected in the field. The travel-time change resulting from thermally induced, longitudinally restrained expansion would be

$$\Delta t_F = t - t_0 - \Delta t_T \tag{9-7b}$$

Using a value of the acoustoelastic constant for the longitudinal waves, selected from Table 9-1, the change in stress may be calculated from

$$d\sigma = \frac{E}{2.45 t_0} \, dt \tag{9-9}$$

where $d\sigma$ = stress change from a reference level
E = Young's modulus
t_0 = travel time in the absence of stress
dt = change in travel time due to stress change

With the travel time t_0 used in Eq. (9-9) now assumed to be equal to t_0 given in Eq. (9-8), the revised form of Eq. (9-9) is

$$\sigma = \frac{E}{2.45 t_0}(t - t_0 - \Delta t_T) \qquad (9\text{-}10)$$

where the $d\sigma$ has been replaced by the actual stress level σ because the right side of the equation may be normalized to a zero reference level.

The temperature effect on wave speed for pearlitic steel has been given in Ref. [15] as

$$\frac{dC}{dT} = 0.55 \ \frac{\text{m}}{\text{s} \ {}^\circ\text{C}} \qquad (9\text{-}11)$$

For the particular probe used in the described experiments, the effect on travel time would be 0.0034 $\mu s/{}^\circ$C. With these values, then, the temperature effect on travel time Δt_T, referred to an arbitrarily chosen reference temperature, could be determined.

A system incorporating many of the features described here has been developed by Polish investigators for the measurement of residual stresses in railroad rail at the steel mill [74].

9-2.6   Summary

While the present discussion has concentrated on the measurement of longitudinal stresses in steel railroad rail, the interest in the measurement of material stresses certainly extends beyond that one example. Unfavorable stress patterns in welded structures, bonded material, ceramics, and cold-worked metals are certainly candidates for investigation by nondestructive means. A number of summary papers [75–80] on the topic of ultrasonic stress measurement in a variety of materials have appeared and the interested reader is directed there for additional information.

9-3   WAVE VELOCITY CHANGES IN TEXTURED CRYSTALLINE MATERIALS

Wave velocities are usually considered to be constant in all directions in any given material; this assumption is adequate for most situations. A material that has the same wave speeds in any direction may be considered to be isotropic with respect to wave propagation. Two particular exceptions to the isotropic assumption occur when a material is stressed or mechanically worked. The effect of applied stresses has been discussed in the preceding section.

Mechanically working a material may affect the wave velocities in several ways. Two of the primary ways are through changes in the residual stress and texture patterns. Residual stresses may be developed in the working process that can cause the wave velocities to vary according to the acoustoelastic effect. Also, material texture (preferred orientation) may occur in the material with a corresponding change in the wave speed. A combination of these two effects may occur. The effects of texture will be discussed in the material to follow.

9-3.1 Effect of Preferred Orientation (Texture) on Wave Speed

While numerous authors have discussed propagation of waves in crystals, a review of the summary presentation on the subject as given by Green will be useful for the present interest [81].

The solutions for wave propagation in bulk, isotropic solids are recalled to be

$$v_1 = [(\lambda + 2\mu)/\rho]^{1/2} \qquad (9\text{-}12a)$$

$$v_2 = v_3 = (\mu/\rho)^{1/2} \qquad (9\text{-}12b)$$

where v_1 = velocity of a pure longitudinal mode having particle motion in the direction of the wave front normal

v_2, v_3 = velocities of pure transverse modes having particle motion perpendicular to the wave front normal and also mutually perpendicular

λ, μ = Lamé constants

ρ = density

In all three wave mode cases, the energy front travels in the same direction as the wave normal.

A general expression for the equation governing the waves propagating in an anisotropic medium, given by Green, is

$$\left| c_{ijkl} l_l l_j - \rho v^2 \delta_{ik} \right| = 0 \qquad (9\text{-}13)$$

where c_{ijkl} = second-order elastic constants

l_l, l_j = direction cosines of the normal to the plane wave

v = velocity (speed)

δ_{ik} = Kronecker delta

Now, let

$$\lambda_{ik} = c_{ijkl} l_l l_j \qquad (9\text{-}14)$$

so that by changing from tensor notation to matrix notation for the elastic constants, assigning $l_1 = l$, $l_2 = m$, and $l_3 = n$, and recognizing certain characteristics of the crystals, one can obtain simplified expressions as follows.

For cubic crystals it has been shown that there are only 12 nonzero elastic constants c_{ij}, i.e.,

$$c_{11} = c_{22} = c_{33}$$

$$c_{12} = c_{21} = c_{13} = c_{31} = c_{23} = c_{32}$$

$$c_{44} = c_{55} = c_{66}$$

where all other c_{ij} are zero. Therefore, Eq. (9-14) becomes

$$\lambda_{11} = l^2 c_{11} + (m^2 + n^2)c_{44}$$
$$\lambda_{12} = \lambda_{21} = lm(c_{12} + c_{44})$$
$$\lambda_{13} = \lambda_{31} = nl(c_{12} + c_{44})$$
$$\lambda_{23} = \lambda_{32} = mn(c_{12} + c_{44})$$
$$l_{22} = (l^2 + n^2)c_{44} + m^2 c_{11}$$
$$l_{33} = (l^2 + m^2)c_{44} + n^2 c_{11} \qquad (9\text{-}14a)$$

Also, by substituting Eq. (9-14) into Eq. (9-13), one can write the matrix equation

$$\begin{vmatrix} \lambda_{11} - \rho v^2 & \lambda_{12} & \lambda_{13} \\ \lambda_{21} & \lambda_{22} - \rho v^2 & \lambda_{23} \\ \lambda_{31} & \lambda_{32} & \lambda_{33} - \rho v^2 \end{vmatrix} = 0 \qquad (9\text{-}15)$$

It can now be shown that the assumption of a set of direction cosines for the wave normal and elastic constants will result in solutions for the three wave velocities.

A plane wave propagating in the [100] direction will have direction cosines $l = 1$ and $m = n = 0$. Substituting these values into Eqs. (9-14a) and (9-15), one obtains the solutions

$$(c_{11} - \rho v^2)(c_{44} - \rho v^2)(c_{44} - \rho v^2) = 0 \qquad (9\text{-}15a)$$

The resulting wave velocities in the [100] direction are, therefore,

$$v_1 = (c_{11}/\rho)^{1/2}$$
$$v_2 = v_3 = (c_{44}/\rho)^{1/2} \qquad (9\text{-}16)$$

With the substitution of Eqs. (9-14a) into an equation for the direction cosines of the particle displacements (eigenvectors) and using direction cosines $l = 1$ and $m = n = 0$, Green showed that the particle motions associated with the velocities in Eq. (9-16) are identical to those assumed for Eqs. (9-12). Here v_2 and v_3 are defined to have particle motions in the [010] and [001] directions, respectively. Also, the energy flux vector can be shown to be in the direction of the wave normal for this case.

For plane waves propagating in the [110] direction, the direction cosines will be $l = m = 1/\sqrt{2}$ and $n = 0$. The resulting velocities are

$$v_1 = [(c_{11} + c_{12} + 2c_{44})/2\rho]^{1/2}$$
$$v_2 = [(c_{11} - c_{12})/2\rho]^{1/2}$$
$$v_3 = (c_{44}/\rho)^{1/2} \qquad (9\text{-}17)$$

An analysis of particle displacements shows that v_1 again is a pure longitudinal

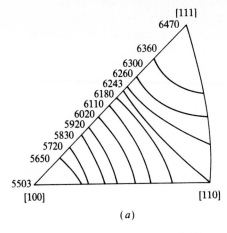

(a)

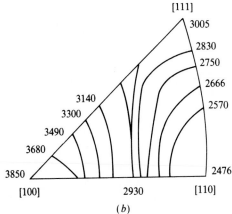

(b)

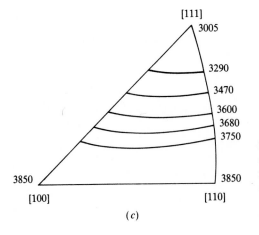

(c)

FIGURE 9-4
(a) Isospeed contours for longitudinal waves in iron, v_1, m/s [42]. (b) Isospeed contours for slow shear waves in iron, v_2, m/s [42]. (c) Isospeed contours for slow shear waves in iron, v_3, m/s [42].

mode, v_2 is a pure transverse mode with particle motion in the [110] direction, and v_3 is also a pure transverse mode but with particle motion in the [001] direction. Moreover, the energy flow is in the same direction as the wave normal.

In the [111] direction, the direction cosines of the wave normal are $l = m = n = 1/\sqrt{3}$ and the velocities are

$$v_1 = \left[\frac{c_{11} + 2c_{12} + 4c_{44}}{3\rho} \right]^{1/2} \qquad (9\text{-}18a)$$

$$v_2 = v_3 = \left[\frac{c_{11} - c_{12} + c_{44}}{3\rho} \right]^{1/2} \qquad (9\text{-}18b)$$

Here again, v_1 is a pure longitudinal mode and v_2 and v_3 are pure transverse

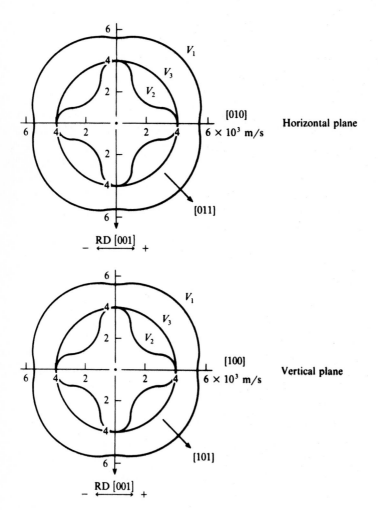

FIGURE 9-5
Wave speeds for (100)[001] texture in iron [42].

modes. For the longitudinal mode, the energy flux is again in the direction of the wave normal, but for the transverse modes, Green showed that the energy flux deviates from the wave normal by an angle

$$\delta = \tan^{-1} \frac{c_{11} - c_{12} - 2c_{44}}{\sqrt{2}\,(c_{11} - c_{12} + c_{44})} \tag{9-19}$$

Wave velocities at various angles can be calculated from Eqs. (9-14a) and (9-15) once a set of direction cosines (l, m, and n) have been assumed. Wave speed variations within crystalline materials are best illustrated using stereographic triangles, which are described by Barrett and Massalski [82]. Stereographic triangles for v_1, v_2, and v_3 in iron are shown in Fig. 9-4(a)–(c), respectively.

Constant-velocity (isospeed) contours for the three types of waves are also shown in an iron crystal in Figs. 9-4(a)–(c). Precise velocities are shown for v_1, v_2, and v_3 along each of the [100], [110], and [111] directions. The velocity of v_1 is seen to be different along each direction. The values for v_2 and v_3 are seen to be equal to each other on the [100] and [111] axes and unequal along the [110] axis. From the isospeed contours, v_3 is seen to be greater than v_2, except at the poles [100] and [111], where they are equal. Figure 9-5 shows a typical velocity contour for a (100)[001] texture in iron.

9-3.2   Ultrasonic Determination of Preferred Orientation

The accurate measurement of ultrasonic transverse wave speeds can be used to determine the existence of preferred orientation in rolled metals. For a material structure having completely random orientation, the two transverse waves would have exactly the same speed when passing through the thickness of the test specimen. On the other hand, if a rolled texture existed in the specimen, a difference in v_2 and v_3 would be noted at certain propagation directions.

Following the work of Sullivan and Papadakis [83], and using Eqs. (9-16), (9-17), and (9-18a, b), a fractional speed change for an assumed (001)[110] texture may be defined as

$$\frac{\Delta v}{v} = \frac{2(S_2 - S_1)}{S_2 + S_1} \tag{9-20}$$

where

$$S_1 = [c_{44}/\rho]^{1/2} \tag{9-21}$$

and

$$S_2 = [(c_{11} - c_{12})/2\rho]^{1/2} \tag{9-22}$$

Here S_1 and S_2 are the shear wave speeds propagated in the [$\bar{1}10$] direction with particle motion polarization in the [100] and [110] directions, respectively.

The percentage of a particular preferred orientation within a rolled specimen can now be measured by noting the difference in travel time of the two transverse waves over the same path length. First, the fast and slow shear wave speeds through the sample are obtained by aligning a contact transverse wave probe so that the propagation is in the rolling plane and at 45° to the rolling direction. The fast and slow travel times t_1 and t_2 are obtained by rotating the shear wave probe on its axis. The time differences for the two polarizations are given by

$$t_1 - t_2 = Lk(S_2 - S_1)/S_1S_2 \tag{9-23}$$

where

$$t_1 = L(1 - k)/v + Lk/S_1 \tag{9-24a}$$

and

$$t_2 = L(1 - k)/v + Lk/S_2 \tag{9-24b}$$

and L is the length and v is the isotropic velocity. Obviously where $S_1 = S_2$, the orientation is random.

The percentage of the preferred orientation being considered is therefore given by

$$k = \frac{S_1 S_2(t_1 - t_2)}{L(S_2 - S_1)} \tag{9-25}$$

Sullivan and Papadakis report the detection of preferred alignment of less than 0.1% in rolled metals. Where the present example has used a (001)[110] texture, the method could be applied to detection of any texture provided that the wave propagation paths were properly defined. Obviously, the accuracy in texture measurement is dependent on the accurate measurement of time and path length for the sample. Several other applications for ultrasonic texture analysis are reported by Green [81].

9-3.3   Summary

The results presented in this section indicate that large variations in wave speeds may be expected for textured conditions that are frequently encountered in wrought metal products. Knowledge of the texture, and hence the degree of anisotropy, may be useful in judging the suitability of materials for particular service. Some applications, such as in electrical transformers, may require a specific texture for optimum performance. A texture in structural steel plate, on the other hand, may render the part unsuitable because of the resulting directional dependence of the elastic properties.

9-4   USING CRITICALLY REFRACTED WAVES FOR NEAR SURFACE INSPECTIONS

Inspection of plates and piping for defects near to the surface may be a difficult task. The use of normal beam probes for this inspection may be unreliable because of the near field effect. Angle-beam shear wave probes may be used, which reflect the ultrasonic pulse off of an opposite surface. These probes too suffer from near field effects as the probe is moved closer to the defect. Another answer is focused probes that use acoustical lenses to concentrate the energy in a small area near to the contact surface. Because of the concentrated energy and the small probe size, however, this type of inspection may be tedious. An alternative to these approaches is the use of critically refracted shear and longitudinal waves that travel along extended paths just below the surface of the material.

9-4.1 Inspecting Layered Material Using Higher-Order Rayleigh Waves

Fundamental mode Rayleigh waves are well known for their characteristic of traveling long distances along surfaces of typical solid-gas interfaces. Less well known are the higher-order mode Rayleigh waves that travel in an upper solid layer when the upper material has a slower shear wave speed than that of the underlying material [19]. These fundamental and first higher-order modes are designated as the M_{11} and M_{21} waves, respectively, where the order is designated by the first subscript. While these wave forms have been studied for a number of years by geophysicists and seismologists, they have only in recent years received any interest in the field of nondestructive evaluation.

At high frequencies where the wavelength of the Rayleigh wave is no greater than one-third to one-fourth of the layer thickness, both the fundamental (M_{11}) and the first higher-order (M_{21}) waves may be excited simultaneously from an outer surface, as shown in Fig. 9-6(a). The model shown depicts a low-speed cold-worked steel layer of thickness H and shear wave speed C_S of 3077 m/s over an isotropic carbon steel that has a bulk shear wave speed of 3257 m/s [14]. For the model shown, the fundamental and higher-order wave forms were excited at the typical 64° Rayleigh wave angle and the wave speeds were experimentally determined to be $C_R = 2888$ and $C_{FR} = 3077$ m/s, respectively, where the speed of the M_{11} wave is designated as C_R and that of the M_{21} is designated as C_{FR}. In this case, C_{FR}, for the speed of the first Rayleigh mode, is the same as C_{ER} that has been used in previous publications to designate this early arriving (first higher-order mode) Rayleigh wave.

The fundamental M_{11} mode excited along the free surface will travel at a speed C_R, while the first higher-order Rayleigh wave mode M_{21} will propagate in the upper layer at a speed approximately equal to the shear wave speed in the upper region C_{FR}. At some distance from the excitation probe, as shown in Fig. 9-6(b), both the M_{11} and the M_{21} may be observed with the M_{21} leading the M_{11} because of the higher speed. Typical arrivals of the two waves are shown in Fig. 9-7. The fact that the higher-order mode Rayleigh wave is essentially captured by the upper layer makes it potentially useful for defect detection there as well as at the interface between the two regions.

An earlier study of the characteristics of these waves was presented by Haüsler [85]. Later, Bray, Egle, and Reiter [86] showed that the wave forms could be easily excited by a surface wave Plexiglas wedge on the surface of used, heavily cold-worked railroad rail, as described in the previous paragraphs. The

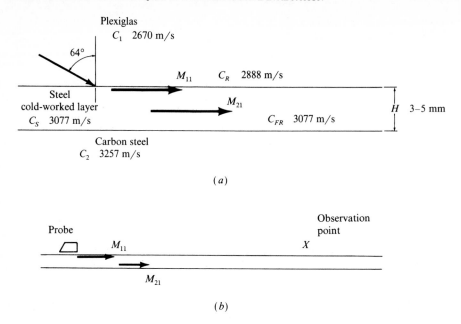

FIGURE 9-6

(*a*) Model of low speed zone over high speed material that excites fundamental (M_{11}) and first higher-order (M_{21}) Rayleigh waves. (*b*) Propagation path for M_{11} and M_{21} waves showing early arrival of the M_{21}.

cold-working resulted in a significant reduction in the shear wave speed in this upper region, creating the physical conditions necessary for the existence of the higher-order Rayleigh wave mode along with the fundamental mode. Later work by Hirao, Kyukawa, Sotani, and Fukuoka [87], also using railroad rail, extended the earlier work of Bray et al. [86] by using carefully controlled excitation frequencies to investigate the propagation characteristics of several of the higher-order Rayleigh wave modes. Their work specifically addressed the potential applications of these wave forms to nondestructive evaluation in railroad rail.

Investigations into the application of the first higher-order Rayleigh wave mode to tasks other than railroad rail inspection led to the possibility of inspecting stainless steel clad material since the shear wave speed in stainless steel material is approximately equal to that of the cold-worked layer in railroad rail. Moreover, thicknesses in the 3–5 mm range that exist for the railroad rail are also found in many stainless steel overlay applications. Figure 9-8 shows a layered model of a stainless steel overlay using wave speeds from typical industrial sources.

The excitation of the first higher-order Rayleigh wave mode in a stainless steel overlay was studied using a 50 mm (2 in) thick carbon steel plate with a stainless steel overlay ~ 9 mm ($\frac{3}{8}$ in) thick, as shown in Fig. 9-9. The overlay

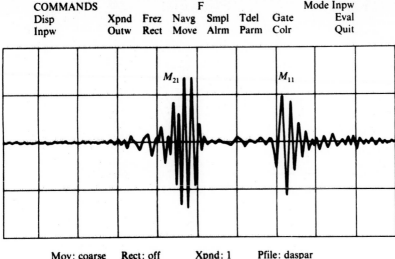

COMMANDS			F				Mode Inpw
Disp	Xpnd	Frez	Navg	Smpl	Tdel	Gate	Eval
Inpw	Outw	Rect	Move	Alrm	Parm	Colr	Quit

Mov: coarse	Rect: off	Xpnd: 1	Pfile: daspar
Del: 140.800	Sdel: 140.800	Tdel: 0.000	Wrt pro: no
Smp: 20.000	Lnth: 36.00	Avrg: 1	Ifil: e:vday005.dat
Gst: 44.800	Glen: 160.000	Pnts: 3200	Ofil: e:data000.dat

FIGURE 9-7
Typical arrivals of M_{21} and M_{11} waves along a surface. Time base 3.6 μs/div. [Full screen time = (screen points × Xpnd)/(sampling rate (MHz)) = 720 × 1(20 × 10)^6 = 36 μs; screen points = the number of horizontal pixels on the display.]

weld material was in rows ~ 25 mm (1 in) apart and the plate was approximately 0.5 m (20 in) square.

The higher-order Rayleigh waves were excited in the overlay using a 64° Plexiglas wedge and a 2.25 MHz commercial transducer. Moreover, weld defects at the interface were found with the technique at the locations marked X on the plate. Thus, the higher-order Rayleigh waves have demonstrated a potential to

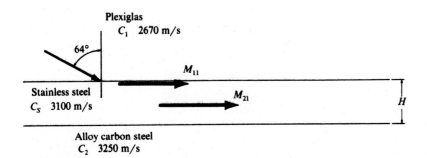

FIGURE 9-8
Stainless steel overlay model that supports the propagation of the M_{21} wave.

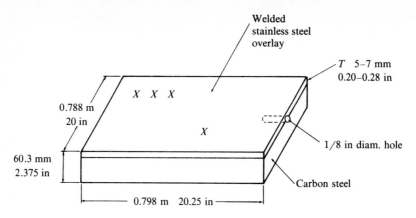

X = defects found with higher-order mode Rayleigh wave

FIGURE 9-9
Carbon steel plate with stainless steel overlay.

inspect stainless steel overlay in a manner far more efficient than the conventional single point contact probe method.

9-4.2 Using Critically Refracted Longitudinal (L_{CR}) Waves for Near Surface Inspections

At the first critical angle, the critically refracted longitudinal mode (L_{CR}) wave is excited. This wave travels at a speed near to the bulk wave speed and just underneath the surface. Where in previous discussion it was presented as being useful for stress measurement, there are also possibilities that it could be used for inspecting for defects near to the surface of pressure vessels, piping, and welds. The unique feature of this wave is that it will be the earliest arrival along its travel path, hence the arrival will not be cluttered with other modes and reflections. The usefulness of this wave form for inspection near to the surface has been discussed by Smith [88].

9-5 GRAIN SIZE DETERMINATION USING ULTRASONIC PULSE SCATTERING

The earlier discussion on ultrasonic pulse attenuation noted that the attenuation rate was a function of the pulse frequency as well as the relative grain size of the material. This has been further quantified by several investigators, for example, Refs. 13, 89, and 90. The authors cite several examples, notably using steel specimens, where correlations were found to exist between attenuation and grain size.

9-6 SUMMARY

It has been the intent of this section to show some possible directions for using ultrasonic analysis for situations other than simple flaw detection. Certainly, stress measurement, texture investigations, and inspection of layered materials represent important applications of nondestructive evaluation. With the material presented here and other sources as listed, the interested reader should be able to investigate any particular application of interest.

PROBLEMS
FOR
PART II

Chapter 5

5-1. The data for elastic constants and wave speed shown in the upper portion of Table 5-1 were obtained from a variety of sources. For several of the materials, use the given values for E, λ, μ, and ρ to calculate the wave speeds. How close is the agreement? What do you expect is the major cause of any disagreements?

5-2. Repeat Problem 5-1, except use C_1, C_2, C_l, C_t, and ρ to calculate E, ν, and k.

5-3. Use Eqs. (5-18), (5-19), and (5-20) to derive Eqs. (5-21), (5-22), and (5-23).

5-4. In Eqs. (5-29a), (5-29b), and (5-29c), the comment was made that the sign of the kx term determines the direction of propagation of the wave front. Show that this is true for both the positive and the negative sign.

5-5. Calculate the pressure reflection and transmission ratios P_r/P_i and P_t/P_i and the power reflection and transmission ratios $\mathrm{Pwr}_r/\mathrm{Pwr}_i$ and $\mathrm{Pwr}_t/\mathrm{Pwr}_i$ for the following interfaces: water/steel, steel/water, copper/steel, and steel/copper.

5-6. The amplitude loss for multiple echos of a 2.25 MHz pulse in aluminum is shown in Fig. 8-18(a). Using the vertical scale, calculate the attenuation coefficient α (dB/m) for this material. Compare this result with the value shown in the top left corner in Fig. 8-18(b). Would you expect a 10 MHz pulse to show any difference in attenuation? Explain your answer.

5-7. An aluminum plate 50 mm thick is to be inspected ultrasonically using a normal beam probe. If the attenuation coefficient in the material is expected to be 85 dB/m, calculate the expected ratio of the pulse–echo heights of two successive pulses.

5-8. The data shown in Table 5-4 indicate that, for the same frequency and the same material, shear waves are more attenuated than are longitudinal waves. Speculate on possible causes for this observed behavior.

5-9. The formation of pulses from superposed, continuous wave forms is shown in Fig. 5-15. Using illustrations, show the effect on the pulse shape of a slight displacement to the right of the lower wave form (B) relative to the upper wave form (A). Discuss how this describes the pulse shape change that may be caused by pulse dispersion. Give one or two examples of material conditions that may cause this type of dispersion. *Note*: The computer program USITUTOR may be used to obtain these illustrations, using slightly differing frequencies.

5-10. The formation of pulses from superposed, continuous wave forms is shown in Fig. 5-15. Using illustrations, show the effect on the pulse shape of a slight reduction in amplitude of the lower wave form (B) relative to the upper wave form (A). Discuss how this describes the pulse shape change that is caused by pulse dispersion. Give one or two examples of material conditions that may cause this type of dispersion. *Note*: The computer program USITUTOR may be used to obtain these illustrations.

Chapter 7

7-1. Calculate the reflection and transmission coefficients for a transducer using lead zirconate titanate (PZT) as the piezoelectric ceramic mounted on a Plexiglas wedge. Assume air to be the backing material.

7-2. Probe ring-down time is an important performance characteristic. Discuss the probable ring-down characteristics of the air-backed transducer described in Problem 7-1 as compared to the same PZT material backed with the $200:100$ tungsten/epoxy mixture. Silk [27] suggests a minimum acoustic impedance of 18.8×10^6 kg/m^2 s as a backing material for PZT. Elaborate on how this higher impedance material would affect probe performance.

7-3. Calculate and sketch the beam divergence angle ϕ and the near field N for the following combinations. Assume longitudinal waves.

(*a*) Material: aluminum
   Probe diameter: 12 mm, 19 mm
   Frequency: 250 kHz, 1 MHz, 2.25 MHz, 10 MHz

(*b*) Material: water
   Probe diameter: as in (*a*)
   Frequency: as in (*a*)

Note: The computer program USITUTOR may be used to perform the calculations and obtain the illustrations.

Chapter 8

8-1. The ultrasonic immersion system shown in Fig. P8-1 is used to inspect the steel billet. Indications are found at beam entrance locations A and B. The probe angle

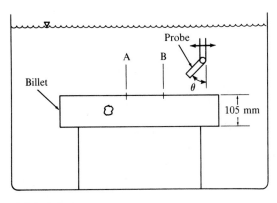
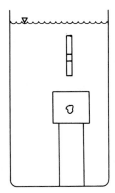

FIGURE P8-1
Immersion inspection arrangement.

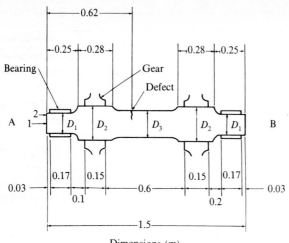

Dimensions (m)

$D_1 = 0.15$ $D_2 = 0.25$ $D_3 = 0.20$

FIGURE P8-2
Power transmission shaft for ultrasonic inspection.

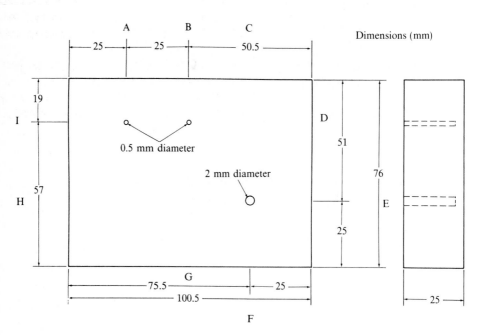

Dimensions (mm)

FIGURE P8-3(a)
Aluminum ultrasonic calibration block.

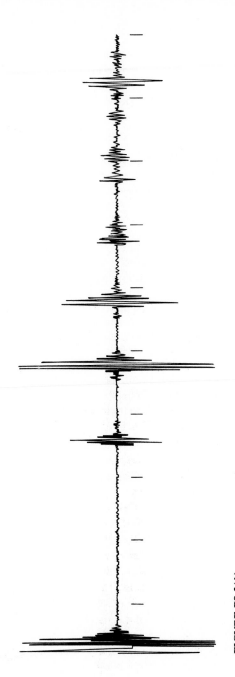

FIGURE P8-3(b)
Digitized ultrasonic A-scan display obtained from aluminum calibration block. The time base is 7.2 μs/div.

$\theta = 18.8°$. What would be the beam entrance angle? Estimate the relative height and the arrival time for the echos obtained for beam entrance locations at A and B. Assume $\alpha = 90$ dB/m.

8-2. The back-echo pattern obtained from an inspection of a diesel engine exhaust valve is shown in Fig. 8-16(a) and (b). A 2.25 MHz probe was used, the valve stem diameter was 14.2 mm, and the length was 128 mm. Identify the back echo and explain the source of the multiple echoes immediately following the back echo.

8-3. The steel shaft for a power transmission assembly as shown in Fig. P8-2 is to be inspected ultrasonically. Normally, the shaft would be inspected from both ends in order to be confident of the results. For the inspection, 19 mm ($\frac{3}{4}$ in) 1.0 and 2.25 MHz probes are to be used. Sketch the expected flaw detector screen appearance for the following conditions:

(a) End A, probe location 1, 1.0 MHz probe
(b) End A, probe location 2, 1.0 MHz probe
(c) End A, probe location 1, 2.25 MHz probe
(d) End A, probe location 2, 2.25 MHz probe
(e) End B, probe location 1, 1.0 MHz probe
(f) End B, probe location 2, 1.0 MHz probe
(g) End B, probe location 1, 2.25 MHz probe
(h) End B, probe location 2, 2.25 MHz probe

8-4. A typical aluminum test block used for calibrating ultrasonic instruments is shown in Fig. P8-3(a). The locations of three drilled holes are indicated. The digitized display in Fig. P8-3(b) shows the ultrasonic signal display obtained with the probe located at one of the locations A through I, on the narrow portion of the block. A 2.25 MHz, 12.7 mm ($\frac{1}{2}$ in) diameter transducer is used. The time base is at 72 μs full screen or 7.2 μs/div. For this problem:

(a) Determine the location of the probe on the test block and indicate this on the diagram.
(b) Identify the back echo. Justify your choice.
(c) Sketch the ray paths for all arrivals between the initial echo and the back echo.

REFERENCES
FOR
PART II

1. Lindsay, R. B., *Mechanical Radiation*, McGraw-Hill, New York, 1960.
2. Kolsky, H., *Stress Waves in Solids*, Dover, New York, 1963.
3. Kinsler, L. E., Krey, A. R., Coppens, A. B., and Sanders, J. V., *Fundamentals of Acoustics*, 3d ed., Wiley, New York, 1982.
4. Hudson, J. A., *The Excitation and Propagation of Elastic Waves*, Cambridge University Press, Cambridge, 1980.
5. Skudrzyk, E., *Simple and Complex Vibratory Systems*, The Pennsylvania State University Press, University Park, PA, 1968.
6. Pierce, A. D., *Acoustics: An Introduction to Its Physical Principles and Applications*, McGraw-Hill, New York, 1981.
7. Graff, K., *Wave Motion in Elastic Solids*, Ohio State University Press, Columbus, OH, 1975.
8. Flinn, R. A., and Trojan, P. K., *Engineering Materials and Their Applications*, 3d ed., Houghton-Mifflin Company, Boston, MA, 1986.
9. Hertzberger, R. W., *Deformation and Fracture Mechanics of Engineering Materials*, Wiley, New York, 1976.
10. Krautkramer, J., and Krautkramer, H., *Ultrasonic Testing of Materials*, 3d ed., Springer-Verlag, Berlin, 1983.
11. Redwood, M., *Mechanical Waveguides*, Pergamon, Oxford, 1960.
12. Kuhn, G. J., and Lutsch, A., "Elastic Wave Conversion at a Solid-Solid Boundary with Transverse Slip," *Journal of the Acoustical Society of America*, vol. 33, no. 7, pp. 949–954, July 1961.
13. Serabian, S., "Frequency and Grain Size Dependence of Ultrasonic Attenuation in Polycrystalline Materials," *British Journal of Nondestructive Testing*, vol. 22, no. 2, pp. 69–77, March 1980.
14. Bray, D. E., Najm, M., and Cornwell, L. R., "Optimization of Ultrasonic Flaw Detection in Railroad Rail," Report DOT/RSPA/DMA-50/83/18, Department of Transportation, Research and Special Programs Administration, NTIS, Springfield, VA, 1983.
15. Egle, D. M., and Bray, D. E., "Nondestructive Measurement of Longitudinal Rail Stresses: Application of the Acoustoelastic Effect to Rail Stress Measurement," Report FRA-ORD-77-341, PB-281164, Federal Railroad Administration, Washington, DC, NTIS, Springfield, VA, January 1978.
16. Cook, E. G., and Van Valkenberg, H., "Surface Waves at Ultrasonic Frequencies," Bulletin No. 198, American Society for Testing and Materials, May 1954.
17. Viktorov, I., *Rayleigh and Lamb Waves*, Plenum, New York, 1967.
18. Bray, D. E., and Finch, R. D., "Flaw Detection in Model Railway Wheels," Report FRA-RT-71-75, PB-199-956, Federal Railroad Administration, NTIS, Springfield, VA, 1971.

19. Sezawa, K., "Dispersion of Elastic Waves Propagating on the Surface of Stratified Bodies and on Curved Surfaces," *Bulletin of the Earthquake Research Institute*, vol. 3, pp. 1–18, 1927.
20. Lehfeldt, W., "Ultrasonic Testing of Sheets with Lamb Waves," *Materialpruf*, vol. 4, no. 9, pp. 331–337, September 1962.
21. Egle, D. M., and Bray, D. E., "Nondestructive Measurement of Longitudinal Rail Stresses," Report FRA-ORD-76-270, PB-272061, Federal Railroad Administrtion, NTIS, Springfield, VA, 1975.
22. Erwin, W. S., "Vital Measurement of Aircraft Parts Made by Supersonic Measurement," *Steel*, vol. 16, no. 10, pp. 131, 188, 190, 192, March 1945.
23. Firestone, F. A., "The Supersonic Reflectoscope for Interior Inspection," *Metal Progress*, vol. 48, no. 3, pp. 505–512, September 1945.
24. Simmons, E., "The Supersonic Flaw Detector," *Metal Progress*, vol. 48, no. 3, pp. 513–516, September 1945.
25. Firestone, F. A., "The Supersonic Reflectoscope, an Instrument for Inspecting the Interior of Solid Parts by Means of Sound Waves," *Journal of the Acoustical Society of America*, vol. 17, no. 3, pp. 287–299, January 1946.
26. Hueter, T. F., and Bolt, R. H., *Sonics*, Wiley, New York, 1955.
27. Silk, M. G., *Ultrasonic Transducers for Nondestructive Testing*, Adam Hilger Ltd., Bristol, 1984.
28. Redwood, M., "A Study of Waveforms in the Generation and Detection of Short Ultrasonic Pulses," *Applied Materials Research*, vol. 2, no. 2, pp. 76–84, April 1963.
29. Egerton, H. B., *Non-Destructive Testing*, Oxford University Press, London, 1969.
30. Washington, A. B. G., "The Design of Ultrasonic Probes," *British Journal of Nondestructive Testing*, vol. 3, no. 3, pp. 56–63, September 1961.
31. Kossoff, G., "The Effects of Backing and Matching on the Performance of Piezoelectric Ceramic Transducers," *IEEE Transactions on Sonics and Ultrasonics*, vol. SU-13, no. 1, pp. 20–30, March 1966.
32. Posakony, G. J., "Influence of the Pulser on the Ultrasonic Spectrum: The Results of an Experiment," *Materials Evaluation*, vol. 43, no. 4, pp. 413–419, March 1985.
33. Kwun, H., Burkhardt, G., and Teller, C., "Ultrasonic Transducer Performance Requirements," *NTIAC Newsletter*, vol. 9, no. 8, Southwest Research Institute, San Antonio, TX, February 1982.
34. Bredael, I., "Characterization of Ultrasonic Transducers," in R. S. Sharpe (ed.), *Research Techniques in Nondestructive Testing*, vol. III, Academic, London, 1977, chap. 5.
35. Papadakis, E. P., "Use of Computer Model and Experimental Methods to Design, Analyze and Evaluate Ultrasonic NDE Transducers," *Materials Evaluation*, vol. 41, no. 12, pp. 1378–1388, November 1983.
36. Cross, B. T., "Sound Beam Directivity: A Frequency Dependent Variable," TR 70-23, Automation Industries, Boulder, CO, April 1970.
37. Hall, K. G., "Visualization Techniques for the Study of Ultrasonic Wave Propagation in the Railway Industry," *Materials Evaluation*, vol. 42, no. 7, pp. 922–929, 933, June 1984.
38. Beissner, R. E., "Electromagneto-Acoustic Transducers: A Survey of the State of the Art," NTIAC-76-1, Nondestructive Testing and Analysis Center, Southwest Research Institute, San Antonio, TX, January 1976.
39. Dobbs, E. R., "Electromagnetic Generation of Ultrasound," in R. S. Sharpe (ed.), *Research Techniques in Nondestructive Testing*, vol. II, Academic, London, 1973, chap. 13.
40. Scruby, C. B., Dewhurst, R. J., Hutchins, D. A., and Palmer, S. B., "Laser Generation of Ultrasound in Metals," in R. S. Sharpe (ed.), *Research Techniques in Nondestructive Testing*, vol. V, Academic, London, 1982, chap. 8.
41. Szilard, J. (ed.), *Ultrasonic Testing, Nonconventional Testing Techniques*, Wiley, Chichester, 1982.
42. Bray, D. E., "Ultrasonic Pulse Propagation in the Cold-Worked Layer of Railroad Rail," Report FRA/ORD-77/09.2, PB-281166, Federal Railroad Administration, NTIS, Springfield, VA, 1978.
43. Vezina, G. A., private communication, 1983.
44. Singh, G. P., "Inspection for Hydrogen Damage in Boiler Waterwall Tubes," *Materials Evaluation*, vol. 43, no. 10, pp. 1164, 1166, September 1985.

45. Silk, M. G., "Ultrasonic Techniques for Inspecting Austenitic Welds," in R. S. Sharpe (ed.), *Research Techniques in Nondestructive Testing*, vol. IV, Academic, London, 1980, chap. 11.

46. Kupperman, D. S., and Reimann, K. J., "Deviation of Longitudinal and Shear Waves in Austenitic Stainless Steel Weld Metal," *Nondestructive Evaluation in the Nuclear Industry—1980*, American Society for Metals, Metals Park, OH, 1980, pp. 239–253.

47. Bray, D. E., "Ultrasonic Angle-Beam Inspection Through the Cold-Worked Layer in Railroad Rail," *NDT International*, vol. 18, no. 3, pp. 139–144, June 1985.

48. Lehfeldt, E., and Höller, P., "Lamb Waves and Lamination Detection," *Ultrasonics*, vol. V, pp. 225–257, October 1967.

49. Silk, M. G., and Lidington, B. H., "A Preliminary Study of the Effect of Defect Shape and Roughness on Ultrasonic Size Estimation," *Nondestructive Testing*, vol. 8, no. 1, pp. 27–31, February 1975.

50. Silk, M. G., "Sizing Crack-like Defects by Ultrasonic Means," in R. S. Sharpe, (ed.), *Research Techniques in Nondestructive Testing*, vol. III, London, Academic, 1977, chap. 2.

51. Cosgrove, D. G., "Check-out of the Delta Technique," Report DP70-001, General Dynamics, Convair Aerospace Division, October 1970.

52. Silk, M. G., and Lidington, B. H., "The Potential of Diffracted Ultrasound in the Determination of Crack Depth," *Nondestructive Testing*, pp. 146–151, vol. 8, June 1975.

53. Thompson, R. B., "Application of Elastic Wave Scattering Theory to the Detection and Characterization of Flaws in Structural Materials," *Wave Propagation in Inhomogeneous Media and Ultrasonic Nondestructive Evaluation*, Applied Mechanics Division (AMD), vol. 62, New York, American Society of Mechanical Engineers, 1984, pp. 61–73.

54. Gruber, G. J., Hendrix, G. J., and Schick, W. R., "Characterization of Flaws in Piping Systems using Satellite Pulses," *Materials Evaluation*, vol. 42, no. 4, pp. 426–432, April 1984.

55. Gruber, G. J., and Jackson, J. L., "Ultrasonic Evaluation of Structural Integrity using Satellite Pulses," *Proceedings of the Fifteenth Symposium on Nondestructive Evaluation*, San Antonio, TX, April 23–25, 1985, pp. 97–101.

56. Special Issue on Imaging, *Materials Evaluation*, vol. 40, no. 1, January 1982.

57. Brunk, J. A., "Application Notes on Ultran's Air/Gas Propagation Transducers," EPN-103, Ultran Laboratories, State College, PA, 1987.

58. Adler, L., Cook, K. V., and Simpson, W. A., "Ultrasonic Frequency Analysis," in R. S. Sharpe (ed.), *Research Techniques in Nondestructive Testing*, vol. III, Academic, London, 1980, chap. 1.

59. Whalen, M. F., Sproules, H. S., Fulghum, C. B., and Gordon, J. H., "An Automated System Using a Multivariate Waveform Classified for Ultrasonic NDE of Heat-Fused PE Pipe Joints," *Materials Evaluation*, vol. 42, pp. 1638–1643, December 1984.

60. Egle, D. M., and Bray, D. E., "Measurement of Acoustoelastic and Third-Order Elastic Constants for Rail Steel," *Journal of the Acoustical Society of America*, vol. 60, no. 3, pp. 741–744, September 1976.

61. King, R. R., Birdwell, J. A., Bray, D. E., Clotfelter, W.N., and Risch, E. R., "Improved Methods for Nondestructively Measuring Residual Stresses in Railway Wheels," *Proceedings of the Ninth Symposium on Nondestructive Evaluation*, San Antonio, TX, April 1973, pp. 91–105.

62. Noronha, P. J., Chapman, J. R., and Wert, J. J., "Residual Stress Measurement and Analysis Using Ultrasonic Techniques," *Journal of Testing and Evaluation*, vol. 1, no. 3, pp. 209–215, May 1973.

63. Noronha, P. J., and Wert, J. J., "An Ultrasonic Technique for the Measurement of Residual Stresses," *Journal of Testing and Evaluation*, vol. 3, no. 2, pp. 147–152, March 1975.

64. Becker, F. L., "Ultrasonic Determination of Residual Stress," Battelle-Pacific Northwest Laboratories, Richland, WA, January 1973.

65. Hsu, N. J., "Acoustical Birefringence and the Use of Ultrasonic Waves for Experimental Stress Analysis," *Experimental Mechanics*, vol. 14, no. 5, pp. 169–176, May 1974.

66. Clotfelter, M. W., and Risch, E. R., "Ultrasonic Measurement of Stress in Railroad Wheels and in Long Lengths of Welded Rail," NASA TM X-64863, NASA Technical Memorandum, July 1974.

67. Hughes, D. S., and Kelly, J. L., "Second-Order Elastic Deformation of Solids," *Physical Review*, vol. 92, no. 4, pp. 1145–1149, December 1953.
68. Murnaghan, D., *Finite Deformations of an Elastic Solid*, Wiley, New York, 1951.
69. Egle, D. M., and Bray, D. E., "Application of the Acoustoelastic Effect to Rail Stress Measurement," *Materials Evaluation*, vol. 37, no. 4, pp. 41–46, 55, March 1979.
70. Thompson, R. B., Smith, J. F., and Lee, S. S., "Effects of Microstructure on the Acoustoelastic Measurements of Stress," in O. Buck and S. Wolf (eds.), *Nondestructive Evaluation: Application to Materials Processing*, The American Society for Metals, Metals Park, OH, 1984, pp. 137–145.
71. Bradfield, G., "Strength, Elasticity and Ultrasonics," *Ultrasonics*, vol. 10, no. 4, pp. 166–172, 1972.
72. Bray, D. E., and Leon-Salamanca, T., "Zero-Force Travel-Time Parameters for Ultrasonic Head Waves in Railroad Rail," *Materials Evaluation*, vol. 43, pp. 854–858, 863, June 1985.
73. Williams, H. D., Armstrong, D., and Robins, R. H., "Application of Ultrasonic Stress Measurements to Problems in the Electric Supply Industry," *Journal of Testing and Evaluation*, vol. 10, no. 5, pp. 217–222, September 1982.
74. Brokowski, A., and Deputat, J., "Ultrasonic Measurements of Residual Stresses in Rails," *Proceedings 11th World Conference on Nondestructive Testing*, vol. 1, American Society for Nondestructive Testing, Columbus, OH, 1985, pp. 592–598.
75. James, M. R., and Buck, O., "Quantitative Nondestructive Measurements of Residual Stresses," *CRC Critical Reviews in Solid State and Material Sciences*, pp. 61–105, August 1980.
76. Aurora, A., and James, M. R., "Ultrasonic Measurement of Residual Stress in Textured Materials," *Journal Testing and Evaluation*, vol. 10, no. 5, pp. 212–216, September 1982.
77. Crecraft, D. I., "Ultrasonic Measurement of Stress," in J. Szilard (ed.), *Ultrasonic Testing*, Wiley, Chichester, 1982, chap. 11, pp. 437–458.
78. Allen, D. R., Cooper, W. H. B., Sayers, C. M., and Silk, M. G., "The Use of Ultrasonics to Measure Residual Stress," *Research Techniques in Nondestructive Testing*, vol. VI, Academic, London, 1982, chap. 4, pp. 151–209.
79. Pao, Y.-H., Sachse, W., and Fukuoka, H., "Acoustoelasticity and Ultrasonic Measurements of Residual Stress," *Physical Acoustics*, vol. XVII, Academic, New York, 1984, pp. 61–143.
80. Rudd, C., and Green, R. E. Jr. (eds.), *Nondestructive Methods for Material Property Determination*, Plenum, New York, 1984.
81. Green, R. E. Jr., *Treatise on Materials Science and Technology: Ultrasonic Investigation of Mechanical Properties*, vol. 3, Academic, New York, 1973.
82. Barrett, C. S., and Massalski, T. B., *Structure of Metals*, 3d ed., McGraw-Hill, New York, 1966.
83. Sullivan, P. F., and Papadakis, E. P., "Ultrasonic Double Refraction in Worked Metals," *Journal of the Acoustical Society of America*, vol. 33, no. 1, pp. 1622–1624, November 1961.
84. Bray, D. E., and Egle, D. M., "Ultrasonic Studies of Anisotropy in Cold-Worked Layer of Used Rail," *Metal Science*, vol. 15, pp. 574–582, November–December 1981.
85. Haüsler, E., "Elastic Waves on the Surface of Solid, Especially Stratified Media, and Their Application to Nondestructive Testing," *Materialprüf*, vol. 2, no. 2, pp. 51–55, 1960.
86. Bray, D. E., Egle, D. M., and Reiter, L., "Rayleigh Wave Dispersion in the Cold-Worked Layer of Used Railroad Rail," *Journal of the Acoustical Society of America*, vol. 64, no. 3, pp. 845–851, 1978.
87. Hirao, M., Kyukawa, M., Sotani, Y., and Fukuoka, H., "Rayleigh Wave Propagation in a Solid with a Cold-Worked Surface Layer," *Journal of Nondestructive Evaluation*, vol. 2, no. 1, pp. 43–49, 1981.
88. Smith, P. H., "Practical Application of 'Creeping' Waves," *British Journal of NDT*, vol. 29, no. 5, pp. 318–321, 1987.
89. Shyne, J. C., Grayeli, N., and Kino, G., "Acoustic Properties as Microstructural Dependent Materials Properties," *Nondestructive Evaluation: Microstructural Characterization and Reliability Strategies*, Warrendale, PA, The Metallurgical Society of AIME, pp. 133–146, 1981.
90. Szilard, J., "Examining the Grain Structure of Metals," in J. Szilard (ed.), *Ultrasonic Testing*, Wiley, Chichester, 1982, chap. 6, pp. 4217–4261.

PART III

MAGNETIC FLUX LEAKAGE TECHNIQUES IN NONDESTRUCTIVE EVALUATION

CHAPTER
10

BASIC
MAGNETISM
FOR NDE

10-1 INTRODUCTION

Although the subject of magnetism is very well documented as advances have been made, advances in the use of magnetic techniques for nondestructive testing and evaluation are to be found only in research articles, despite the fact that such techniques are widely used. In principle, magnetic nondestructive testing consists of magnetizing a test part, generally a ferromagnetic material, and scanning its surface with some form of flux-sensitive sensor [1]. It is therefore essential that the practitioner have a clear understanding of both the magnetizing techniques used to produce magnetic flux leakage (MFL) from surface and subsurface flaws, and how commonly used MFL sensors react to such fields.

Commonly used magnetizing methods consist of applying current directly to the part, if permitted by the relevant inspection specifications, or the inducement of magnetic flux in the part due to the proximity of some magnetizing agent, such as a permanent magnet or electrical current through a conductor. Once maintained at some relatively high flux density level, the variety of sensors that may be used include coils, C-core yokes, solid state magnetic sensors, either alone or in arrays, and magnetic powder. Historically, the use of magnetic particles has been termed magnetic particle inspection (MPI), while the use of other forms of flux-sensitive sensor is known as magnetic flux leakage inspection (MFL). MPI is a small subset of magnetic flux leakage testing.

Much of the development of magnetism in the first four chapters of this part of the book is more general than might be considered necessary for the study of magnetic testing at the technical level. What distinguishes the early material from a physics text is that applications and questions are taken from NDE. Later chapters consider in detail the reaction of flaws in steel to the application of magnetic fields and give examples of commonly found magnetic techniques for product evaluation. A short chapter on demagnetization ends this part.

10-2   GENERAL PHYSICAL UNITS

Two common systems of units are used in this text [2]. This occurs because while scientists work in the International System (SI) of units and, therefore, use the ampere per meter for the magnetic field intensity (H) and the tesla (or weber per square meter) for the flux density (B), much of the available equipment is calibrated in oersteds (for H) and gauss (for B). Where possible, conversions between the two sets of units are given.

As magnetism historically evolved separately from electricity, different systems of units also evolved that seemed to be best fitted to the particular subject. In the first major step toward unification of units, the CGS system, length was measured in centimeters, mass in grams, and time in seconds. These lead, in mechanics, to such derived units as the dyne for force (1 dyn = 1 g cm s^{-2}) and the erg for energy (1 erg = 1 dyn cm = 1 g cm^2 s^{-2}). Unfortunately, however, if the system is developed to include electrical units, the common practical units such as the ampere for current, the volt for electrical pressure, and the ohm for electrical resistance do not follow naturally. If, on the other hand, it is required that the system of units in which we work include the practical electrical units, then the SI system must be used. In this system, length is measured in meters, mass in kilograms, time in seconds, and current in amperes. Derived units include force in newtons (1 N = 1 kg m s^{-2}), energy in joules (1 J = 1 N = kg m^2 s^{-2}), and work rate in watts (1 W = 1 J/s = 1 kg m^2 s^{-3}).

At this point, requiring that the magnetic permeability of vacuum in the CGS system be unity eliminates the use of common practical units for electromagnetism. However, by demanding that practical units appear within the framework of the SI system, the compromise that must be made is that the permeability of vacuum is numerically

$$\mu_0 = 4\pi \times 10^{-7} \text{ H/m (henrys/meter)} \tag{10-1}$$

So defined, this rationalization process within the meter-kilogram-second-ampere system permits electrical potential to be measured in volts, resistance in ohms, and inductance in henrys. The intensity of a magnetic field is measured in amperes per meter and the flux density is in tesla or webers per square meter.

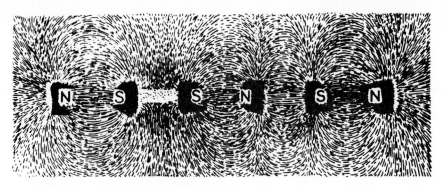

FIGURE 10-1
Fields of bar magnets revealed by iron filings. (*Reproduced with permission from "Magnetism" in Encyclopaedia Britannica, 14th edition, © 1973 by Encyclopaedia Britannica, Inc.*)

10-3 MAGNETIC POLE

When magnetism was first studied, it was found convenient to introduce the concept of the magnetic pole. Such poles are often used in magnetic NDE for modeling flux leakage fields from flaws (see Chaps. 16 and 17). In this chapter, the pole is used to develop some of the more elementary concepts of magnetism.

The region outside of a bar or other magnet, in which the magnet exerts some influence, is known as the magnetic field. Its intensity is defined shortly. Such fields can be detected by their influence on magnetic particles and measured with a gaussmeter. Figure 10-1 shows such fields in the vicinity of three bar magnets. It is relatively easy to see why the concept of the magnetic pole arose; the magnet can be replaced by two point sources from which the external field is considered to have arisen.

These point sources are the north pole, from which field lines are considered to emerge, and the south pole, into which they are absorbed. The strength of individual poles is given by the coulombic force between them,

$$\overline{F} = \left(\mu_0 p_1 p_2 / 4\pi r^2\right)\hat{r} \tag{10-2}$$

where p_1 and p_2 are the pole strengths, r is the distance between the poles, and $\hat{r}$ is a unit vector in the r direction. North poles have positive pole strengths while south poles have negative pole strengths. This permits repulsive forces between like poles and attractive forces between unlike poles.

In the SI system, the force F is in newtons, the distance r is in meters, and the pole strengths are in ampere meters.

10-4 LINES OF FORCE

Field lines are also known as lines of force, perhaps because of the way in which magnetic particles join together to make the magnetic field visible. The following

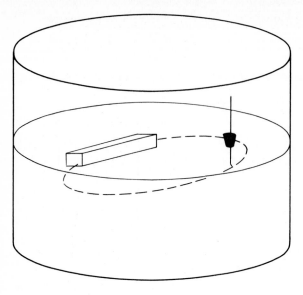

FIGURE 10-2
Flotation method for magnetic field lines.

properties are given to such lines:

1. They begin and end on poles.
2. They appear to repel each other.
3. Their direction is that in which a free north pole would travel, if allowed to do so.

Property 3 can be demonstrated by sticking a magnetized needle through a cork as shown in Figure 10-2 so that it will float from one pole of a magnet to the other.

10-5 MAGNETIC FLUX DENSITY

If, in Eq. (10-2), the pole strength p_2 is set at 1 unit, then the force becomes the force per unit pole, and defines the magnetic flux density B:

$$\bar{B} = \left(\mu_0 p_1/4\pi r^2\right)\hat{r} \qquad (10\text{-}3)$$

The units are those of F/p, i.e., newtons per ampere meter or webers per square meter.

10-6 MAGNETIC FIELD INTENSITY

Outside of the magnet, the strength of the magnetic field in air is related to the flux density through the relation

$$\overline{H} = B/\mu_0 \tag{10-4}$$

From Eq. (10-3), it can be seen that

$$\overline{H} = \left(p_1/4\pi r^2 \right)\hat{r} \tag{10-5}$$

This defines the strength of the magnetic field, which is also known as the magnetic field intensity, the magnetizing force, or just simply, the field. It is clear that, since the units of H are those of $F/\mu_0 p$, i.e., amperes per meter, H is not a force. In this text, H will be referred to as the magnetic field intensity. Further, the term *magnetic field* will not be used in other than the general sense.

In the CGS system, the unit of field intensity is the oersted and that of flux density is the gauss. In air, oersteds and gauss are numerically equal. The value of the earth's field intensity is about 0.2 Oe (16 A/m), while that a short distance from the pole of a magnet might be 1000 Oe (79,580 A/m). Small changes in the earth's magnetic field strength, such as are of importance to geophysicists, are measured in gammas, 1 gamma $= 10^{-5}$ Oe (8×10^{-4} A/m).

10-7 MAGNETIC FIELD INTENSITY DUE TO A BAR MAGNET

Equation (10-5) is important because it can be generalized so as to develop the magnetic field intensity from a collection of poles, i.e.,

$$H = (1/4\pi)\left(\sum p_i/r_i^2 \right)\hat{r}_i \tag{10-06}$$

where p_i is the pole strength of the ith pole, r_i is its distance from the field point, and the $\hat{r}_i$ are the associated unit vectors.

The simplest case is that of a bar magnet that has two poles of equal and opposite strength [3]. As shown in Fig. 10-3, the total force at the point Q is the vector resultant of forces due to the north $(+p)$ and south $(-p)$ poles separated by the distance $2d$. A unit pole at Q is influenced by forces

$$F_{1a} = \left(\mu_0 p/4\pi r_a^2 \right)\hat{r}_a$$

$$F_{1b} = \left(-\mu_0 p/4\pi r^2 \right)\hat{r}_b$$

as shown. It is relatively easy to show that the resultant field intensity at Q is given by

$$H_Q = \begin{cases} \dfrac{F_{1x}}{\mu_0} = \dfrac{p(d+x)}{\left[y^2 + (d+x)^2 \right]^{3/2}} + \dfrac{p(d-x)}{\left[y^2 + (d-x)^2 \right]^{3/2}} \\[4mm] \dfrac{F_{1y}}{\mu_0} = py\left| \dfrac{1}{\left[y^2 + (d+x)^2 \right]^{3/2}} + \dfrac{1}{\left[y^2 + (d-x)^2 \right]^{3/2}} \right| \end{cases} \tag{10-7}$$

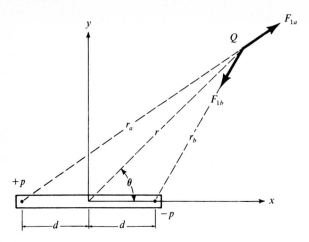

FIGURE 10-3
Forces at Q due to poles $+p$ and $-p$ as shown.

where the x and y components are found separately. It is clear that if many poles contribute to the final field intensity, this approach becomes unwieldy. Equation (10-7) provides a good idea of what the leakage field from a defect might look like, if modeled by two such poles.

Equation (10-7) simplifies considerably under two circumstances:

1. When Q lies on the y axis, putting $x = 0$ gives

$$H_x = 2pd/(y^2 + d^2)^{3/2}$$
$$H_y = 0 \tag{10-8}$$

and as d is made smaller,

$$H_x \rightarrow 2pd/y^3 \tag{10-9}$$

2. When Q lies on the x axis, putting $y = 0$ gives

$$H_x = -2(2pd)x/(x^2 - d^2)^2$$
$$H_y = 0 \tag{10-10}$$

and again, as d is made smaller,

$$H_x \rightarrow -4pd/x^3 \tag{10-11}$$

These equations illustrate the way in which the field intensity falls off along the two axes defined in Fig. 10-3.

10-8 POLE CONCEPT IN NDE

The field external to magnets, or to cracks or voids within ferromagnetic parts, can be modeled by assuming the existence of poles, or various densities of poles.

The pole concept has been used extensively in the USSR to model defect leakage fields; some of these ideas are further discussed in later chapters. The major problem that arises with such modeling is exactly how to relate the pole strengths and their densities to the level of magnetization around the flaw within the inspected part.

10-9 MAGNETIC MOMENT

If a small magnet is suspended in the field B of another magnet, coil, etc., then a couple acts on the suspended magnet that tends to turn it into the external field direction. Figure 10-4 shows how this couple acts. Its moment about the suspension point is the sum of the moments acting on each pole, i.e.,

$$\Gamma = (pB)\, d \sin \theta + (pB)\, d \sin \theta$$

$$= (2pd)\, B \sin \theta \tag{10-12}$$

The quantity $m = 2pd$ is the magnetic moment of the magnet. The more general vector relation is

$$\overline{\Gamma} = \overline{m} \times \overline{B} \tag{10-13}$$

where the magnetic moment of the magnet is a vector quantity, the direction of which is that of the magnet axis running from the S pole toward the N pole within the magnet. Magnetic moment is an important quantity in magnetism, and is more fundamental than this simple theory shows. It can be measured accurately for atomic magnets, where p and d lose their individual meanings.

The work done in turning the magnet through a small angle, $d\theta$ at some ambient angle θ, by the external field is given by

$$dE_p = 2(pB \sin \theta)(d)(d\theta)$$

$$= mB \sin \theta\, d\theta$$

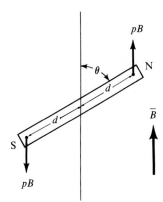

FIGURE 10-4
Forces on a suspended magnet constituting a couple.

From this, the total energy change of the magnet is

$$E_p = \int mB \sin \theta \, d\theta$$

if it is assumed that $E_p = 0$ at $\theta = 90°$. Integration gives

$$E_p = -mB \cos \theta \tag{10-14}$$

From Eq. (10-14), it can be seen that:

1. $E_p = -mB$ if the magnet is parallel to the external field.
2. $E_p = 0$ if the magnet is perpendicular to the external field.
3. $E_p = +mB$ if the magnet is antiparallel to the external field.

In vector notation, Eq. (10-14) is written as

$$E_p = -\overline{m} \cdot \overline{B} \tag{10-15}$$

The couple of Eq. (10-13) is responsible for rotating magnetic particles in the vicinity of a defect into the leakage field direction. In this case, the defect field first magnetizes the particles, inducing N and S poles in them, which then interact with the leakage field by rotating into alignment. The induced N and S poles also attract other particles, so long chains of particles are formed and the leakage field is made visible. In the process, their energy is minimized.

10-10   MAGNETIC DIPOLE

In many cases, the length of the magnet can be allowed to shrink, and equations for the external field intensity can be derived that are simpler than those that occur when $2d$ is kept relatively large. In the following derivation, the magnetic moment m will be used for $2pd$. First, the magnetic scalar potential at the point Q of Fig. 10-3 that is due to $+p$ is stated as $\Omega = p/4\pi r_a$ and that due to $-p$ is $-p/4\pi r_b$. The total potential is then

$$\Omega = (p/4\pi)(1/r_a - 1/r_b) \tag{10-16}$$

since potentials are additive. Now, expressing r_a and r_b in terms of r, d, and θ, it can be seen that

$$r_a^2 = r^2 \left[1 + 2(d/r)\cos \theta + (d/r)^2\right]$$

$$r_b^2 = r^2 \left[1 - 2(d/r)\cos \theta + (d/r)^2\right]$$

by using the cosine rule and anticipating that $r \gg d$. Using these relations in Eq. (10-16) yields

$$\Omega = (p/4\pi r)\left[\{1 + 2(d/r)\cos \theta + (d/r)^2\}^{-1/2}\right.$$
$$\left. - \{1 - 2(d/r)\cos \theta + (d/r)^2\}^{-1/2}\right]$$
$$= (p/4\pi r)\left[\{1 - (d/r)\cos \theta - 1/2(d/r)^2 \cdots \}\right.$$
$$\left. - \{1 + (d/r)\cos \theta - 1/2(d/r)^2 \cdots \}\right]$$

using the binomial expansion.

$$\Omega = -(2pd)\cos\theta/4\pi r^2$$

keeping only the first term of the expansion, i.e.,

$$\Omega = -(m/4\pi r^2)\cos\theta \tag{10-17}$$

Note that the terms in $(d/r)^2$ cancel, so that the next term in the expansion will be of the order $(d/r)^3$. With $r \gg d$, this and higher terms are ignored. The remarkably simple Eq. (10-17) is used to find the magnetic field intensity of the pair of poles, known as a dipole, from $\overline{H} = -\nabla\Omega$. The gradient function ∇ is given in rectangular, cylindrical, and spherical coordinates, respectively, by

$$\nabla = \hat{i}\partial/\partial x + \hat{j}\partial/\partial y + \hat{k}\partial/\partial z$$

$$\nabla = \hat{r}\partial/\partial r + (\hat{\phi}/r)\,\partial/\partial\phi + \hat{z}\partial/\partial z$$

$$\nabla = \hat{r}\partial/\partial r + (\hat{\theta}/r)\,\partial/\partial\theta + \hat{\phi}(1/r\sin\theta)\,\partial/\partial\phi \tag{10-18}$$

In order to obtain H, the vector components of Eq. (10-17) must be found in the coordinate system that is most suited to the problem at hand. For example, in spherical coordinates,

$$H = \left[\hat{r}\partial/\partial r + (\hat{\theta}/r)\,\partial/\partial\theta + (\hat{\phi}/r\sin\theta)\,\partial/\partial\phi\right](m\cos\theta/4\pi r^2)$$

$$= -(2m/4\pi r^3)\cos\theta\hat{r} - (m/4\pi r^3)\sin\theta\hat{\theta} + 0 \tag{10-19}$$

since there is no $\hat{\phi}$ term.

The value of H at any point can be found from $H = (H_r^2 + H_\theta^2)^{1/2}$, i.e.,

$$H = (m/4\pi r^3)(3\cos^2\theta + 1) \tag{10-20}$$

This method eliminates the use of vector quantities in the geometrical part of the computation; only relatively simple derivatives are required, the vector relationships being carried by the gradient function. This is a powerful technique and widely used.

The angle that the magnetic field intensity vector makes with the direction of r is given by

$$\tan\beta = H_\theta/H_r = 0.5\tan\theta \tag{10-21}$$

If Eq. (10-17) is written in vector form, then it becomes

$$\hat{\Omega} = -\overline{m}\cdot\overline{r}/4\pi r^2 \tag{10-22}$$

A volume distribution of such dipoles will have a magnetic scalar potential of

$$\Omega = \sum(\overline{m}\cdot\overline{r}/4\pi r^2)\,dv \tag{10-23}$$

Many defect fields can be modeled very simply by the correct positioning of such a dipole or a collection of dipoles.

10-11 MAGNETIC EFFECTS OF STEADY CURRENTS

So far, simple concepts of magnetostatics have been outlined in order to visualize what defect leakage fields might look like. The issue of how test parts are magnetized was avoided. In the next few paragraphs, the magnetic effects of electric current in conductors are derived, leading to formulas that are applicable to commonly used magnetizing methods.

10-11.1 Magnetic Field Intensity Due to a Current in a Straight Wire

The general formula [4] for the magnetic field intensity, $\overline{H}$ at some point P, due to a current I in an element of conductor $\overline{dl}$ is given (without proof) by

$$\overline{H} = \frac{I}{4\pi} \int \frac{\overline{dl} \times \hat{r}}{r^2} \tag{10-24}$$

This convenient equation indicates that H:

1. is proportional to the current I that causes it
2. is related to the geometry of the full circuit that carries the current, and
3. is in a direction that is perpendicular to both the filament $\overline{dl}$ and the unit vector $\hat{r}$

Equation (10-24) is the starting point for the computation of H for a variety of simply and not-so-simply shaped conductors. For the straight wire of Fig. 10-5, which is not infinitely long, it can be seen that

$$z = a \cot \theta$$
$$dl = dz = d(a \cot \theta) = -a \csc^2\theta (d\theta)$$
$$r = a \csc \theta$$
$$\hat{r} = \bar{r}/|\bar{r}|$$

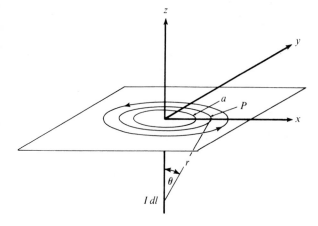

FIGURE 10-5
Coordinate axes for the computation of the magnetic field intensity at point P caused by current I in a long straight wire.

These values in Eq. (10-24) yield

$$\overline{H} = \frac{I}{4\pi} \int \frac{(-a \csc^2\theta \, d\theta)(r \sin \theta)(-\hat{j})}{r^2}$$

$$= (I\hat{j}) \int \sin \theta \, d\theta$$

$$= (I/4\pi a)(\cos \theta_1 - \cos \theta_2] \, \hat{j} \tag{10-25}$$

For an infinitely long, straight conductor, θ_1 can be approximated by 0 and θ_2 by π, so that Eq. (10-25) becomes

$$\overline{H} = (I/2\pi a) \, \hat{j} \tag{10-26}$$

In order to use this relation generally, note that there is an obvious symmetry about the z axis. From this symmetry, it can be seen that lines of constant H generate cylinders of constant radius about the conductor. Further, the vector cross product of Eq. (10-24) yields the familiar right hand rule for the direction of H (i.e., if the thumb of the right hand is placed along the conductor in the direction of the current, then the fingers curl around the conductor in the direction of H).

Expressed in CGS units, Eq. (10-26) becomes

$$H = 2I/10a \text{ Oe} \tag{10-27}$$

Equations (10-26) and (10-27) can be used to show that the connection between the ampere per meter and the oersted is

$$1 \text{ A/m} = 4\pi \times 10^{-3} \text{ Oe}$$

10-11.2 Axial Magnetic Field Intensity Due to a Current Loop

Considering a one-turn current loop in the (x, y) plane, as shown in Fig. 10-6, the element of the loop can be written as

$$\overline{dl} = a(-\hat{i} \sin \theta + \hat{j} \cos \theta) \, d\theta$$

and the vector $\bar{r}$ from the loop element to some point on the z axis is given by

$$\bar{r} = -\hat{i} \cos \theta - \hat{j}a \sin \theta + \hat{k}z$$

so that Eq. (10-24) becomes

$$\overline{H} = \frac{I}{4\pi} \int \frac{(\hat{i}za \cos \theta + \hat{j}za \sin \theta + \hat{k}a^2) \, d\theta}{(z^2 + a^2)^{3/2}}$$

The first two terms integrate to 0 and the third term gives

$$\overline{H} = \frac{I}{2} \frac{a^2}{(z^2 + a^2)^{3/2}} \hat{k} \tag{10-28}$$

which is along the z axis.

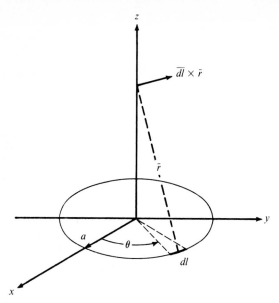

FIGURE 10-6
The vectors used in the computation of the magnetic field intensity along the axis of a coil.

10-11.3 Magnetic Field Intensity on the Axis of a Coil

For a coil of N turns, wrapped tightly together, ignoring the physical dimensions of the conductor, Eq. (10-28) yields

$$H = \frac{NI}{2} \frac{a^2}{\left(z^2 + a^2\right)^{3/2}} \tag{10-29}$$

Coils that roughly obey this relation are commonly used in magnetic NDE to magnetize parts either prior to or during inspection.

Equation (10-29) only gives the field intensity on the axis of the coil. The value of H at the general nonaxial point is much more difficult to compute, but it is not generally necessary to compute it, because it can be measured with a gaussmeter.

The field intensity at the center is found by putting $z = 0$ to give

$$H_c = NI/2a \text{ A/m} \tag{10-30}$$

That is, the field intensity at the center depends linearly upon the product of number of turns and the current, and inversely upon the coil radius or diameter. The product NI is known as the magnetomotive force of the coil and is measured in ampere turns.

The CGS formula equivalent to Eq. (10-30) is

$$H_c = 2\pi NI/10a \text{ Oe}$$

10-11.4 Magnetic Field Intensity Due to a Helmholtz Pair

A commonly used coil configuration is that of two coils of the same radius and number of turns on a common axis and separated by a distance that makes the second derivative of H zero on the axis midway between the two coils (Fig. 10-7). From Eq. (10-29), it can be deduced that the magnetic field intensity at the axial point P is given by

$$H_p = \frac{NIa^2}{2}\left[\frac{1}{(z^2 + a^2)^{3/2}} + \frac{1}{\{(2b - z)^2 + a^2\}^{3/2}}\right]$$

The reader should compute dH_p/dz and d^2H_p/dz^2, and show that (i) at $z = b$, the first derivative vanishes, and (ii) the second derivative also vanishes when $a = 2b$. At this separation,

$$H_c = 8NI/125^{1/2}a \tag{10-31}$$

The reader can also show that the first nonvanishing derivative is d^4H_p/dz^4 and also that, by expanding H_p as a Taylor series about $z = a/2$, the magnetic field intensity can be written as

$$H = H_c\left[1 - (144/125)\left(z - \frac{1}{2}a\right)/a^4\right] \text{A/m} \tag{10-32}$$

The importance of Helmholtz coil pairs is that very uniform fields can be obtained.

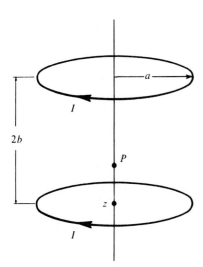

FIGURE 10-7
Double coil system. This is a Helmholtz system when $a = 2b$.

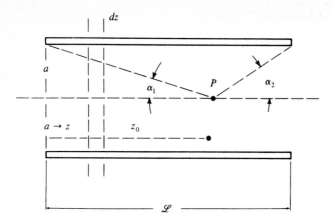

FIGURE 10-8
The evolution of a long solenoid from a series of short coils of length dz.

10-11.5   Magnetic Field Intensity of a Solenoidal Coil

A solenoid is made by winding a few layers of wire uniformly on an elongated tube, as shown in Fig. 10-8. The axial value of H is found by dividing the solenoid up into small length elements dz using Eq. (10-28) for each element and then taking the sum over the length of the coil. If the solenoid has N turns altogether, which are wound evenly over a length $\mathscr{L}$, then each element contains $N\,dz/\mathscr{L}$ turns and the magnetic field intensity at the point P, due to one such element, is given by

$$h_p = \frac{NI\,dz}{2\mathscr{L}} \frac{a^2}{\left[(z_0 - z)^2 + a^2\right]^{3/2}}$$

Integrating over the length of the coil gives

$$H_p = \frac{NIa^2}{2\mathscr{L}} \int \frac{dz}{\left[(z_0 - z)^2 + a^2\right]^{3/2}}$$

$$= \frac{NI}{2\mathscr{L}}[\cos \alpha_1 + \cos \alpha_2] \tag{10-33}$$

where α_1 and α_2 are the angles from P to the ends of the coil, as shown in Fig. 10-8. From this relation, it can be seen that:

1. At the center of the solenoid,

$$H_c = NI \cos \alpha/\mathscr{L} \text{ A/m}$$

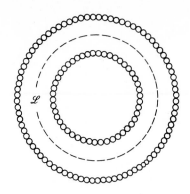

FIGURE 10-9
Ring solenoid.

2. At each end of the solenoid,

$$H_e = NI \cos \alpha / 2 \mathscr{L} \, \text{A/m}$$

3. If the solenoid is made infinitely long, which is generally considered to be the case when the length is greater than at least seven times the diameter of the windings, then the field intensity at the center is

$$H_c(\infty) = NI/\mathscr{L} \qquad (10\text{-}34)$$

Inspection of Eq. (10-33) indicates that as the point P passes outside the confines of the solenoid, the angle α_1 becomes more nearly equal to the complement of α_2, so that the cosine term in brackets becomes smaller and the magnetic field intensity decreases. Use is made of this when demagnetizing elongated parts; the coil is energized with ac to a field level that is sufficient to saturate the part, and as it is withdrawn from the coil, it experiences a smaller and smaller rapidly reversing field strength. This decreasing ac field acts to demagnetize the material, as long as it can penetrate into the material.

10-11.6 Ring Solenoid

If the solenoid of Fig. 10-8 is bent into a ring, then the ends are eliminated and the cosine terms are no longer present in Eq. (10-33). It simply becomes identical with Eq. (10-34). Further, this value of H exists at every point around the dashed circle of Fig. 10-9.

The importance of Eq. (10-34) and the ring solenoid is that it permits computation of the field intensity within ring-shaped specimens of magnetic materials. Now, if the value of the flux density within the material can be measured, the $B\text{-}H$ properties of the material can be investigated.

CHAPTER
11

MAGNETISM IN MATERIALS

11-1 INTRODUCTION

In Chap. 10, some of the elementary concepts relating to the nature of the field around a magnet were discussed. The units commonly used and the use of electric current to produce the field intensity necessary to magnetize the parts to be inspected were also discussed. Several simple formulas that are useful in magnetic NDE were derived. However, nothing has yet been introduced regarding the effects of such externally applied magnetic fields upon real flawed and unflawed materials. In this chapter, how external fields interact with different materials, which leads to their classification as dia-, para-, ferri-, ferro-, and antiferromagnetics, is discussed.

While magnetic NDE is limited to ferromagnetic materials because of their common use in industry, some of the sensors that are used to detect MFL fields contain materials (ferrites) that are more properly classed as ferrimagnetic. Most other common materials, which are generally referred to as being nonmagnetic, are in fact very weakly magnetic and classed according to whether they are repelled or attracted by a strong magnetic field, i.e., dia- or paramagnetic.

An overview of dia- and paramagnetism in materials is given before concentrating on those materials that are most used in magnetic NDE.

11-2 MAGNETIZATION

Since all matter consists of atoms and atoms contain electrons in motion, large numbers of orbital electrons exist in all materials. In substances that are good electrical or thermal insulators, these circulating electrons remain bound to their parent nuclei, but their circulatory motion permits them to be treated as currents, and so they produce magnetic fields. The basic difference between such insulating substances and those that are good conductors is that some of the outer shell electrons are lost from their original parent atoms when the substance condenses and become virtually free to wander about within the material. Application of a voltage to such conductors drives these "free" electrons preferentially in one direction, the resulting charge transport being an electric current. Charge transport motion produces magnetic fields, the strengths of which are proportional to the charge transport rates.

Atomic currents, as opposed to the electronic current, are due to bound electrons circulating around the nucleus; such circulating currents are magnetic dipoles and the external field that they create can be determined by specifying the magnetic dipole moment $\overline{m}$. For a collection of many dipoles in a volume ΔV, the macroscopic magnetization is defined as [3]

$$\overline{M} = \lim_{\Delta V \to 0} \frac{1}{\Delta V} \sum \overline{m}_i \tag{11-1}$$

where $\overline{m}_i$ is the moment of the ith dipole. How the summation occurs determines the magnetic nature of the material. The magnetization M gives a macroscopic description of the effect of all the atomic currents, since it measures their number per unit volume, multiplied by the average magnetic moment of each. In this way, it is possible to proceed from the atomic concept of magnetism to the more measurable macroscopic magnetization with which the inspector is familiar.

11-3   DIAMAGNETISM

If a magnetic field intensity is applied to a material, it modifies the atomic currents in a manner that tends to weaken the field strength [5]. Before the field is applied, in classical circular orbit theory, the electrostatic force of attraction between an electron at distance R from a nucleus is balanced by the centripetal force. After the field is applied, if the electron stays in its orbit, the orbit becomes distorted and the angular velocity of the electron changes by

$$\Delta\omega = \pm eB_m/2m_e$$

where e = the charge on the electron
  m_e = the electron mass
  B_m = the microscopic flux density at the position of the electron

It either speeds up or slows down depending upon the relative direction of the applied magnetic force in relation to the coulombic centripetal force. The orbital magnetic moment, which is defined as the product of the area of the current loop and the current circulating in the loop, is generally $-e\omega R^2/2$; thus a small change $\Delta\omega$ in ω causes a small change in the moment of

$$\Delta m = \pm\left(e^2R^2/2m_e\right)B_m$$

The magnetization is the sum of all such moment changes per unit volume, so for a large number N of similar atoms or molecules per unit volume,

$$\overline{M} = -\left(Ne^2/2m_e\right)\overline{B}_m\sum_i R_i^2$$

where the sum is taken over all of the electrons in the atom or molecule. In a volume where N is large, all orientations of R with respect to B_m will be present and so R_i^2 must be averaged over all such orientations. This averaging yields $r_i^2 = (2/3)R_i^2$ and so the magnetization becomes

$$M = -\left(Ne^2/6m_e\right)B_m\sum_i r_i^2$$

where r_i is the radius of an electron orbit, which can take on all orientations.

The magnetization per unit field intensity is known as the magnetic susceptibility and is given by

$$\chi_m = M/H \simeq M/H_m$$

or
$$\chi_m = -\left(\mu_0 Ne^2/6m_e\right)\sum_i r_i^2 \tag{11-2}$$

This result was derived in the classical manner by Langevin. A more complex approach replaces the classical electron orbits with Schrödinger wave functions, quantizes the orientations of the orbits, but yields essentially similar results.

Equation (11-2) indicates that diamagnetism must be present in all materials and that since none of the above quantities are temperature dependent, the diamagnetic susceptibility must also be virtually independent of temperature.

Diamagnetism is particularly prominent in materials with atoms or ions that have closed shells, for, in these cases, the diamagnetism is not masked by stronger paramagnetic effects or ferromagnetic behavior. Typical examples are monatomic rare gases, polyatomic gases with filled shells, ionic salts with filled shells, and covalent bonded elements, e.g., C, Si, and Ge.

11-4 PARAMAGNETISM

In addition to orbital electron motion, the nuclei and electrons in atoms and molecules also have their own axial spins. Thus orbital and spin moments occur, and must be added vectorially to give the total magnetic moment $\overline{m}_i$. Paramagnetism results when the spins tend to align in the external magnetic field. The magnetic energy associated with a magnetic moment that is aligned at angle θ to such a field, as previously given by Eq. (10-15), is

$$E_p = -\mu_0 mH\cos\theta \tag{10-15}$$

A system of many such moments at absolute temperature T possesses thermal and magnetic energies, which are continually exchanged. That is, transitions occur from one precessional state to another of different inclination.

Temperature randomizes such orientations, but those that are along (or nearly along) the field direction have a lower magnetic energy and are therefore preferred. In order to find the average over all orientations, Maxwell–Boltzmann statistics are employed, the result being

$$\langle m \cos \theta \rangle = \mu_0 m \left[\coth(\mu_0 mH/kT) - (kT/\mu_0 mH) \right]$$

For a system of N molecules per unit volume, this term becomes the magnetization M, so

$$M = \mu_0 Nm \left[\coth(\mu_0 mH/kT) - (kT/\mu_0 mH) \right] \qquad (11\text{-}3)$$

where k is the Boltzmann constant $(1.38 \times 10^{-23} \text{ J/K})$. This expression was first derived by Langevin. At relatively small field values, for which $\mu_0 mH/kT < 1$, the term in brackets approximates to $\mu_0 mH/kT$ and the magnetization becomes

$$M = \left(\mu_0 Nm^2/3kT \right) H$$

The low-field paramagnetic susceptibility (M/H) is given by

$$\chi_m = \mu_0 Nm^2/3kT \qquad (11\text{-}4)$$

It can be seen from Eq. (11-4) that χ_m is inversely proportional to T. This was first reported experimentally by Pierre Curie (1895), who defined χ_m as C/T. The Curie constant, is therefore given by $\mu_0 Nm^2/3k$.

This theory was originally derived for gases, where the individual magnetic moments do not interact. As atoms or molecules are brought closer together, such interactions become more important and must be included. One way to do this is to postulate a molecular field that is proportional to the magnetization M, i.e., $H' = \gamma M$. Including this molecular field, the total field intensity acting in the vicinity of an atom or molecule is now $H + H'$ and Curie's rule becomes

$$\chi_m = M/(H + \gamma H') = C/T$$

Solving for M yields

$$M = CH/(T - \gamma C)$$

The paramagnetic susceptibility becomes

$$\chi_m = C/(T - \gamma C)$$

or
$$\chi_m = C/(T - \theta) \qquad (11\text{-}5)$$

where γ is a measure of the strength of the atomic interactions. Equation (11-5) is a more general form of Curie's Law and is known as the Curie–Weiss Law (1907).

Both positive and negative values of θ have been found, although a typical value is about 10 K. A positive value for θ (and, therefore, the molecular field) indicates that it aids the applied field, while a negative value indicates the opposite.

The predictions of the Curie, Curie–Weiss, and Langevin (diamagnetism) relations for the variation of the magnetic susceptibility with temperature are shown in Fig. 11-1.

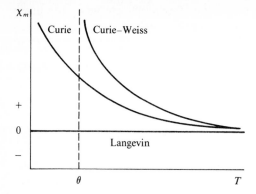

FIGURE 11-1
Temperature dependence of the magnetic susceptibility as predicted by the Curie, Curie–Weiss, and Langevin relations.

Examples of paramagnetic substances are transition metal and rare earth ions with incomplete inner shells, salts of transition elements [e.g., $KCr(SO_4) \cdot 12H_2O$], salts and oxides of rare earths.

11-5 TYPICAL MAGNETIC SUSCEPTIBILITIES

As previously outlined, there exists a large class of materials for which the magnetization, weak as it is, obeys the relation $M = \chi_m H$, where χ_m, the magnetic susceptibility, is positive for paramagnetics and negative for diamagnet-

TABLE 11-1
Magnetic susceptibility of some paramagnetic and diamagnetic materials at room temperature

Material	χ_m	$\chi_{m,\,mass}$ $(m^3 / kg\,m)$
Aluminum	2.3×10^{-5}	0.82×10^{-8}
Bismuth	-1.66×10^{-5}	-1.70×10^{-8}
Copper	-0.98×10^{-5}	-0.11×10^{-8}
Diamond	-2.2×10^{-5}	-0.62×10^{-8}
Gadolinium chloride ($GdCl_3$)	276.0×10^{-5}	114.0×10^{-8}
Gold	-3.6×10^{-5}	-0.19×10^{-8}
Magnesium	1.2×10^{-5}	0.69×10^{-8}
Mercury	-3.2×10^{-5}	-0.24×10^{-8}
Silver	-2.6×10^{-5}	-0.25×10^{-8}
Sodium	-0.24×10^{-5}	-0.25×10^{-8}
Titanium	7.06×10^{-5}	1.57×10^{-8}
Tungsten	6.8×10^{-5}	0.35×10^{-8}
Carbon dioxide (1 atm)	-0.99×10^{-8}	-0.53×10^{-8}
Hydrogen (1 atm)	-0.21×10^{-8}	-2.47×10^{-8}
Nitrogen (1 atm)	-0.50×10^{-8}	-0.43×10^{-8}
Oxygen (1 atm)	209.0×10^{-8}	155.0×10^{-8}

Source: From Reitz/Milford, Foundations of Electromagnetic Theory, © 1967, Addison-Wesley Publishing Co., Inc., Reading, Massachusetts. Reprinted with permission.

ics. Tabulations generally list the mass or molar susceptibilities, i.e.,

$$\chi_m(\text{mass}) = \chi_m/\rho$$

$$\chi_m(\text{molar}) = \chi_m A/\rho \tag{11-6}$$

where ρ is the density and A is the atomic (or molecular) mass. Typical values are given in Table 11-1. It is clear from the table that susceptibilities are very small.

11-6 FERROMAGNETISM

Ferromagnetic materials are paramagnetics with large molecular fields. A very high flux density is induced in such materials for a relatively low field intensity. Further, the materials exhibit hysteresis, as will be shown. In order to determine the relationship between the magnetization of such materials and the applied magnetizing field intensity, it is necessary to consider first the classic experiment of Rowland.

11-6.1 Rowland Ring Method

In order to investigate the magnetic properties of ferromagnetic materials, Rowland performed experiments on ring samples of such materials. As shown in Chap. 10, ring samples are particularly easy to analyze, since theoretically there are no poles created by surfaces that might be perpendicular to the direction of the magnetic field intensity H. The field intensity along the average radius of the ring is easily calculable from Eq. (10-34). In practice, the number of turns around the sample that are required to produce a field intensity of at least 16,000 A/m (200 Oe) are required for adequate magnetization of common steels. Low permeability materials or high field tests will require more.

The effect of applying such field intensities via the coil (ab) as shown in Fig. 11-2 must be detected; this is accomplished by wrapping a coils of N turns

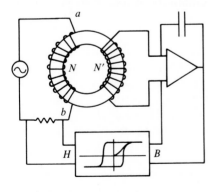

FIGURE 11-2
Schematic for the measurement of the magnetic properties of ring samples. (*From the Nondestructive Testing Handbook, Vol. 4, 2d ed. Reprinted with permission.*)

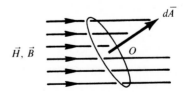

FIGURE 11-3
The vector nature of the magnetic field, flux density, and area element.

over a small section of the ring and connecting its output to an integration circuit. Early forms of this particular measurement employed a ballistic galvanometer. The essence of the investigation is that a controlled variation of H produces a measurable change in the magnetic flux in the ring and that this flux can be detected by the N'-turn coil/integrator system.

In the modern concept of this measurement [6], H is varied slowly (over perhaps several seconds) from a value sufficient to saturate the material in one direction to a value sufficient to saturate it in the other direction. This is accomplished with a low frequency generator driving a power supply. Induced flux is measured as indicated, its density being computed as follows. The magnetic flux ϕ is defined by

$$\phi = \overline{B} \cdot \hat{n}\, dA = (B \cos \theta)\, dA$$

as shown in Fig. 11-3. The angle θ is between the vectors of B and $\hat{n}\, dA$. In this type of measurement, it is arranged that the vector of B is parallel to that of $\hat{n}\, dA$, so that $\overline{B} \cdot \hat{n}\, dA = B\, dA$ and $\theta = 0$.

The results of such measurements, made with a computer-controlled system, may be as shown in Fig. 11-4.

11-6.2   Hysteresis Phenomenon

Beginning with the material in the unmagnetized state (i.e., $H = 0$ and $B = 0$), a small increase in H raises the flux density B in a nonlinear fashion along OA (Fig. 11-4). This parabolic increase of B with H is known as the Rayleigh region, after its discoverer. For many industrial steels, in this region, the slope of OA is relatively small, perhaps $(20-100) \times \mu_0$. This is the region that is scanned by eddy currents when such NDE techniques are used on unmagnetized ferromagnetic materials. It extends for a few tens of amperes per meter (i.e., a few oersteds) on either side of the origin.

At larger values of H, there follows a region (AC) in which the flux density rises rapidly for a small change in H. It is intuitively obvious that the internal process that gives rise to the magnetization along OA is joined by a second process along AC.

The region AC is followed by a region CD in which it is clear that the flux density is beginning to level off, implying that perhaps one of the magnetization processes is beginning to saturate. This is followed by a region DE in which the second process is saturating; above E, the increase of B with H is linear.

If H is reduced to zero, the path followed by the material is not $EDCOA$ but along EB_r to the remanence value, i.e., the saturation flux density at $H = 0$. This value varies widely for materials upon which magnetic NDE is performed, depending upon the material chemistry, the internal stress state, and the number of dislocations per unit volume within the material.

When H is reversed, a situation first exists in which H and B are in opposite directions within the material. The flux density follows the path $B_r F H_c E'$, i.e., it is remagnetized in the opposite direction. The value of H at H_c

is that which is required to hold the ring sample at zero flux density, i.e., the coercivity. (Should the field intensity be held at this level and then removed, the material will again assume a positive value of B and is certainly not demagnetized.)

Further increase in H beyond H_c in the reverse direction causes saturation at E'; returning H to the positive direction reproduced the rest of the loop symmetrically. Since, if H is swung first positive and then negative, the value of B seems to lag in direction behind that of H, the material exhibits magnetic hysteresis and the full B-H loop is known as the hysteresis loop for that material. The same material might also exhibit a stress–strain hysteresis loop.

Should the material not be taken to saturation on the initial magnetization, but rather to a value such as given by the point D, then it may follow a minor loop. In this case, the flux density for $H = 0$ is known as the residual magnetism. Obviously the maximum value of the residual magnetism is the remanence B_r. Further, the magnetizing field strength needed to hold the material at zero induction under these circumstances is known as the coercive force. Inner loops for various maximum values of H are shown in Fig. 11-4.

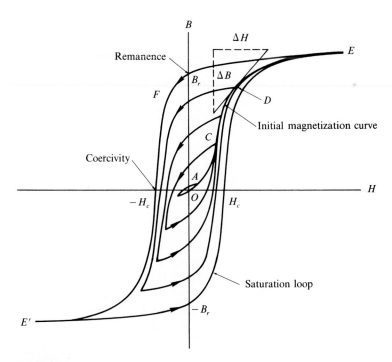

FIGURE 11-4
B-H curve for a typical steel, showing the outer saturation hysteresis loop, the positions of the remanence and coercivity, the initial (virgin) magnetization curve, and some minor (internal) hysteresis loops. (*From the Nondestructive Testing Handbook, Vol. 4, 2d ed. Reprinted with permission.*)

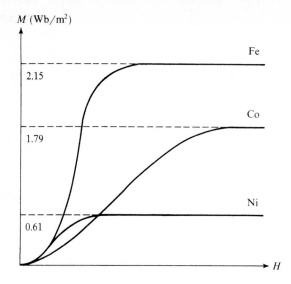

M (Wb/m^2)

Fe

2.15

Co

1.79

Ni

0.61

H

FIGURE 11-5
Magnetization versus H at room temperature for Fe, Co, and Ni. (*Cullity, Introduction to Magnetic Materials, © 1972, Addison-Wesley Publishing Co., Inc., Reading, Massachusetts. Reprinted with permission.*)

The Rowland ring experiment permits not only the B-H hysteresis loop to be measured, but also allows the virgin (initial) magnetization curve to be recorded. Further, removing the effect of H from B, curves that represent the variation of the magnetization M with H can be drawn. Figure 11-5 shows the M-H variations for the $3d$ transition metals iron, cobalt, and nickel, which are all ferromagnetic. Each curve shows that by its shape and by the value of H at which saturation occurs, the saturation magnetization is structure-dependent.

In explaining magnetization curves, it is necessary to explain both the way in which the saturation magnetization (M_s) is reached and, also, its final value. The first problem is relatively simple if two bold assumptions are made, but the latter is more difficult to explain. The saturation magnetization of iron, for example, is 2.15 Wb/m^2 (21.5 kG), which is at least six orders of magnitude greater than that for paramagnetic substances.

It was discovered early that above a certain temperature (the Curie temperature T_c), ferromagnetic materials become paramagnetic and in the paramagnetic state, these otherwise ferromagnetic materials follow the Curie–Weiss Law with $\theta \sim T_c$. The implication of this is that for these materials, θ is large and positive, so that the molecular field coefficient is also large and positive. The molecular field aids the applied field in magnetizing the material.

Weiss postulated that the molecular field acts below the Curie temperature and is so strong that it magnetized the material to saturation even when there is no applied field. In effect, H' takes over almost completely. Second, in order to overcome the objection that it is relatively easy to demagnetize a ferromagnetic material, Weiss also postulated that even in the demagnetized state, the material is divided into very many small domains, each of which is spontaneously magnetized to its saturation value M_s, but that the domains have a random orientation.

11-6.3 Domain Theory

At high magnification, domains were eventually observed. Figure 11-6 is an example of five domains of size about 10^{-2} cm on a side in an iron crystal. The upper left diagram shows the orientation of the magnetization M_s in the five domains in zero external applied field, while the diagram at the upper right is a microphotograph. Because of the sizes of the domains, the net magnetization is upward.

As the applied field is raised, in the low-field region of the *B-H* curve, the domain walls move so that those domains that have a magnetization in the direction of the external field grow at the expense of those that have an opposite magnetization. As Fig. 11-6 shows, stronger fields cause more domain wall movement and a reversal of the field causes the opposite situation to occur.

At higher external fields, the domain boundary displacements become irreversible, and at even higher fields, the magnetization vectors of the domains that were not originally in the direction of the applied field are rotated gradually so as to lie parallel to the external field direction. Figure 11-7 shows the regions of the *B-H* curve where these phenomena occur. In the region where irreversible domain wall motion occurs, some of the domains, notably those that are not oriented with the external field, are gradually reduced in size and eventually disappear. Thus, during magnetization, the number of domains gradually falls. The ease with which a high level of magnetization can be induced by a relatively low level of applied field is controlled by how easy it is to magnetize the crystals that make up the material along certain crystal axis directions and the density of microscopic dislocations in the sample. Domain walls can pin themselves to such dislocations.

The presence of domain walls can be verified using extremely fine magnetic particles, since there is some MFL at the surface of the material from one domain to another.

11-6.4 Domain Arrangements at Various States of Magnetization

The relative number and orientation of domains in a ferromagnetic material at various points on the *B-H* curve is shown in Fig. 11-8. At *O*, there are a large number of domains with zero net magnetization. At *B*, the external field has eliminated those domains that were oriented in a direction opposite to that of *H*, but many domains still remain. At *C*, only a few domains still remain, their orientations generally being along *H*. When *H* is removed, some domains are reestablished by strains and dislocations within the material, so that the net magnetization at *D* is somewhat less than at *C*. At *E*, the coercivity point, some of the domains that were easy to reorient with the reserved field have done so, and the net magnetization is zero. This is also the case at *E'*.

How and under what conditions the domain walls move is quite complicated. The crystal structure of the material, the density of impurities, and the

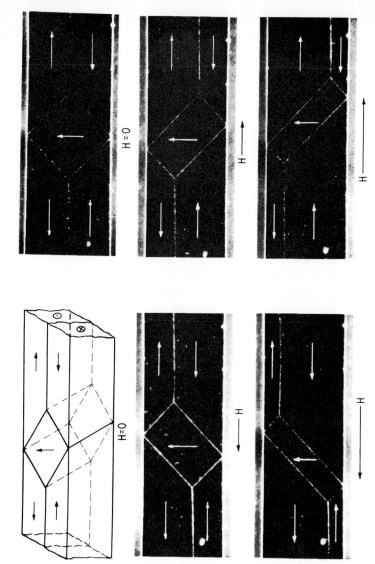

FIGURE 11-6
Reversible domain wall motion in an iron crystal. The maximum value of H in either direction is about 800 A/m (10 Oe). (*Reproduced with permission from R. W. De Blois and C. D. Graham, "Domain Observations in Iron Whiskers," J. Appl. Phys. 29, 1958.*)

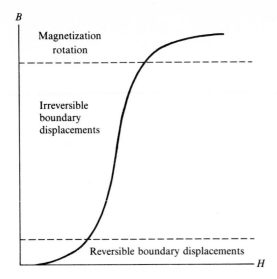

FIGURE 11-7
Three contributions to magnetization.
(*Reprinted from Introduction to Solid State
Physics, 3rd ed., by Charles Kittel with
permission from John Wiley & Sons, Inc.*)

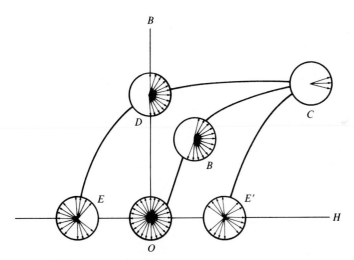

FIGURE 11-8
Relative numbers of domains and their directions around the *B-H* curve. (*Cullity, Introduction to
Magnetic Materials,* © *1972, Addison-Wesley Publishing Co., Inc., Reading, Massachusetts. Reprinted
with permission.*)

density of dislocations between crystal grains all affect the net value of flux
density for a given external applied field intensity.

11-6.5   Typical Values for Saturation
Magnetization

Table 11-2 gives values for the saturation magnetization (in gauss) at room
temperature and also at 0 K, the number of Bohr magnetons per magnetic atom

TABLE 11-2
Number n_B of Bohr magnetons per magnetic atom and data on saturation magnetization and Curie points*

| Substance | Saturation magnetization M_s (G) | | n_B (0 K) per formula unit | Ferro-magnetic Curie temper-ature (K) |
	Room temperature	0 K		
Fe	1707	1740	2.22	1043
Co	1400	1446	1.72	1400
Ni	485	510	0.606	631
Gd†	...	2010	7.10	292
Dy†	...	2920	10.0	85
Cu_2MnAl	500	(550)	(4.0)	710
MnAs	670	870	3.4	3.18
MnBi	620	680	3.52	630
Mn_4N	183	...	1.0	743
MnSb	710	...	3.5	587
MnB	152	163	1.92	578
CrTe	274	...	2.5	339
$CrBr_3$†	...	...	...	37
CrO_2	515	...	2.03	392
$MnOFe_2O_3$	410	...	5.0	573
$FeOFe_2O_3$	480	...	4.1	858
$CoOFe_2O_3$	400	...	3.7	793
$NiOFe_2O_3$	270	...	2.4	858
$CuOFe_2O_3$	135	...	1.3	728
$MgOFe_2O_3$	110	...	1.1	713
UH_3†	...	230	0.90	180
EuO	...	1920	6.8	69
$GdMn_2$	...	215	2.8	303
$Cd_3Fe_5O_{12}$	0	605	16.0	564
$Y_3Fe_5O_{12}$ (YIG)	130	200	5.0	560

*Data selected with the assistance of R. M. Bozorth. General references: *AIP Handbook*, sec. 5g, 1963; *Landolt-Bornstein*, vol. 2, 6th ed., pt. 9, 1962.
†Materials having magnetic properties only below room temperature (300 K).

(explained later), and the Curie temperatures (T_c) for a variety of ferromagnetic substances. Some of the materials (as indicated by the †) are only magnetic below room temperature (300 K). Many are binary compounds of otherwise nonmag-netic elements and many are the ferrite oxides.

The important point for NDE is that all of the materials that are commonly held to be ferromagnetic do stay highly ferromagnetic up to about $0.6T_c$, with the magnetization falling off rapidly only in the region of $T > 0.8T_c$. In some steel mills where the product must be inspected at elevated temperatures, it is necessary to use electromagnetic techniques such as eddy currents or electromagnetic acoustics.

11-6.6   Origin of the Molecular Field

The hypothesis that the molecular field (H') is proportional to the magnetization (M) implies that the phenomenon that produced it is cooperative in nature. In 1928, Heisenberg showed that it was due to quantum mechanical exchange forces that occur between neighboring ions. Being an entirely nonclassical force, the theory that underlies ferromagnetism is not based on any classical physics. The exchange interaction occurs between the spins of neighboring ions. Figure 11-9 illustrates the classical and quantum-mechanical pictures that lead to magnetism in a chain of atoms. In Fig. 11-9(a), the classical case, the net spins (S) on each atom in the linear chain all point in the same direction. One possible excitation is for one reversed spin [Fig. 11-9(b)], but a lower excitation energy occurs when all the spins precess [Fig. 11-9(c)]. An expanded version is shown in Fig. 11-9(d), with successive spins advanced in phase by a constant amount and the relative spin orientations over one full wavelength drawn out. In Fig. 11-9(e) the tips of the projections of the spin vectors on the bases of the cones sweep out one wavelength of the spin wave or magnon. The nearest-neighbor interaction energy

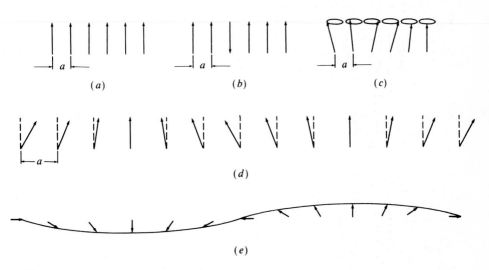

FIGURE 11-9
(a) Classical picture of the ground state of a simple ferromagnet, with all the atomic spins parallel.
(b) A possible excitation, with one spin reversed. (c) An elementary excitation known as a spin wave.
(d) Expanded picture of a spin wave. (e) Spin waves from above, showing one wavelength.
(*Reprinted from Introduction to Solid State Physics, 3rd ed., by Charles Kittel with permission from John Wiley & Sons, Inc.*)

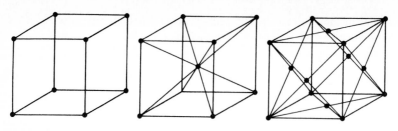

FIGURE 11-10
Ion locations for three types of cubic crystal structure. Cubic *I* is known as body-centered cubic and cubic *F* is known as face-centered cubic.

associated with the spin wave is given by

$$E_{ex} = -2J_{ex}\bar{S}_i \cdot \bar{S}_j \tag{11-7}$$

where $\bar{S}_i$ and $\bar{S}_j$ are the spin angular moments of neighboring atoms (i, j) and J_{ex} is an exchange integral. This energy is much less than that required to flip one spin into the condition shown in Fig. 11-9(*b*). Such exchange energies are an important part of the total energy of many molecules and also of the covalent bond.

If J_{ex} is positive, E_{ex} is minimum when S_i is parallel to S_j; if it is negative, then the reverse is true. Ferromagnetism, which is due to the alignment of these spins, requires a positive value of J_{ex}, which is quite rare. For cubic crystal structures (Fig. 11-10) there may be six nearest neighbors (cubic *P*), or eight [cubic *I*, or body-centered cubic (BCC)], or only four [cubic *F*, or face-centered cubic (FCC)]. The BCC structure supports ferromagnetism, with eight nearest-neighbor interactions. Since such exchange interactions fall off quite rapidly with distance, generally only the nearest- and next-nearest-neighbor interactions are considered.

In the solid state, the 3*d* electrons dominate the interatomic interactions for Fe, Co, and Ni, and so a plot of J_{ex} versus r_a/r_{3d} (where r_a is the radius of the atom and r_{3d} is that of the 3*d* atomic subshell) is quite instructive. Figure 11-11

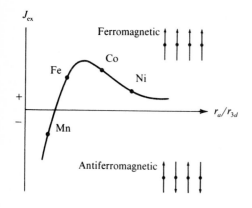

FIGURE 11-11
Bethe–Slater curve for J_{ex} versus r_a/r_{3d}. (*Cullity, Introduction to Magnetic Materials,* © *1972, Addison-Wesley Publishing Co., Inc., Reading, Massachusetts. Reprinted with permission.*)

shows this Bethe–Slater curve. It is seen that when r_a/r_{3d} is large, J_{ex} is small and positive. As it decreases, the $3d$ electrons become closer, the value of J_{ex} becomes larger, and the requirement of ferromagnetism for almost parallel spins becomes more favorable. At smaller values of r_a/r_{3d}, J_{ex} becomes smaller and eventually negative, and other forms of magnetism occur. Manganese, for example, exhibits a condition known as antiferromagnetism below 100 K. In this condition, the spins are antiparallel and spin waves are still possible.

The Bethe–Slater curve also suggests why some elements are not ferromagnetic, but their alloys can be. The value of r_a/r_{3d} must be maintained large enough for J_{ex} to be positive. This occurs in MnBi.

11-6.7 Equivalence of the Molecular Field and Exchange Forces

Equating the potential energy of an atom in the molecular field with the exchange energy yields an expression for the exchange integral, i.e.,

$$J_{ex} = 3kT_c/2zS(S + 1) \qquad (11\text{-}8)$$

where k is Boltzmann's constant, T_c is the Curie temperature, z is the number of nearest neighbors, and S is the spin quantum number. For BCC iron with $z = 8$ and $S = 1/2$, $J_{ex}(\text{Fe}) = 0.25kT_c = 3.6 \times 10^{-21}$ J.

11-7 BAND THEORY IN RELATION TO MAGNETIC MATERIALS

Since exchange forces exist between spin states of neighboring atoms, ferromagnetism is a collective phenomenon rather than an atomic one. In solids, however, the individual atoms do not retain sharply defined atomic energy levels. Rather, the outer levels, which contain the valence electrons, form bands. The Pauli Exclusion Principle, which requires that no two electrons can have the same four quantum numbers, governs energy bands in solids. Atomic shells and subshells, along with the maximum permitted number of electrons in each, are given in Table 11-3.

The inner (K) shell has two electrons with spin quantum numbers $m_s = \pm 1/2$. The L shell has up to eight electrons in two subshells ($2s$ and $2p$). The M

TABLE 11-3
Atomic shells and subshells with maximum number of permitted electrons

Shell	K		L		M			N		
Subshell	$1s$	$2s$	$2p$	$3s$	$3p$	$3d$	$4s$	$4p$	$4d$	$4f$
Number of electrons	2	2	6	2	6	10	2	6	10	

TABLE 11-4
Transition metals

Number	Ca	Sc	Ti	V	Cr	Mn	Fe	Co	Ni	Cu
$3d$	0	1	2	3	5	5	6	7	8	10
$4s$	2	2	2	2	1	2	2	2	2	1
$3d + 4s$	2	3	4	5	6	7	8	9	10	11

shell has up to 18 electrons in three subshells ($3s$, $3p$, and $3d$). In the transition metals, the K and L shells are filled. Proceeding up the Periodic Table, as shown in Table 11-4, the $3d$ and $4s$ levels do not fill one after the other; rather, they overlap in energy and the lowest available energy state is filled first. In vanadium, for example, a $3d$ level has a lower energy than a $4s$ level, and so is filled. In chromium, the opposite is true.

Bringing together millions of atoms to form a solid causes the $4s$ and $3d$ levels to broaden out into bands, which overlap. Figure 11-12 illustrates this overlapping. When atoms are far apart (large d) sharp energy levels (E) exist for the $1s$, $2s$, $2p$, $3d$, and $4s$ states. Electronic transitions between these states result in x-ray and optical line spectra. However, as the atoms approach each other and the solid state forms, the first levels to overlap in the transition metals series are the $4s$ levels, and so these form an energy band. They are closely followed by overlapping of the $3d$ levels and then the $4s$ and $3d$ bands also overlap.

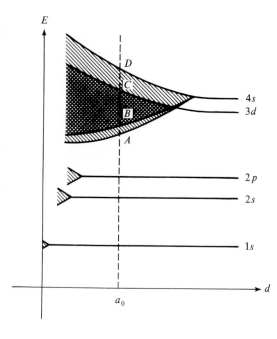

FIGURE 11-12
The broadening of single energy states (at relatively large interatomic distances d) into energy bands (at relatively small interatomic distances, as are found in the solid state). (*Cullity, Introduction to Magnetic Materials,* ©*1972, Addison-Wesley Publishing Co., Inc., Reading, Massachusetts. Reprinted with permission.*)

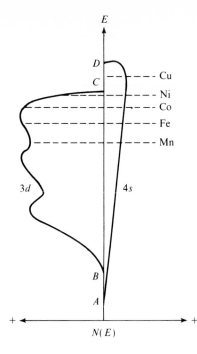

FIGURE 11-13

A rigid band model for the $3d$ (left) and $4s$ (right) densities of states. The transition metals fill the states in the bands up to the levels shown. (*Cullity*, *Introduction to Magnetic Materials*, © *1972*, *Addison-Wesley Publishing Co.*, *Inc.*, *Reading*, *Massachusetts. Reprinted with permission.*)

Transition metals have interatomic distances in the region of the dashed line at $d = d_0$. Here, the $4s$ and $3d$ levels have banded and overlapped, while the $2p$, $2s$, and $1s$ levels still have sharply defined energies. In effect, in these metals, the $4s$ and $3d$ electrons have been donated by their original parent atoms to the conduction band, while the $2p$, $2s$, and $1s$ electrons are still localized around their parent nuclei.

This band idea also accounts for the difference between electrical conductors and insulators. Conductors have a partially filled conduction band; insulators have completely filled bands, so that there are no electrons available for charge transport. For the transition metals, Fig. 11-12 shows that the $4s$ electrons have energies that range from A to D, while the $3d$ electrons have energies that range from B to C. The uppermost energy within the band for each metal is shown in Fig. 11-13.

The bands are actually very closely spaced levels, as required by the Pauli Principle. The density of these levels $N(E)$ for the transition metals is shown in Fig. 11-13. The $3d$ states are on the left and the $4s$ states are on the right of the center line. The $3d$ state density is greater than the $4s$ state density because there are five $3d$ levels but only one $4s$ level per atom. In the model shown in Fig. 11-13 the density of states does not vary between elements, and the area between A and the horizontal dashed lines $[N(E)\,dE]$ measures the total number of electrons in the conduction band for each element. The dashed lines represent the maximum energy an electron can have within the band, irrespective of its "character," and is known as the Fermi energy.

The rigid band model shows that for copper, the $3d$ band is filled, so that the conduction electrons show only $4s$ character. Filled energy bands, like filled electronic shells, cannot contribute to the magnetic moment (because there are just as many spins of one orientation as there are of the other) means copper cannot be paramagnetic. It is, in fact, diamagnetic.

11-7.1 Spin Inbalance

Since there are more $3d$ than $4s$ states, spin inbalance occurs in the transition metals. Ferromagnetism is due to this spin inbalance, which is maintained by the exchange force.

11-7.2 Ferromagnetic Alloys

Ferromagnetism is found in binary and ternary alloys of Fe, Ni, and Co with one another, in their alloys with other elements, and in alloys that do not contain any of the ferromagnetic elements. It is impossible to make hard and fast rules, as both M_s and T_c are unpredictable. Detailed explanations are given in two excellent texts by Cullity [5] and Bozorth [7].

11-8 ANTIFERROMAGNETISM

Antiferromagnetism occurs at low temperatures in some materials that otherwise are paramagnetic. Below a transformation temperature (T_N) the molecular field causes spontaneous magnetization of two sublattices within the material. The net magnetization is the difference between the two sublattice magnetizations, so the final level is quite low.

11-9 FERRIMAGNETISM

Originally thought to be ferromagnetic, the materials that exhibit ferrimagnetism consist of saturated domains and show both saturation and hysteresis effects. Above a Curie temperature, they become paramagnetic. Known as ferrites, they are very important from an engineering viewpoint because they are poor electrical conductors. Useful ferrites fall into two classes: cubic and hexagonal.

11-9.1 Cubic Ferrites

Cubic ferrites have the general formula $MO \cdot Fe_2O_3$, where M is a divalent metal ion (e.g., Mn, Fe, Ni, Co, or Mg). Of these, $Co \cdot Fe_2O_3$ is magnetically hard, i.e., it has a high coercivity H_c, while the rest are magnetically soft. Commercial ferrites commonly have two kinds of divalent ions, i.e., $(Ni, Zn)O \cdot Fe_2O_3$. Their crystal structure is complex, with 56 ions per unit cell. The larger O^{++} ions (radius 1.3×10^{-10} m) are closely packed in a FCC structure, with the smaller metal ions (radius 0.8×10^{-10} m) between them. Sites available within the

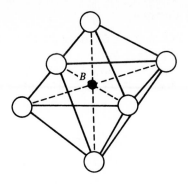

FIGURE 11-14
Tetrahedral (A) and octahedral (B) sites that are available to metal ions for the formation of ferrimagnetism in ferrites.

structure for the metal ions are either (1) tetrahedral sites (A), of which one-eighth are occupied or (2) octahedral sites (B), of which one-half are occupied. These sites are shown in Fig. 11-14. Exchange interactions occur between the A-A sites, the A-B sites, and the B-B sites, all of which are negative, the A-B interaction being the strongest. The effects of these interactions are that the A moments are all parallel and antiparallel to the B moments, and the magnitudes of the A and B sublattice saturation magnetizations are not equal. This is the cause of ferrimagnetism and it must obviously be weaker than ferromagnetism. For example, in $NiO \cdot Fe_2O_3$, there is one Ni^{++} ion with moment $2\mu_B$ and two Fe^{3+} ions with moment $5\mu_B$, so that if the ferrite were ferromagnetic, the total moment per molecule would be $12\mu_B$. However, the saturation magnetization is only $2.3\mu_B$, which is evidence for the antiparallel nature of the spins.

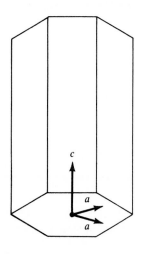

FIGURE 11-15
The directions of the a and c axes in a hexagonal crystal.

11-9.2   Hexagonal Ferrites

With hexagonal ferrites, there are 64 ions in a unit cell, which is longer in the c direction (23.2×10^{-10} m) than in the a direction (5.9×10^{-10} m). The a and c axes of hexagonal crystals are shown in Fig. 11-15. Commercially important hexagonal ferrites are $BaO \cdot 6Fe_2O_3$ and $SrO \cdot 6Fe_2O_3$. Typical saturation magnetization is 0.48 Wb/m^2.

11-9.3   Importance of Ferrites in NDE

Since the $3d$ electrons are relatively tightly bound, they are poor electrical conductors. They are used as cores for pickup coils and as such, they effectively amplify relatively small MFL fields, without the induction of eddy currents.

11-10   MEASUREMENT OF THE MAGNETIC PROPERTIES OF FERROMAGNETIC MATERIALS

Ferro- and ferrimagnetic materials support a relatively large flux density (B) for a relatively small applied field intensity (H) and exhibit hysteresis. The method used to measure the flux density is important because it can be used throughout magnetic NDE for the detection of magnetic flux and, therefore, finds application in both magnetization and demagnetization techniques.

Modern equipment for the measurement of B (Fig. 11-2) employs an integrating fluxmeter, shown schematically in Fig. 11-16, which must remain stable during one oscillation of the magnetic field intensity H. As the magnetic flux through the pickup coil (N' turns) changes, an electromotive force (EMF) is induced in the coil according to Faraday's Law of Induction, i.e.,

$$e = -N' \, d\phi/dt \tag{11-9}$$

where $d\phi/dt$ is the time rate of change of flux. B-H measurements are generally

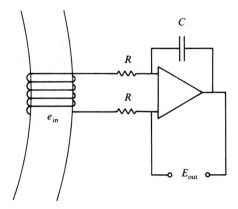

FIGURE 11-16
Schematic of integrator method for B.

made with the coil wrapped tightly around the sample to exclude flux in the air, so that (for large rings of small cross-sectional area) $\phi = BA$ and changes in flux are given by $d\phi = A\,dB$. This EMF is the input, e_{in}, to the integration circuit. Its output EMF, (E_{out}), is given by

$$E_{out} = -(k/RC)\int e_{in}\,dt \qquad (11\text{-}10)$$

where RC is the time constant of the circuit and k is a constant that depends on its design. Using Eq. (11-9) in Eq. (11-10) gives

$$E_{out} = (kN'/RC)\int d\phi = (kN'A/RC)\int dB \qquad (11\text{-}11)$$

When this output, which is proportional to changes in B, is fed to the y plates of an oscilloscope or the vertical deflection of a chart recorder and the voltage across a resistor in the H supply circuit is fed to the x plates or horizontal deflection, the result is a B-H curve for the material. Adjustment of the maximum current in the magnetizing circuit permits inner hysteresis loops to be traced, and increasing the frequency of power supply enables the shielding effects of eddy currents in the sample to be seen.

The virgin curve can be obtained by joining the turning points of several inner loops or using a suitably programmed power supply. Further, since the flux pickup coil (N') is wound inside the field coil (N), the values of B so obtained include the ambient value of H. This can also be removed, so that only the magnetization M is recorded.

11-11 MAGNETIC PERMEABILITY

For weakly magnetic materials, the susceptibility is defined as $M = \chi_m H$. This implies that there is also a linear relation between B and H, i.e., $B = \mu H$. For the ferro- and ferrimagnetic materials, the flux density is related to the field intensity through the magnetic permeability, defined from the relation

$$B = \mu_0\mu_r H \qquad (11\text{-}12)$$

In Eq. (11-12), the magnetic permeability is $\mu_0\mu_r$, where μ_r is the relative permeability and $\mu_0 = 4\pi \times 10^{-7}$ H/m. From Fig. 11-4, it is clear that μ_r is a function of B or H. Figure 11-17 illustrates both the B-H variation exhibited by a common pipe steel [Fig. 11-17(a)] for the virgin curve and also how the relative permeability μ_r varies. The virgin curve OQ reaches 1.5 T (15 kG) after application of 4775 A/m (60 Oe). The dashed line below the B-H curve is the variation of the residual magnetism with H and shows that the material is effectively saturated after application of 4000 A/m (50 Oe). (In order to obtain such data, the material is taken from 0 to some point such as N and the field H is then removed. The material follows a path such as NN' along an inner loop.)

The relative permeability, found from B/H, has an initial value of about 70, rises to a maximum value of 420 at 1750 A/m (22 Oe), and then falls. This is not a particularly large value, but is typical of an industrial steel. In some special magnetic materials, the maximum relative permeability can reach values as large as $10^5\mu_0$.

Another useful quantity is the incremental permeability, defined from

$$\mu_d = (1/\mu_0)(dB/dH) \tag{11-13}$$

For the steel of Fig. 11-17, μ_d has a maximum value of $800\mu_0$ at $H = 1330$ A/m (15 Oe), and falls off rapidly for applied field values higher than this. At saturation, when M can increase no further, the limiting value of μ_d is unity.

11-12 RANGE OF MAGNETIC PARAMETERS IN NDE

The low carbon steels used in the automotive, construction, and oil industries are all ferromagnetic, the parameters of saturation (B_s), remanence (B_r), and coercivity (H_c) depending on some or all of the following: chemical composition, inclusion density, residual stress state, and heat treatment. The basic prerequisite is BCC phase of iron, the 400 series of stainless steels and many specialty steels. The 300 series of stainless steels, which contain roughly 22% of chromium and some nickel, crystallize in the FCC configuration and are not ferromagnetic. (These steels contain small pockets of ferrite and are known to age into a mildly magnetic condition.)

An example of the range of magnetic properties encountered within oil field tubular materials is shown in Fig. 11-18. The data show the virgin curve, the hysteresis and inner loops, the residual flux density variation, and the incremental permeabilities for a specialty steel for use in deep H_2S gas wells (DSG-90) and a common well casing material (K-55). The higher strength material exhibits a very square loop, with a saturation flux density at 10,000 A/m (120 Oe) of $B_s = 1.7$ T (17 kG), a remanence of $B_r = 1.49$ T (14.9 kG), and a coercivity of $H_c = 1150$ A/m (14.5 Oe). For the lower strength material, $B_s = 1.5$ T (15 kG), $B_r = 0.94$ T (9.4 kG), and $H_c = 880$ A/m (11 Oe). It also has a more rounded loop. The knee in the second quadrant of the higher strength material B-H curve is typical of materials that have been quenched and tempered, so that good stress relief has occurred.

The lower plots show that both materials are effectively saturated after application of 3200 A/m (40 Oe), the wide disparity between the values of B_r being attributed mainly to the internal stress state of the ring samples used. Data such as these are useful in setting applied field intensity levels for adequate magnetization prior to performing inspection in the residual induction.

Typical values for industrial steels are 1.4–1.8 T for B_s, 240–1600 A/m (3.5–20 Oe) for H_c, and 0.3–1.5 T for B_r. The careful experimenter will doubtless encounter values outside these ranges.

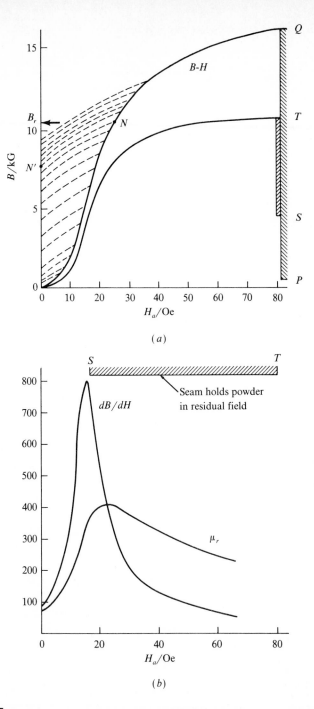

FIGURE 11-17
Initial magnetization data for a typical pipe steel. (a) Initial B-H and remanent flux density from 0–80 Oe. In this sample, a tight seam holds magnetic powder as shown on the right of the figure in (i) active (PQ) and (ii) residual fields (ST). (b) The values of μ_r and $(1/\mu_0)(dB/dH)$ for the same sample. The ability of the flaw to hold magnetic powder is indicated along the top of the figure (*Courtesy of American Society for Nondestructive Testing, Columbus, Ohio, 1980. From the Nondestructive Testing Handbook, Vol. 4, 2d ed. Reprinted with permission.*)

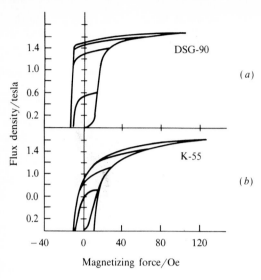

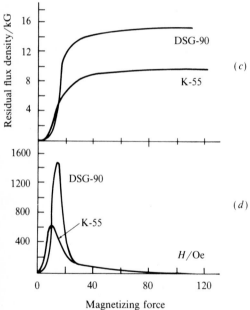

FIGURE 11-18
B-H data for a high grade (DSG-90) and a low grade (K-55) OCTG. (*a*) Initial curve and hysteresis loops for DSG-90. (*b*) Same for K-55. (*c*) Residual flux density after application of field *H* for these materials. (*d*) Derivatives of curves in (*c*) with respect to *H*. Note that the materials are effectively saturated after $H = 40$ Oe. [*Reproduced with permission from Electromagnetic Methods of NDT, W. Lord (ed.), 1985, published by Gordon and Breach, NY 1985.*]

11-13 MAGNET MATERIALS AND TRANSFORMER STEELS

Transformer and electromagnetic steels are made so that the area enclosed by the *B-H* curve is small and a small field intensity causes a high flux density. This helps maximize energy transfer by reducing core losses, since the energy needed to take a magnetic material around its *B-H* loop is proportional to $\int \overline{H} \cdot d\overline{B}$.

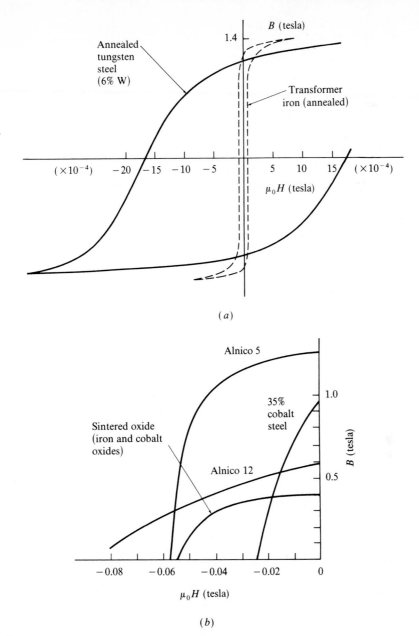

FIGURE 11-19

(a) Comparison of the hysteresis curves of several materials. (Note that $\mu_0 H$ is plotted along the abscissa instead of just H. $\mu_0 = 4\pi \times 10^{-7}$ H/m). (*Data from R. M. Bozorth, Ferromagnetism, Van Nostrand, New York, 1951.*) (b) Hysteresis curves of permanent magnet materials. (*Reitz/ Milford, Foundations of Electromagnetic Materials, © 1972, Addison-Wesley Publishing Co., Inc., Reading, Massachusetts. Reprinted with permission.*)

Figure 11-19(a) compares typical B-H loops for transformer steel and an annealed tungsten steel. Coercivities are 80 and 1350 A/m, respectively, while the remanences and saturation flux densities are roughly equal.

In designing permanent magnets, the loop area is important in that, in the second quadrant, the designer aims to maximize the product BH. This ensures a high value for B and a high coercivity H_c, so that the magnet has both a strong external field and the ability to retain its magnetization against abuse. Figure 11-19(b) illustrates second quadrant B-H data for typical magnet materials. Alnico 5, for example, shows a B_r of 1.27 T and a coercivity of 43,800 A/m (550 Oe). State-of-the-art magnet material research has provided rare-earth-based materials such as samarium–cobalt and neodymium–iron that exhibit coercivities in the millions of oersteds. Demagnetization other than by raising the materials above their Curie temperatures is virtually impossible.

CHAPTER 12

MACROSCOPIC
FIELD
RELATIONS

12-1 INTRODUCTION

It was seen in Chap. 11 that by defining the magnetization of a material in terms of individual atomic moments, dia- and paramagnetism could be explained. Ferro- and ferrimagnetism required the introduction of the Heisenberg quantum-mechanical exchange force. In this chapter, the classical macroscopic approach to electromagnetism is used to derive relations that are useful in NDE. The approach leads directly to relations between the three important magnetic vectors $\overline{H}$, $\overline{B}$, and $\overline{M}$ and to the boundary conditions that are obeyed at the interfaces between magnetic and nonmagnetic materials.

The macroscopic approach assumes that magnetic phenomena can be explained in terms of circulating (amperian) current distributions, without relating them to electronic and atomic magnetic moments [1]. Such current distributions are shown in Fig. 12-1. If they are evenly distributed within the material, these currents tend to cancel out. However, if they are not uniformly distributed, regions will exist in which this cancellation does not occur.

12-2 MAGNETIZATION AND AMPERIAN
CURRENT DISTRIBUTIONS

When applied to a nonuniform current distribution such as Fig. 12-1(b), vector calculus yields the result

$$\overline{J}_M = \text{curl } \overline{M} \tag{12-1}$$

217

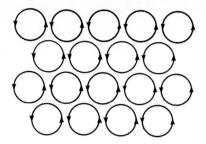

Uniform

(a)

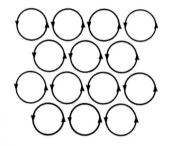 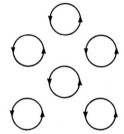

Nonuniform

(b)

FIGURE 12-1
(a) Amperian currents uniformly distributed in a material and (b) nonuniformly distributed.

where $\bar{J}_M$ is the amperian current density that provides the magnetization $\overline{M}$. In rectangular coordinates, the curl function is given by

$$\text{curl } \overline{M} = \begin{vmatrix} \hat{i} & \hat{j} & \hat{k} \\ \partial/\partial x & \partial/\partial y & \partial/\partial z \\ M_x & M_y & M_z \end{vmatrix} \qquad (12\text{-}2)$$

It is emphasized that $\bar{J}_M$ is not a transport current density.

12-3 MAGNETIC FIELD PRODUCED BY A MAGNETIC MATERIAL

From Eq. (11-1), the magnetization of a small volume of magnetic material located at the point (x', y', z') can be written as

$$\Delta \overline{m} = \overline{M}(x', y', z') \, \Delta v'$$

where $\Delta v'$ is the volume element that has a magnetic moment of $\Delta \overline{m}'$. The

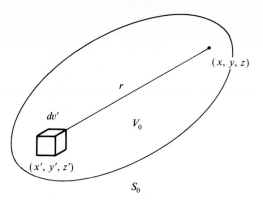

FIGURE 12-2
The magnetic field at (x, y, z) is the sum of contributions from small volume elements such as the one at (x', y', z') taken over the volume V_0 of the material.

magnetic effect at some other point (x, y, z) is an integral over all of the contributions from all such $\Delta \overline{m}\,s$ within the material (see Fig. 12-2). At this point it is expedient to introduce the magnetic vector potential that is created by the effect of the $\Delta \overline{m}\,s$, as

$$\vec{A}(x, y, z) = \frac{\mu_0}{4\pi} \int_{v_0} \frac{\overline{M}(x', y', z') \times \vec{r}}{r^2}\, dv'$$

This is an integral of $\overline{M}$ over the volume of the material. It is widely used in electromagnetic theory because it is found easier to compute such integrals than to compute $B(x, y, z)$ directly. The flux density will follow from the vector relation $\overline{B} = \text{curl }\overline{A}$. From the identities of vector calculus, $\overline{A}$ splits into two terms:

$$\overline{A}(x, y, z) = \frac{\mu_0}{4\pi}\left[\int_{v_0} \frac{\text{curl }\overline{M}}{|\vec{r}|}\, dv' + \int_{S_0} \frac{\overline{M} \times \hat{n}}{|\vec{r}|}\, da' \right] \tag{12-3}$$

where the first term is the contribution from the volume V_0 and the second is that from the surface S_0. S_0 bounds V_0 and $\hat{n}\, da'$ is the outward unit vector to this surface. The vector potential is thus represented by a volume integral and a surface integral. Postulating a surface current density that is given by

$$\overline{j}_M = \overline{M} \times \hat{n} \tag{12-1a}$$

and using Eq. (12-1), the vector potential at the field point $[\vec{r} = (x, y, z)]$ is

$$\overline{A}(\vec{r}) = (\mu_0/4\pi)\left[\int_{v_0} (\overline{J}_M/|\vec{r}|)\, dv' + \int_{S_0} (\overline{j}_M/|\vec{r}|)\, da' \right] \tag{12-3a}$$

Now the vector potential at $\vec{r}$ is caused by amperian current distributions both inside and on the surface of the material.

In order to find the more practical quantity $\overline{B}(\vec{r})$, the curl must be taken. The result of this vector calculus exercise is

$$\overline{B}(\vec{r}) = (\mu_0/4\pi) \int_{V_0} \left[\text{curl }\overline{M} \times \vec{r}/r\right] dv'$$

or
$$(\mu_0/4\pi)\left[-\int_{V_0} \text{grad}(\overline{M}(\vec{r}) \cdot \vec{r}/|r^2|)\, dv' + 4\pi\overline{M}(\vec{r})\right]$$

which again has two terms. It is seen that $\overline{B}(\vec{r})$ is caused by two contributions, one of which is proportional to the local magnetization at $\vec{r}$. The first term can be explained by postulating that it is caused by a magnetic scalar potential, given by

$$\Omega(\vec{r}) = (1/4\pi)\int_{V_0} \left(\overline{M}(\vec{r}) \cdot \vec{r}/|r^2|\right) dv' \qquad (12\text{-}4)$$

so that the equation for $\overline{B}(\vec{r})$ finally becomes

$$\overline{B}(\vec{r}) = -\mu_0\left[\text{grad}\,\Omega(r) - \overline{M}(\vec{r})\right] \qquad (12\text{-}5)$$

I.e., inside the material, $\overline{B}$ is given by the gradient of a scalar potential and the local magnetization. Outside the material, where there is no magnetization, the value of $\overline{B}$ is just that of the scalar potential. This is given by

$$\overline{H}(\vec{r}) = -\text{grad}\,\Omega(\vec{r}) \qquad (12\text{-}6)$$

which was used in Chap. 10 to compute the magnetic field intensity in the vicinity of a magnetic dipole.

12-4   MAGNETIC POLE DENSITY

It is encouraging when it can be shown that vector calculus will yield familiar concepts. In Chap. 10, the concept of the magnetic pole was introduced. Such poles emerge from vector calculus as follows. The right hand side of Eq. (12-4) is written as

$$\overline{M}(\vec{r}) \cdot \vec{r}/|\vec{r}^2| = \overline{M} \cdot \text{grad}(1/|\vec{r}|)$$
$$= \text{div}(\overline{M}/|\vec{r}|) - (1/|\vec{r}|)\text{div}\,\overline{M}(\vec{r})$$

both of which follow from standard vector calculus results. The result for $\Omega(\vec{r})$ is

$$\Omega(\vec{r}) = (1/4\pi)\left[\int_{S_0} \frac{\overline{M}\cdot\hat{n}}{|\vec{r}|}\, da' - \int_{V_0} \frac{\text{div}\,\overline{M}}{|\vec{r}|}\, dv'\right] \qquad (12\text{-}7)$$

That is, the scalar potential consists of two terms, one of which contains $\overline{M}\cdot\hat{n}$, the other div $\overline{M}$. Poles are introduced by defining a volume pole density $\rho_M(\vec{r})$ and a surface pole density $\sigma_M(\vec{r})$:

$$\rho_M(r) = -\text{div}\,\overline{M}(\vec{r})\ \text{A/m}^2$$
$$\sigma_M(r) = \overline{M}(\vec{r}) \cdot \hat{n}\ \text{A/m} \qquad (12\text{-}8)$$

The volume pole density is just the divergence of $\overline{M}$, or $\partial M_x/\partial x + \partial M_y/\partial y + \partial M_z/\partial z$. Under conditions of uniform magnetization, these derivatives are all zero, so that $\rho_M = 0$. Should any of the derivatives be finite, then in effect, poles exist within the material. The surface pole density is the scalar product of the magnetization and the unit normal to the surface; thus, wherever field lines emerge from the surfaces of magnetized materials, in effect, surface poles are present. In reality, poles exert a demagnetizing influence, which destroys the uniformity of M inside the material, so that ρ_M is no longer zero. The combination of surface and volume poles is often taken as one pole, placed inside the ends of magnetized material, as shown in Fig. 10-3.

12-5 MAGNETIC FIELD INTENSITY

Generally, both conventional and amperian currents might be present in a material. Both contribute to the flux density and under such circumstances, Eq. (12-5) must be written

$$\bar{B}(\bar{r}) = \mu_0\left[\frac{I}{4\pi}\int_V \frac{\overline{dl} \times \hat{r}}{r^2} - \text{grad }\Omega(\bar{r}) + \overline{M}(\bar{r})\right] \qquad (12\text{-}9)$$

where the first term is just Eq. (10-24) and V extends over all current-carrying regions. In order to bring this equation in line with measured quantities, the magnetic field intensity is taken as

$$\overline{H}(\bar{r}) = \frac{I}{4\pi}\int_V \frac{\overline{dl} \times \hat{r}}{r^2} - \text{grad }\Omega(\bar{r}) \qquad (12\text{-}10)$$

and Eq. (12-9) becomes

$$\bar{B}(\bar{r}) = \mu_0\left[\overline{H}(\bar{r}) + \overline{M}(\bar{r})\right] \qquad (12\text{-}11)$$

As Eq. (12-10) shows, the field intensity consists of two parts, one of which is an integral over a geometrical distribution of transport currents, the other being the gradient of a scalar potential. The former term is used to compute field intensities that are due to currents in wires, coils, etc., while the latter is used where no transport currents are involved, such as close to a magnet.

12-6 AMPERE'S CIRCUITAL LAW

In the absence of magnetic materials and replacing $I\,\overline{dl}$ as $\bar{J}\,dv'$, Eq. (10-24) may be written as

$$\bar{B}(\bar{r}) = \frac{\mu_0}{4\pi}\int \frac{\bar{J} \times \hat{r}}{r^2}\,dv' \qquad (12\text{-}12)$$

Taking the curl of this relation results in the simple relation

$$\text{curl }\bar{B}(\bar{r}) = \mu_0\bar{J} \qquad (12\text{-}13)$$

which is known as the differential form of Ampere's Circuital Law. By the use of Stokes' Theorem (another well-known vector calculus relation, which states that for any vector $\overline{Q}$, its line integral around a closed contour C is given by $\oint_c \overline{Q} \cdot d\bar{s} = \int \text{curl }\overline{Q} \cdot \hat{n}\,da$), Eq. (12-13) is transformed to

$$\oint_c \overline{B} \cdot d\bar{s} = \int \text{curl }\overline{B} \cdot \hat{n}\,da = \mu_0\int \bar{J} \cdot \hat{n}\,da \qquad (12\text{-}14)$$

That is, the line integral of B around a closed path is equal to the product of μ_0 and the total current $\int \bar{J} \cdot \hat{n}\, da$ flowing through that path. This is a very powerful relation, because for currents flowing down wires, $\int \bar{J} \cdot \hat{n}\, da$ is just the current I, and so Eq. (12-14) gives

$$\oint_c \bar{B} \cdot d\bar{s} = \mu_0 I \qquad (12\text{-}15)$$

Example 12-1. For a long straight wire carrying a current I, the contour C is taken as a circle of radius a, centered on the wire (Fig. 10-5). The line integral becomes $\oint \bar{B} \cdot d\bar{s} = B(2\pi a)$, from which

$$B = \mu_0 I / 2\pi a \qquad (12\text{-}16)$$

This result falls out of Ampere's Law in a natural way because most of the mathematics has been performed within the vector calculus operations that produced Eq. (12-15).

12-6.1 Magnetic Field Intensity Inside and Outside a Straight Wire

So long as the wire is not magnetic, Eq. (12-16) gives the field intensity H outside the wire. Using $B = \mu_0 H$ yields

$$H = I / 2\pi a$$

The simplest way to find the field intensity inside the conductor is to take a closed contour in the form of a circle centered on the axis of the conductor, as shown in Fig. 12-3. In this case, the flux density at the point P is given by

$$B_p = \mu_0 I' / 2\pi r$$

where I' is the current flowing through the circle of radius r. For direct current, the density of the current is constant across the conductor, so that

$$I' = \left(\pi r^2 / \pi R^2 \right) I$$

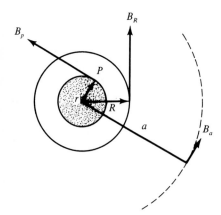

FIGURE 12-3
Flux densities B_p (inside conductor), B_R (on surface), and B_a (in air) outside the conductor.

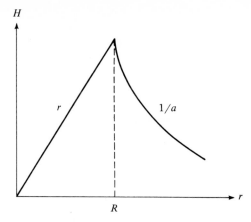

H

r

1/a

R

r

FIGURE 12-4
The linear rise of the field intensity with radius inside the conductor and the hyperbolic decay outside.

and the result for B_p is

$$B_p = \left(\mu_0 r/2\pi R^2\right) I$$

Thus

$$H_p = rI/2\pi R^2 \tag{12-17}$$

Equation (12-17) shows that the field and flux density inside the conductor rise from zero at the center in a linear manner with the radius until the outside surface is reached. It then falls off outside the conductor as $1/a$ [Eq. (12-16)]. The variation of H along any radius is shown in Fig. 12-4.

12-6.2 Modification to Include Magnetic Material

In the presence of magnetic materials, the amperian current densities must be included, and so Eq. (12-13) becomes

$$\text{curl } \bar{B} = \mu_0\left(\bar{J} + \bar{J}_M\right)$$

$\bar{J}_M$ is eliminated using Eq. (12-1) so as to give

$$\bar{J} = \text{curl}\left(\bar{B}/\mu_0 - \bar{M}\right)$$

$$= \text{curl } \bar{H} \tag{12-18}$$

using Eq. (12-11). Equation (12-18) indicates that the magnetic field intensity is related to the transport current density $\bar{J}$ through its curl. Finally Stokes' Theorem ($\oint_c \bar{H} \cdot d\bar{s} = \int \text{curl } \bar{H} \cdot \hat{n}\, da$) is used to give

$$\oint_c \bar{H} \cdot d\bar{s} = \int \bar{J} \cdot \hat{n}\, da = I \tag{12-19}$$

In effect, the line integral of the tangential component of $\bar{H}$ around a closed path is equal to the current within that path. This, again, is a very powerful relationship.

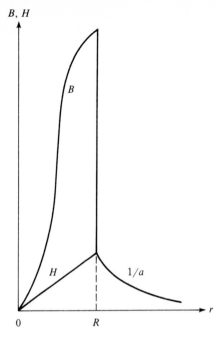

FIGURE 12-5
The variation of flux density B inside and field H outside of a cylindrical conductor carrying a current.

A common NDE application is to pass current along a steel bar and inspect the surface with magnetic particles. Considering a cylindrical steel bar, Eq. (12-19) indicates that when direct current is used, the field intensity at distance r from the center is given by

$$H_p = rI/2\pi R^2$$

so that it is identical with the nonmagnetic case (Fig. 12-5). The flux density must be introduced through the B-H curve for the material. Because H is zero at $r = 0$, then B will also be zero at the center of the bar. The flux density level at the surface requires knowledge of the B-H properties of the steel. As shown in Chap. 11, a value of I such that the surface value of H is 3200–4000 A/m (40–50 Oe), which can easily be measured with a gaussmeter, will ensure that many industrial steels are magnetized almost to saturation in this region.

Outside of the bar, while the current is flowing, the field intensity falls off as $1/a$, as before.

12-6.3   Magnetization of Tubes by the Current Method

Another common NDE application is to pass current along a tube and inspect the outside surface with magnetic particles. For the inner surface, Eq. (12-19) indicates that $H = 0$ and so also $B = 0$. The bore of the tube merely shifts the origin of Fig. 12-5 to the center of the tube.

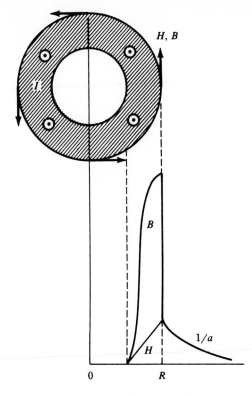

FIGURE 12-6
Variation of B and H inside and outside a tube carrying a direct current I.

Figure 12-6 illustrates the situation in which a dc current I is passed along a tube. Note the continuity of the tangential value of H and the discontinuity of the tangential value of B at the outer surface. Boundary conditions are derived later in this chapter. The ac case is left as an exercise.

12-7 SUSCEPTIBILITY AND PERMEABILITY

The three relations $B = \mu H = \mu_0 \mu_r H$, $B = \mu_0(H + M)$, and $M = \chi_m H$ can be manipulated to show that

$$\mu_r = \mu/\mu_0 = 1 + \chi_m \qquad (12\text{-}20)$$

Equation (12-20) shows that magnetic susceptibility measures how minutely different the permeabilities of dia- and paramagnetic materials are from unity.

12-8 MAGNETIC BOUNDARY CONDITIONS

At the interfaces between magnetic and nonmagnetic materials, the following boundary conditions apply. They often enable the inspector to take measurements external to the part and deduce the internal magnetic condition.

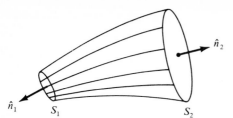

FIGURE 12-7
A tube of magnetic flux, illustrating div $B = 0$.

12-8.1   Continuity of Flux

One of Maxwell's laws is that the divergence of the flux density is everywhere zero, i.e., div $\overline{B} = 0$. From the divergence theorem, this is equivalent to $\int \overline{B} \cdot \hat{n} \, da = 0$, i.e., the total magnetic flux through any closed surface is zero. In effect, incoming and outgoing fluxes must balance. As shown in Fig. 12-7, which represents a tube of flux, this equation gives

$$\overline{B}_1 \cdot \hat{n}_1 \, da = \overline{B}_2 \cdot \hat{n}_2 \, da$$

or $$\phi(S_1) = \phi(S_2) \tag{12-21}$$

Taking the argument one step further, flux lines can never terminate but rather form closed loops. In effect, lines of B are continuous.

12-8.2   Discontinuous Nature of H

Since $\overline{B} = \mu_0(\overline{H} + \overline{M})$ and div $\overline{B} = 0$, then div $\overline{H} = -\text{div}\,\overline{M}$. As Eq. (12-8) shows, $-\text{div}\,\overline{M}$ is just the volume charge density ρ_M, so that when the divergence theorem is applied to a flux tube (Fig. 12-7), the result is

$$\overline{H}_1 \cdot \hat{n}_1 \, da - \overline{H}_2 \cdot \hat{n}_2 \, da = \rho_M \, dv \tag{12-22}$$

This result means that there is a discontinuity in the field intensity that is determined by the total pole strength intercepted by the tube. As an example [2], the way in which B and H change for a cylindrical bar of length/diameter ratio 5, uniformly magnetized, is as shown in Fig. 12-8. Although M is uniform, neither B nor H are uniform. B has its maximum value at the center (away from the demagnetizing effects of the poles), but is continuous across the ends [Eq. (12-21)]. H is discontinuous at the ends and obeys the relation

$$H_{\text{inside}} - H_{\text{outside}} = \sigma_s$$

since the right hand side of Eq. (12-22) becomes $\sigma_s \cdot \hat{n} \, da$ at the interface.

12-8.3   Continuity of Normal Component of B

Consideration of a pillbox volume at the interface between two media, as shown in Fig. 12-9, by use of Eq. (12-21) gives

$$\overline{B}_2 \cdot \hat{n}_2 \, \Delta S + \overline{B}_1 \cdot \hat{n}_1 \, \Delta S = 0$$

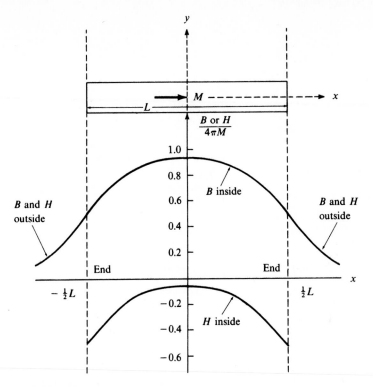

FIGURE 12-8
Axial B and H inside and outside a uniformly magnetized bar of $L/D = 5$. Note the continuity of B at the ends, the discontinuity in H at the ends, and B inside is in the opposite direction to H.

Since $\hat{n}_1 = -\hat{n}_2$, this reduces to

$$\left(\bar{B}_2 - \bar{B}_1\right) \cdot \hat{n}_2 = 0$$

or
$$B_{1,n} = B_{2,n} \tag{12-23}$$

i.e., the normal components of the flux density are continuous across an interface. Thus if a gaussmeter is used to measure the flux density emerging from the end of

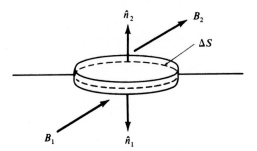

FIGURE 12-9
The use of an elementary pillbox, which is used to deduce Eq. (12-23).

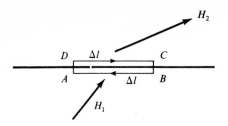

FIGURE 12-10
Illustrating the development of the boundary condition of Eq. (12-24).

a magnetized material, the flux density just inside the material will have the same value. As shown in Fig. 12-8, the flux density will rise with distance from the interface.

12-8.4   Tangential Surface Boundary Condition

Consideration of a closed path $ABCD$ in an interface, as shown in Fig. 12-10, by the use of Ampere's Circuital Law ($\oint \overline{H} \cdot d\overline{l} = I$) gives

$$H_2 \cdot \Delta l + H_1 \cdot (-\Delta l) = I$$

so long as AD and BC can be ignored. If surface currents are present, then the preceding above relation can be written as

$$H_{2,t} - H_{1,t} = \hat{j}_s \cdot \hat{l}_0 \tag{12-24}$$

i.e., the tangential components of H are related to any surface current density j_s, which may be present, and the unit vector $\hat{l}_0$ in the direction of Δl.

In testing with stationary fields, j_s is zero and this boundary condition simplifies to

$$H_{1,t} = H_{2,t} \tag{12-25}$$

As an example, Fig. 12-11 shows a typical method for the magnetization of ferromagnetic parts using a coil. If the part is stationary and the field is caused by dc, Eq. (12-25) is obeyed and so the tangential field intensity at the surface of the part, which can be measured with a gaussmeter, has the same value as the field intensity just inside the surfce. If, however, the part passes quickly through the coil, or ac excitation is used, then Eq. (12-24) is applicable because surface eddy currents are induced in the part.

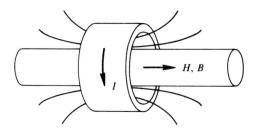

FIGURE 12-11
Magnetization of bars, tubes, etc., according to Eqs. (12-24) and (12-25).

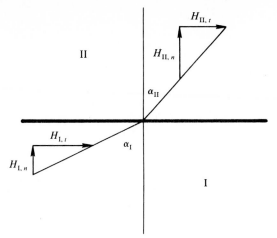

FIGURE 12-12
Illustrating the combination of earlier boundary conditions into the Eq. (12-26).

12-8.5 Static Permeability Relation

In the static case, two of the above relations can be combined to provide information about how field lines emerge from magnetic materials. Considering Fig. 12-12, with medium I ferromagnetic and medium II air, then Eq. (12-23) gives $\mu_r H_{I,n} = H_{II,n}$ and Eq. (12-25) gives $H_{I,t} = H_{II,t}$. Division gives

$$\mu_r \cot \alpha_I = \cot \alpha_{II} \tag{12-26}$$

As an example, with $\mu_I = 100$ and $\alpha_I = 89°$, Eq. (12-26) shows that 100 cot 89° = cot α_{II}, i.e., $\alpha_{II} = 30°$. With $\mu_I = 1000$, the value of α_{II} falls to 3.3°. Thus, in the infinitely high permeability case, the field lines leave the surface normally. This occurs in Förster's theory of flux leakage (Chap. 16). In practice, however, high applied field intensities imply relatively low permeabilities, such as 100–200.

CHAPTER
13

THE MAGNETIC CIRCUIT

13-1 INTRODUCTION

In much of magnetic NDE, the part undergoing inspection forms part of a magnetic circuit, which may contain one or more air gaps. For example, in the case of a short part placed inside a coil so as to establish longitudinal magnetization (Fig. 13-1), the flux generated by the coil exists mainly in air. Further, it is easy to observe with a gaussmeter that the field intensity at the point R is greater when the part is removed. The part appears to be responsible for a field intensity that opposes that of the coil. This field, known as the demagnetization field, is discussed in this chapter, along with methods that are used in magnetic NDE for reducing it.

13-2 TOROID OF DIFFERING MATERIALS

If the ring of Fig. 10-9 is broken up into materials of differing permeabilities (Fig. 13-2), Ampere's Circuital Law gives $\oint \overline{H} \cdot \overline{dl} = NI$ since the current surrounds the sample N times [8]. Writing the magnetic flux as $\phi = \mu_0 \mu_r H A$, where A is the cross-sectional area of the material (assumed constant) within the toroid, then Ampere's Law becomes

$$\oint \phi \, dl / \mu_0 \mu_r A = NI$$

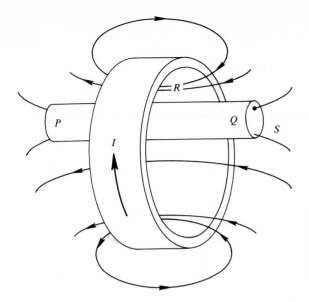

FIGURE 13-1
Relatively large air component of flux loop for an encircling coil. (*From the Nondestructive Testing Handbook, Vol. 4, 2d ed. Reprinted with permission.*)

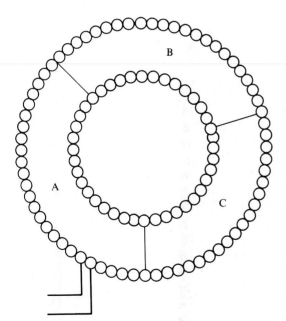

FIGURE 13-2
Toroid made from several different materials.

and taking ϕ as essentially constant within the toroid winding,

$$(\phi/\mu_0)\oint dl/\mu_r A = NI \tag{13-1}$$

Now, by analogy with Ohm's Law, it is possible to define:

1. a magnetomotive force (MMF), which is equal to NI ampere turns
2. a magnetic reluctance of $R = \oint dl/\mu_0 \mu_r A$ (which is analogous to electrical resistance, defined as $R = \int dl/\sigma A$, where σ is the electrical conductivity of the material)

Equation (13-1) becomes

$$\phi = \text{MMF}/R \tag{13-2}$$

where the total reluctance is defined as

$$R = \sum_i dl_i/\mu_0 \mu_{ri} A_i \tag{13-3}$$

It is relatively easy to show that reluctances in parallel obey the same law that is obeyed by resistances in parallel. Despite the implication of constant permeability contained within these equations, they provide a reasonable approximation to the magnetic state of different materials when brought into close proximity with each other [9]. Some examples of flux loop parameters that are calculated by this method are given later in this chapter. Modern techniques would employ iterative computer programs to allow for the problems introduced by nonlinearity in the respective permeabilities.

13-3   MAGNETIC CIRCUIT OF A PERMANENT MAGNET

Consider the magnetic circuit shown in Fig. 13-3, which consists of a permanent magnet (XY) with soft-iron flux concentrators. The soft iron contributes no flux of its own. In this situation, Ampere's Circuital Law gives $\oint \overline{H} \cdot d\overline{l} = 0$ or

$$\left[\int_x^y \overline{H} \cdot d\overline{l}\right]_{\text{magnet}} = -\left[\int_x^y \overline{H} \cdot d\overline{l}\right]_{\text{elsewhere}}$$

$$= \int_x^y dl/\mu_0 \mu_r A$$

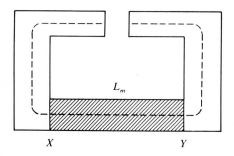

FIGURE 13-3
Permanent bar magnet with soft iron flux concentrators.

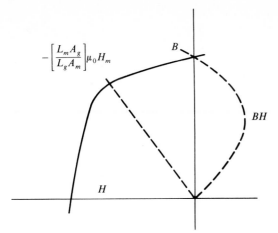

FIGURE 13-4
Second quadrant hysteresis curve for a permanent magnet showing the demagnetizing field line and the BH product.

Now, since $\phi = [BA]_{mag}$ for the magnet and the left hand side is just $H_m L_m$ for the magnet, then the preceding equation becomes

$$H_m L_m = -[BA]_{mag} R \qquad (13\text{-}4)$$

In order to solve this equation, the second-quadrant B-H curve for the magnet material is needed. Such a curve is shown in Fig. 13-4. For the soft-iron pole pieces and air gap, the reluctance is given by

$$(L/\mu_0 \mu_r A)_{Fe} + L_g/\mu_0 A_g \qquad (13\text{-}5)$$

where the first term is that for the iron and the second is that for the air gap. In the approximation in which $\mu_r \gg 1$, the soft-iron term is neglected and the result of combining Eqs. (13-4) and (13-5) is

$$B_{mag} = -[L_m A_g/L_g A_m]\mu_0 H_m \qquad (13\text{-}6)$$

Equation (13-6) represents a straight line in the second quadrant that intersects the B-H curve for the magnet at its operating point.

In this simple theory, A_g is generally taken as the pole face area in order to arrive at a first approximation of a real life situation. In reality, the flux leakage from the pole pieces must be accounted for. It is clear that in electrical theory, current is constrained to flow within conducting media; in magnetostatics, there is no such constraint on the flux, so we should not expect too much accuracy from the simple electrical analogy once we acknowledge the presence of leakage flux.

13-4 RESIDUALLY MAGNETIZED SLIT RING

As an illustrative example of the preceding theory, the field intensity within the gap of a slit ring can be computed [10]. Figure 13-5 illustrates a magnetized ring of length L_m, with a slot of width L_g. The field intensity in the ring is H_m while

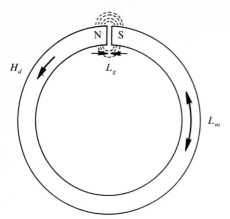

FIGURE 13-5
Residually magnetized ring.

that in the gap is H_g. Ampere's Circuital Law gives

$$H_m L_m = -H_g L_g \qquad (13\text{-}7)$$

which indicates that (1) the direction of the field intensity in the gap is the opposite of that in the ring and (2) as the gap is made wider, the two field intensities (H_m and H_g) will depend upon how H_m changes with B. The approximation used is

$$B = 0.96(H + 10)$$

where B is in tesla and H is in oersted. [This is typical of an industrial carbon steel and is a straight line drawn through the coercivity (10 Oe) and remanence (0.96 T) as shown in Fig. 13-6.] Three gap widths are investigated, with μ_r assumed constant at 250 and L_m is taken as 500 mm. At a gap width of 0.5 mm,

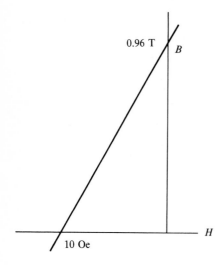

FIGURE 13-6
Approximation to the second quadrant B-H curve for an industrial carbon steel ($B_r = 0.96$ T and $H_c = 10$ Oe).

the reluctance of the ring is given by

$$R_{0.5} = \left(L_m/\mu_r A_m + L_g/A_g \right)/\mu_0 \quad \text{from Eq. (13-3)}$$
$$= 2.5/\mu_0 A_m \quad \text{with the approximation } A_m = A_g$$

Inspection of the magnitudes of the two terms shows that the steel has the larger reluctance and cannot be ignored. Using $[BA]_m R_{0.5} = -H_m L_m$ gives

$$B_m(2.5/\mu_0) = 500 H_m$$

so that $B_m = -200\mu_0 H_m$

This is a line in the second quadrant of the B-H curve for the ring. The linear approximation is now used to give

$$H_m = -0.048 \text{ Oe}$$

which is obviously a very small demagnetizing field. The same equation also gives

$$B_m = 0.955 \text{ T} = 9.55 \text{ kG}$$

which is fractionally less than the remanence value. Finally, Eq. (13-7) is used to give

$$H_g = 47.8 \text{ Oe}$$

Thus the field intensity in the gap is much larger than the demagnetizing field H_m in the ring.

Increasing the gap width to 1 mm raises the reluctance to $3/\mu_0 A_m$, and performing the suggested calculations yields $B_m = -167\mu_0 H_m$, $H_m = -0.057$ Oe, and $H_g = 28.6$ Oe. Thus an increase in the gap width also causes an increase in the gap field intensity. Finally, increasing the gap width to 2 mm yields the following results: $R = 4/\mu_0 A_m$, $B_m = -131\mu_0 H_m$, $H_m = -0.076$ Oe, and $H_g = 19.1$ Oe. Equation (13-3) shows in this case that the reluctance of the air gap is the same as that of the steel.

13-4.1 Comparison with Experiment

The values found in the preceding example are tabulated in Table 13-1 from which it is seen that while the value of H_g falls as the gap width rises, the value of the product $H_g L_g$ also rises. Experimental data indicate that the value of $H_g L_g$ rises to a maximum value at a certain slit width and then falls. Values of H_g,

TABLE 13-1
Magnetic field characteristics for increasing gap width (L_g)

L_g / mm	B_M / T	H_m / Oe	H_g / Oe	$H_g L_g$
0.5	0.955	0.048	47.8	23.9
1.0	0.955	0.057	28.6	28.6
2.0	0.953	0.076	19.1	38.1

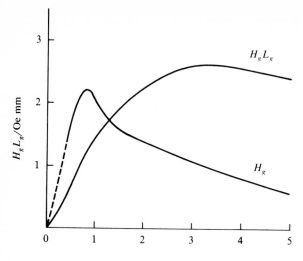

FIGURE 13-7
Typical variation of the field intensity (H_g) and product $H_g L_g$ for a through slit in a residually magnetized ring.

taken with a gaussmeter inside the slot, first rise from zero, then peak at a specific slit width, and then fall (Fig. 13-7). The dashed lines are extrapolations of the data that are required because of the finite thickness of the Hall element used to take these data.

It is clear that the constant permeability approximation predicts the general trend in the data for relatively wide gaps. At small gap widths the theory is not valid, and modern techniques involving computer solutions that employ finite element and difference techniques must be employed.

Some interesting differences exist between the leakage fields from through slots and those from slots that do not pass all the way through the material, especially in residual induction [10, 31]. These are discussed in Chap. 15.

13-5   DEMAGNETIZATION FIELD

The presence of a break in a ring of ferromagnetic material can be thought of as setting up poles in the vicinity of the break, which in turn are responsible for the creation of a demagnetizing field intensity within the ring. This phenomenon occurs wherever field lines leave a magnetized material. In most cases, however, the size of the demagnetizing field is difficult, if not impossible, to calculate, but for some simple geometries it can be computed. Figure 13-8 illustrates a uniformly magnetized sphere of magnetization M [5]. The number of poles created on the annular strip shown (radius r, width dr) is $\overline{M} \cdot \hat{n}\, da = 2\pi M\, dr$. The same number of poles exist on the curved surface of the sphere, shown shaded in Fig. 13-8, which has an area of $da = (2\pi a \sin\theta)(a\, d\theta)$, so that the surface pole density is given by

$$\sigma_s = (2\pi M r\, dr)/(2\pi a^2 \sin\theta\, d\theta)$$

Since $r = a \sin\theta$, where a is the radius of the sphere, so that $d\theta = a\cos\theta\, d\theta$, the

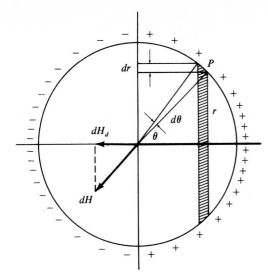

FIGURE 13-8
Topography for the calculation of the demagnetizing field due to uniformly magnetized sphere.

surface pole density becomes

$$\sigma_s = M \cos \theta$$

This is not constant, but has larger values at smaller values of θ. Now a small element of surface charge at the point P causes a field intensity at the center of the sphere that obeys the inverse square law, and when the element is rotated around the axis to sweep out the shaded area, the field component that is perpendicular to the axis cancels out and what is left is an element of the demagnetizing field of strength

$$dH_d = (2\sigma_s)(2a^2 \sin \theta \, d\theta)(\cos \theta)/4\pi a^2$$
$$= (\sigma_s/\pi) \sin \theta \cos \theta \, d\theta$$

The total demagnetizing field intensity H_d, found by integration of dH_d from $\theta = 0\text{--}90°$, is just

$$H_d = M/3 \tag{13-8}$$

In this case, the demagnetization field is proportional to the magnetization. More generally, it is often assumed that for other shapes,

$$H_d = N_d M \tag{13-9}$$

where N_d is known as the demagnetization factor. The demagnetization factor for the sphere is $1/3$.

13-6 DEMAGNETIZATION FACTOR CURVES

It is clear from the foregoing argument that demagnetization factors can be computed only for those shapes that retain uniform magnetization. Values of N_d have been computed for spheroids; in the case of the prolate spheroid in which

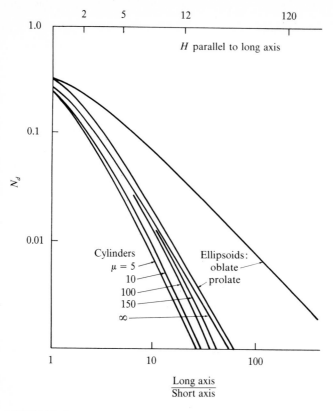

FIGURE 13-9
Demagnetization factor versus axial ratio for short parts with the applied field parallel to the long axis
of the part. (*Courtesy R. M. Bozorth, Ferromagnetism.*)

the major axis (c) is considerably longer than the minor axis (a), so that the
elongated shape approximates that of a long rod of length $2c$, radius a, the
result is

$$N_d = (a/c)^2[\ln(2c/a) - 1] \qquad (13\text{-}10)$$

For example, with $c/a = 10$, the demagnetization factor is 0.02.

A set of curves has been provided by Bozorth for parts to which axial ratios
can be assigned and for which the field intensity is parallel to the longer axis.
These curves are presented in Fig. 13-9, but when using them, it should be
recalled that samples that are commonly inspected by magnetic NDE cannot
generally be uniformly magnetized. In effect, the demagnetization factor varies
along the length of a part. The curves show the values of $N_d/4\pi$ for oblate and
prolate ellipsoids and for cylinders of materials of different permeability. They
indicate that N_d decreases with part length and increases with part permeability.

Solenoid

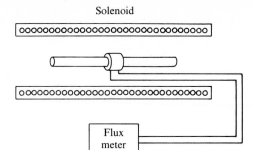

Flux meter

FIGURE 13-10
Method for *B-H* parameters of a long sample.

13-7 APPARENT PERMEABILITY

One method for performing a *B-H* investigation of an elongated sample is to use a long solenoid for *H* and a pickup coil/integrating fluxmeter system for the measurement of *B*, as shown in Fig. 13-10. The applied field intensity is given by

$$H_a = NI/L$$

from Chap. 10, or *nI*, where *n* is the number of turns per meter on the solenoid. This value must be corrected for the demagnetization factor of the sample so as to obtain the true field intensity within the material. This is given by

$$H = H_a - N_d M \tag{13-11}$$

but since $B = \mu_0(H + M)$, Eq. (13-11) can be rearranged to give

$$B = \mu_0\left[H_a + (1 - N_d)M\right]$$

From this equation, *M* is obtained as $[(B - \mu_0 H)/\mu_0(1 - N_d)]$ and returned to Eq. (13-11) to give

$$H = H_a - \frac{N_d(B - \mu_0 H)}{\mu_0(1 - N_d)} \tag{13-12}$$

which provides the correction for the applied field to yield the true field intensity in the specimen.

The apparent permeability of the specimen is found by division by *B*, i.e.,

$$\frac{H}{B} = \frac{H_a}{B} - N_d(B - \mu_0 H)/\mu_0 B(1 - N_d)$$

or $\quad (\mu_0\mu_{\text{true}})^{-1} = (\mu_0\mu_{\text{app}})^{-1} - N_d/\mu_0(1 - N_d) \quad$ since $B - \mu_0 H \simeq B$

$$1/\mu_{\text{true}} = 1/\mu_{\text{app}} - N_d/(1 - N_d) \tag{13-13}$$

An approximation to Eq. (13-13) may be found by noting that since N_d is usually very small, then $N_d/(1 - N_d) \rightarrow N_d$ and the result is

$$1/\mu_{\text{true}} = 1/\mu_{\text{app}} - N_d \tag{13-13a}$$

Thus in this approximation, the true and apparent permeabilities are connected very simply through the demagnetization factor.

TABLE 13-2
Comparison of measured and calculated effective permeabilities

| | Measured μ_{app} (G / Oe) | | | Calculated |
| | Test piece diameter (in) | | | μ_{app} |
L / D	0.5	1.0	1.5	(G / Oe)
3	11.7	13.6	12.7	13
5	23.3	25.2	23.3	25
8	45.0	45.4	43.2	43
10	57.0	58.5	56.0	55
12	69.4	68.8	68.7	67

Source: Reproduced from ref. 11 by permission of the American Society of Nondestructive Testing.

13-8   APPARENT PERMEABILITY OF VERY SHORT PARTS

As the axial ratio of the part decreases, the demagnetization factor increases [11]. Equation (13-13a) shows that at some point, for a short part with a large value of μ_{true}, the limit is approached where

$$\mu_{app} \rightarrow 1/N_d \qquad (13\text{-}14)$$

The implication of Eq. (13-14) is that for very short parts, the permeability that they exhibit is more dependent upon the demagnetization factor, which is a function of the part geometry, than upon the true material permeability. This is reflected in the use of formulas such as

$$\mu_{app} = 6(L/D) - 5 \qquad (13\text{-}15)$$

for short parts within specific axial ratios. Many similar formulas appear in experimentally produced inspection specifications.

A comparison of measured values of apparent permeability for L/D ratios of cylindrical samples of bar stock in the range 3–12, and values calculated from Eq. (13-15) has been made by McClurg [11]. This is presented in Table 13-2. Within this range of L/D ratios, Eq. (13-13) shows that the two terms on the right hand side of the equation are very similar in magnitude.

13-9   EFFECT OF THE DEMAGNETIZING FIELD

In order to illustrate the effect of the demagnetizing field, caused by the presence of the ends of a bar sample, Cullity [5] has presented the following data for a 24 cm long cold-worked iron rod of $L/D = 35$. For this value, the Bozorth curves give $N_d = 0.002$. The values of H_a, the applied field intensity, and B are taken

TABLE 13-3
Demagnetization characteristics for rods with $L / D = 35$

H_a / Oe	B / kG	$(B - \mu_0 H_a)$ / kG	M / kG	H_{demag} / Oe	H / Oe
8.1	1.08	1.07	1.07	2.1	6.0
16.2	3.85	3.83	3.83	7.7	8.5
26.9	7.91	7.88	7.89	15.8	11.1
35.0	10.08	10.04	10.07	20.1	14.9
43.0	12.42	12.38	12.40	24.8	18.2
53.9	14.86	14.81	14.83	29.6	24.3
80.7	18.22	18.14	18.18	36.3	44.4

Source: Reproduced by permission of Addison-Wesley.

with an experimental system such as Fig. 13-10; the rest of Table 13-3 is constructed using the preceding equations. It is clear that while the values of B and M do not differ by much, the demagnetizing field intensity is considerable, being of the order of one-half of the applied field strength. Note that the last column of the table gives the calculated true field strength within the rod.

When plotted on the same scale, the B-H and B-H_a curves show the effect of the demagnetizing field (Fig. 13-11). The B-H curve, which is that which would be obtained from a ring sample, without the complication of the poles at the ends of the sample, lies entirely to the left of the B-H_a curve. The data for $H_a = 26.9$ Oe, for example, give $\mu_{true} = 713$ and $\mu_{app} = 294$. (The reader should verify that the difference between the reciprocals of these values is close to N_d.)

The effect of high demagnetization fields is to both push the B-H_a curve closer to the field intensity axis and to make the curve more linear. Under such circumstances, the use of a low, but constant, value for the apparent permeability is more justified.

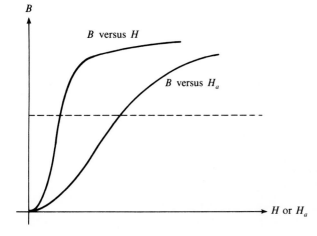

FIGURE 13-11
Flux density in a bar sample plotted against the applied field (H_a) and the internal field in the material (H).

13-10 MAGNETIZATION OF SHORT PARTS

Consider, for example, the magnetization of a 15 cm (6 in) long bar of diameter 2.5 cm (1 in), with a coil as shown in Fig. 13-1. For a relative permeability of 100, the Bozorth curves give $N_d = 0.03$. Table 13-3 shows that, for a ring of soft, cold-worked iron, 880 A/m (11.1 Oe) are required to raise the flux density to 7.9 kG. For a bar of $L/D = 35$, 2150 A/m (27 Oe) are required for the same value of B. Now, with $N_d = 0.03$, the demagnetizing field strength at the same flux level is $N_d M = 21,000$ A/m (236.7 Oe), and so the applied field strength needed to cause this level of magnetization is 19,700 A/m (247.8 Oe). The relative permeability at this point is 31.

From Eq. (10-30), it is seen that large numbers of ampere turns are required of coils in order to produce such high values of H_a. It is known, however, that the field strength on the inner diameter (ID) of a coil is greater than at its center, does not vary much over one-tenth of the diameter of the coil, and, for the same number of ampere turns, does not vary much with coil diameter. In effect, H near to the ID of a coil is not seriously affected by current flowing in the diametrically opposite part of the coil, so that coil diameters are relatively unimportant if magnetization of short parts is to be accomplished by placing them against the ID of the coil. Under these circumstances, in order to obtain a flux density of 10.85 kG in a sample of L/D between 2 and 15, the number of ampere turns on the coil should be

$$NI = 45,000/(L/D) \tag{13-16}$$

The value of flux density quoted is such as to give adequate MFL from surface breaking transverse flaws in the part in residual induction for the application of wet fluorescent magnetic particles. At such high flux density levels, the application of dry particles causes "furring" along the emergent field lines (Fig. 13-12) with the possibility that tight flaws may be obscured by the particles themselves.

FIGURE 13-12
Dry magnetic particles "furring" along external field lines at the end of a tube.

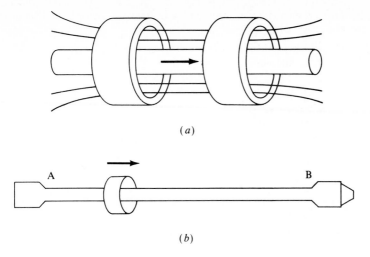

FIGURE 13-13
Two common methods for longitudinal external magnetization of elongated parts: (*a*) part moving;
(*b*) magnetizing coil moving.

13-11 MAGNETIC CIRCUITS USED IN INSPECTION

The rest of this chapter is devoted to a description of some of the magnetic circuits that are used to establish flux in parts that are inspected by MFL. Not all of the many possibilities are given.

13-11.1 Elongated Tubes and Rods

Much tubular material and rods are inspected for transverse flaws, both in the new and used condition, by magnetizing them longitudinally to saturation and scanning the outer surface with some form of MFL sensor. Typical discontinuities include:

1. New: quench cracks, seams, overlaps, pits, slugs
2. Used: fatigue cracks, pits, eroded areas

Figure 13-13 shows two common situations: (*a*) a tube passed slowly through one or two coils and (*b*) a coil and detector assembly passed slowly along the tube. In either case, the material is more-or-less centered within the coil and magnetized to saturation by direct current. Experience indicates that coil fields of $H = 16,000$ A/m (200 Oe), measured at the surface of the tube with a gaussmeter, provide sufficient flux within the material to give adequate MFL from inner-surface flaws such as fatigue cracks and pits. The double coil system is used to minimize the L/D ratio of the material, since in this case, L is the average distance between where the flux lines enter and leave the material. For tubes, D can be taken as twice the wall thickness.

Where only outer diameter (OD) surface-breaking cracks are required, ac can be used in the coil(s). Sensors are commonly coils, Hall elements, and magnetodiodes, which ride at about 1 mm from the OD surface. In the case of drill pipe, regions A and B [Fig. 13-13(b)] are also inspected with magnetic particles because the sensors do not ride the tube well in these areas, which also contain the heat-affected zone between the tube and the tool joints.

13-11.2 Wire Rope Inspection

Figure 13-14 shows typical cross sections of wire rope. New rope is examined to ensure that there are no breaks in the individual strands by magnetizing the rope to saturation with the yoke technique shown in Fig. 13-15, while used rope is examined for cross-section loss that might be due to elongation, outer surface wear, and breakage of strands [12]. The use of electromagnet (a) or permanent magnet (b) yokes depends to some extent upon the accessibility of the rope.

The flux loop is derived as follows. If there are N electromagnet cores, each of length L_M, cross section A_M, and effective relative permeability μ_M, and the two end pieces of the loop have an effective length L_E and permeability μ_E, then the total magnetic reluctance is given by

$$R = (1/\mu_0)(L_M/\mu_M N A_M + 2L_E/\mu_E A_E + 2L_A/A_A + L_R/\mu_R A_R) \quad (13\text{-}17)$$

where the air gaps have an effective length of L_A and area A_A and the rope has an effective length in the instrument of L_R and a cross-sectional area A_R. Further, if there are n turns on each electromagnet core, each carrying I amperes, then

$$nNI = \phi/\mu_0 R = \oint \overline{H} \cdot d\overline{l} \quad (13\text{-}18)$$

By making suitable estimates, a reasonably good estimate of the flux in the rope can be found, and from it, the flux density.

Where permanent magnets are used, the soft iron of the electromagnet is replaced by cylindrical magnets. If the rope is moved relatively slowly, so as to minimize the effect of motionally induced eddy currents, Eq. (13-4) defines the magnetic circuit.

Recalling that the boundary condition on B is that it is continuous throughout the flux loop, its value in the air gap can easily be measured with a Hall element gaussmeter. Since discontinuities can exist deep within the rope, a high value of B is needed to cause MFL at the surface. An additional problem is that the instrument must permit the passage of splices, so the sensors must be maintained at some distance from the surface of the rope. Commonly used sensors are Hall elements and encircling coils.

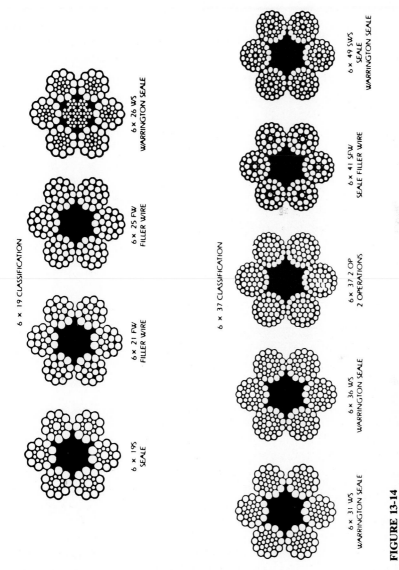

FIGURE 13-14

Cross-sectional area configurations for various type of wire rope. (*From Wire Rope Corporation of America. Reproduced with permission from ASNT Handbook, Vol. IV, 1986.*)

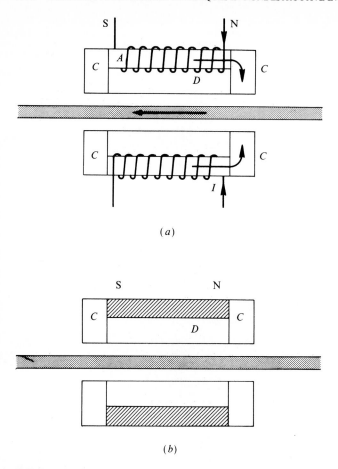

FIGURE 13-15
Setup for magnetic flux leakage testing of elongated parts: (*a*) flux loop for wire rope inspection with an electromagnet; (*b*) with permanent magnets. (*From the Nondestructive Testing Handbook, Vol. 4, 2d ed. Reprinted with permission.*)

13-11.3 Internal Casing, Tubing, and Line Pipe Inspection

In the case of down-hole casing, underground gas transmission lines, and installed ferromagnetic tubing in heat exchangers, inspection is performed with MFL from the inside. Typical flaws include pitting, holes, splits, erosion of either ID or OD, hydrogen and temperature cycling damage, and various forms of cracking.

Figure 13-16 illustrates one form of pipeline inspection "pig," which consists of a drive/cleaning section, the flux loop/detector section, and a recorder package, universally jointed to permit negotiation of bends [13–15].

Cleaning package   Flux loop   Recorder package

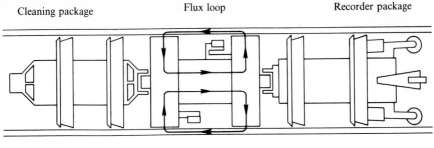

Sensors

FIGURE 13-16
Schematic diagram of a magnetic flux leakage pipeline inspection system. (*From the Nondestructive Testing Handbook, Vol. 4, 2d ed. Reprinted with permission.*)

The magnetization system must be as efficient as possible, to keep the mass of the pig down. This is accomplished by considering the lengths and areas of the flux loop components and using steel brushes between the pole pieces and the pipe so as to reduce the air-gap reluctance. The flux loop equation is

$$NI/\phi = \left[L_c/\mu_c A_c + 2t_g/A_g + L_p/\mu_p A_p \right]/\mu_0 \qquad (13\text{-}19)$$

where the subscripts represent the core (c), the air gap (g), and the pipe (p). The value of A_g, the area of the air gap, is given by the product of the circumference of the pole piece and its width. Shims or wire brushes are used on pole pieces to minimize the air gap. Computation of the three contributions to the reluctance indicate that no component can be ignored (see Problem 8, Chap. 13).

Figure 13-17 shows an almost identical system used for the inspection of installed well casing [16]. The drive system is provided by a work-over rig and the MFL signals are sent by telemetry up the wire line to the logging truck. Any down-hole electronics must be designed to withstand elevated temperatures, e.g., 200°C. In one application, the core is hollow so as to reduce A_c and made from a low permeability material so as to reduce μ_c. This is done to raise the core reluctance, which is controllable, and so effectively reduce the effect of variations of t_g and A_p, which are not.

The MFL sensors in both applications ride in spring-loaded shoes. The flux density in the tube wall is as high as possible in order to provide sensitivity to outer surface pitting. (If the tube wall is not saturated, flux diversion from an OD surface discontinuity merely drives the level of flux density in the wall beneath the pit to a higher level.)

In some systems, the sensors detect pits but not elongated regions of erosion. For this reason, eddy current sensors are added, which respond to the effects of lift-off, wall loss, and ID changes. Since the inspection systems may not move at constant speed, it is essential to provide sensors that are not speed-dependent.

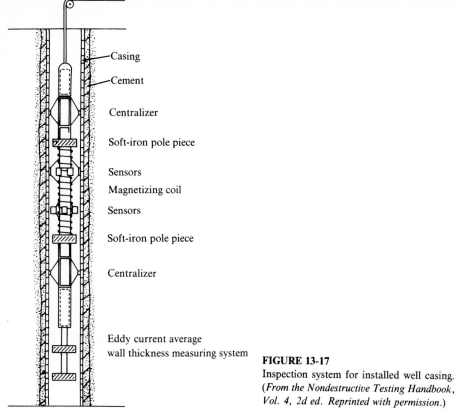

Casing

Cement

Centralizer

Soft-iron pole piece

Sensors
Magnetizing coil
Sensors

Soft-iron pole piece

Centralizer

Eddy current average
wall thickness measuring system

FIGURE 13-17
Inspection system for installed well casing.
(*From the Nondestructive Testing Handbook,
Vol. 4, 2d ed. Reprinted with permission.*)

For the assessment of damage to the heat exchanger and the chemical plant tubing up to 5 cm ID, permanent magnet flux loops can be employed. Samarium–cobalt and neodymium–iron are well suited to this application.

13-11.4   Internal End Area Inspection for Drill Pipe

Transverse fatigue cracking sometimes occurs in the region where a tool joint meets the tube in drill pipe. One form of internal inspection [17] is to magnetize from the OD and use magnetic particles and an optiscope on the ID. This is tedious, but can be automated by the use of a ring of coil or Hall element sensors, as shown in Fig. 13-18. The field intensity needed from the coil can be found with a Hall element gaussmeter placed inside the tube and simulated test flaws.

13-11.5   Ends of Tubes and Rods

Tube and rod ends are often inspected for transverse cracks both prior to and after upsetting and threading. Magnetization is accomplished with a coil [18], the

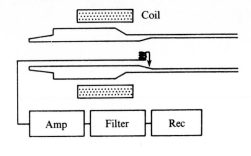

FIGURE 13-18
Drill pipe internal inspection system with external coil excitation.

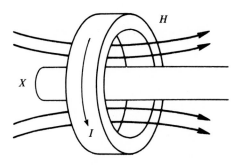

FIGURE 13-19
Coil magnetization of the end of a tube or rod. (*From the Nondestructive Testing Handbook, Vol. 4, 2d ed. Reprinted with permission.*)

criterion being that the end of the material must be as saturated as possible if inspection is to be performed for internal and external transverse cracks in residual induction (Fig. 13-19). Higher air fields are required of the coil than is the case for the elongated tube of Fig. 13-13, because of the physical presence of the end of the material. The effective L/D ratio is smaller and the demagnetization field is larger.

A good experimental method is to test the external longitudinal field intensity at X (Fig. 13-19) with a gaussmeter after applying increasing values of I with the coil. At some current value, which depends on the number of turns on the coil, the diameter or thickness of the material, and the length of material undergoing magnetization, the material will saturate. At this point, because of large external fields, wet fluorescent MPI should be used.

The threaded connections of drill pipe tool joints and drill collars are often inspected in this manner. The coil is often several turns of welding cable wrapped tightly around the end of the part and connected to a capacitor discharge magnetizer (Chap. 14). Several "shots" are required, the particles being applied between shots.

13-11.6 Local Yoke Magnetization

Yokes with adjustable legs (Fig. 13-20) are commonly used in magnetic NDE. The flux loop is given by [19]

$$NI/\phi = (1/\mu_0)\left[L_c/\mu_c A_c + 2t_g/A_g + L_p/\mu_p A_p\right]$$

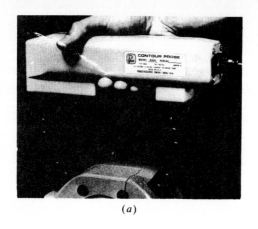

(a)

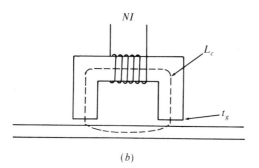

(b)

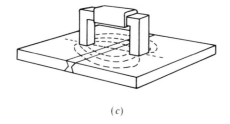

(c)

FIGURE 13-20
(a) Hand held prove-up yoke with adjustable legs. (*Courtesy Parker Research, Dunedin, Florida.*) (b) Flux loop parameters. (c) Spreading of flux lines into a part. (*From the Nondestructive Testing Handbook, Vol. 4, 2d ed. Reprinted with permission of ASNT.*)

as before, where the subscripts represent the core (c), air gap (g), and part (p). Adjustable legs minimize t_g. In most applications, the reluctance of the part is unknown because the flux lines that are generated by NI spread out into the part [Fig. 13-20(c)] in an uncontrolled manner. However, industrial yokes are designed to provide 30–40 Oe tangentially at the surface of the part, midway between the legs. This and the region of effectiveness of the surface field can be checked with a Hall element gaussmeter. Equation (12-25) is applicable.

Yokes are designed to provide 50–60 Hz ac for the detection of surface-breaking cracks and rectified ac for prove-up of subsurface flaws. The field intensity at any point below the surface of the inspected part is given by

$$H = H_0 \exp\left(-d\sqrt{\pi f \mu \sigma}\right) \sin\left(2\pi f t - d\sqrt{\pi f \mu \sigma}\right) \qquad (13\text{-}20)$$

where H_0 = the surface field intensity
d = the depth into the part
f = the applied frequency
μ = the permeability of the local region of the part
σ = its electrical conductivity

The $d\sqrt{\pi f \mu \sigma}$ term within the argument of the sine expression gives the phase difference between the field at the surface and at depth d and is unimportant in MPI. The argument of the exponential term may be written as d/δ, where δ is known as the skin depth, i.e.,

$$\delta = 1/\sqrt{\pi f \mu \sigma}$$

$$= 10^4/2\pi\sqrt{10 f \mu_r \sigma} \qquad (13\text{-}21)$$

in SI units. At one skin depth, the field intensity is $1/e$ ($0.37\times$) of its surface value. Skin depths for typical steels are about 1 mm. At five skin depths, the field intensity is essentially zero, showing the confinement of the field, and the resultant flux density within the part, to the surface. With dc fields, penetration is much deeper into the part.

One rough method for assessing the ability of such a yoke to adequately magnetize a part is to specify its ability to hold a mass of steel against gravity. The approximate formula for the force that holds two magnetic materials together is

$$F = \tfrac{1}{2} B_a^2 A_a / \mu_0 \qquad (13\text{-}22)$$

where B_a is the flux density, assumed constant, within the air gap of area A_a. With two such forces (one for each leg) lifting a mass M,

$$2F = B_a^2 A_a / \mu_0 = Mg$$

or, since the boundary condition on the flux density is that B_m in the material is equal to B_a and so

$$B_m = \mu_0 Mg / A_a \qquad (13\text{-}23)$$

For example, if it is specified that the yoke lift 18 kG (40 lb), and its pole pieces have an area of 6.45×10^{-4} m^2 (1 in^2), then Eq. (13-23) gives $B_m = 0.6$ T.

13-11.7 YOKES IN RAIL INSPECTION [20]

In order to illustrate the effects of eddy currents, which are induced whenever there is relative motion between the magnetizing agent and the part, an example

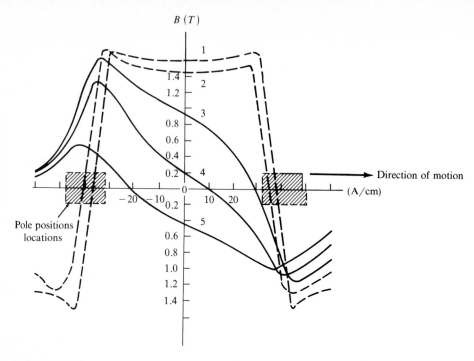

FIGURE 13-21

Distribution curves for the longitudinal component of magnetic induction in the rail head in the surface layer and in its central section. (*Reprinted from "Flaw Detection in Rails," Report FRA/ORD/77/10. U.S. Department of Transportation, Federal Railroad Administration 1978.*)

is given from the inspection of railroad rails. Figure 13-21 shows the velocity dependence of the flux density in typical rail, with the yoke moving at 41 km/h. Curve 1 shows the zero velocity situation, in which there is a tangential field of 1.7 T under each pole and 1.6 T over a 35 cm region between the poles. Curve 2 shows the effect of the motion on the flux density at the surface, which is not great. Curves 3, 4, and 5 show the lowering of the flux density deeper into the rail. The eddy currents shield the inner regions of the rail in a nonlinear manner and curves 3–5 suggest that the best place for MFL sensors is just in front of the following pole.

13-12   OTHER FLUX LOOP APPLICATIONS

Flux loops are used in many applications. Some typical example are:

1. *Determination of the thickness of steel strips [19].* For thin strips, e.g., razor blade steel or recording tape, the reluctance equation can be used to determine a magnetic thickness gauge or a permeability change indicator. Variations in thickness of ferric or chromium oxide on recording tape have been monitored in such a manner.

2. *Determination of anisotropy in steel.* If a second coil is wound onto a magnetizing ac yoke, then its output will reflect changes in the permeability of the part (thickness variations are eliminated via the skin depth effect). Many steels are rolled to have different grain sizes in different directions; this affects the average permeability along and perpendicular to the rolling direction, which can be detected by such a yoke method.

3. *Coupling locators.* Since steel cable (wire line) stretches as tools are lowered down the bore of casing in a hydrocarbon well, the couplings that are used to join the casing sections together form convenient depth markers. Their position is found by lowering a centralized bar magnet down the well and detecting the change in the flux within the flux loop, which consists of the magnet, the bore of the casing, and the surrounding steel, with an encircling coil as the tool passes a coupling. The additional steel of the coupling lowers the flux loop reluctance and so raises the total flux in the magnet and its surrounding coil.

CHAPTER
14

MAGNETIZATION IN THE CIRCUMFERENTIAL DIRECTION

14-1 INTRODUCTION

Many inspected parts have an axis of symmetry and can be easily magnetized in the circumferential direction so as to produce MFL from longitudinally oriented flaws. Parts are also sometimes remagnetized circumferentially so as to remove the high external fields that might be caused at their ends when the methods of Chap. 13 are used. Three methods are commonly used. Where the part is solid, or only an outer surface inspection is needed, and the specifications permit, the passage of current along the part is often used. Where a more uniform flux density is required in hollow parts, an internal conductor method may be used, with either direct or pulse current [21, 22]. Finally, for round bars and tubes, rotating yokes are sometimes used. These techniques are discussed in this chapter.

14-2 CURRENT APPLIED TO THE PART

In this method (Fig. 14-1) the part is clamped between copper mesh electrodes and current is applied at a level that gives a certain minimum surface field intensity, according to $\oint \overline{H} \cdot d\overline{l} = I$. It is then inspected with MPI. The pads minimize arc burns, which should be removed prior to part use. Some inspection specifications do not permit current to be applied to the part for this reason.

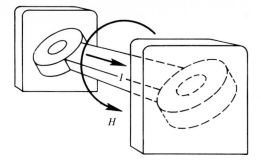

FIGURE 14-1
Circumferential magnetization of elongated parts by the use of a current I along the part.

If the part is a round bar or tube, Ampere's Circuital Law gives

$$H = I/2\pi a \ (\text{A/m}) \tag{14-1}$$

where a is the part radius. As an example, the dc current required to produce a surface field intensity of 4000 A/m (50 Oe) for a 7.6 cm diameter (3 in) bar is

$$I = 2\pi a H = (2\pi)(0.076 \text{ m})(4000 \text{ Am})$$
$$= 960 \text{ A}$$

It should be noted that the electrical resistance of the part being inspected is given by $R = L/\sigma A$ (L is the part length, A is the cross-sectional area, σ is the electrical conductivity), so that part dimensions affect the resistance presented to the supply circuit. It is prudent to ensure that the power supply is capable of providing sufficient current to produce 3200–4000 A/m at the part surface for all the dimensions of parts to be inspected.

If the part has an irregular cross section, a Hall element gaussmeter should be used to determine the external field strength at the region of maximum cross section. Obviously if this material is saturated, then the surface material at any point of lesser cross section also will be saturated.

14-3 INTERNAL CONDUCTOR METHOD

The presence of a bore in the part indicates that an internal conductor can be used to establish magnetization. Two techniques are commonly used. In the first, direct current is passed along a centered conductor so as to achieve a flux density that is directed parallel to the inner and outer surfaces of the part. In the second, an off-centered rod connected to a pulsing system is used. Uniform flux densities are also established by this technique, since the pulse induces eddy currents into the material, which have the effect of smearing out any nonuniformity of the field.

14-3.1 Central Conductor Method

Figure 14-2 illustrates the field intensity and flux density for a noneccentric tube that is magnetized by the central conductor method. The field intensity (H) is

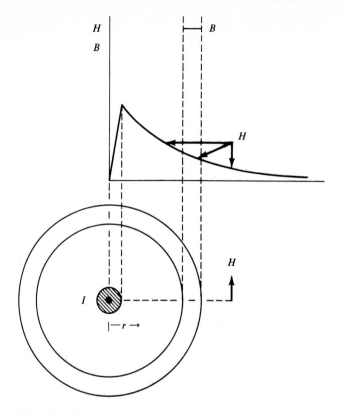

FIGURE 14-2
Field (H) and flux density (B) variations for the magnetization of a steel tube by a centered conductor carrying dc.

highest at the surface of the conductor and falls off as $1/r$ outside the conductor. In order to raise the flux density within the inspected part to a high level, the value of H at the OD of the part should be 3200–4000 A/m (40–50 Oe), so that this requirement governs the central conductor current. When this field intensity is removed, the material should be at its remanent value (B_r) and if the material is perfectly concentric and has uniform permeability, there should be no external flux. For active field inspection with wet particles, lower values are used.

A comparison between the field intensity (H) and flux density (B) for the applied current and dc central conductor methods is shown in Fig. 14-3 for a typical tube wall [23]. The figure shows that the current used provides the same field intensity at the OD surface, irrespective of the method used, but the flux densities are quite different.

14-3.2 Capacitor Discharge Magnetization

A more complex magnetization method is that in which a pulsed current is used, the duration of the pulse being of the order of 10–100 ms. This method is

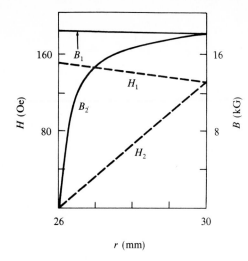

FIGURE 14-3

Comparison between the central conductor and the applied current methods for the circumferential magnetization of tubes. The subscript 1 denotes central conductor with dc; the subscript 2 denotes applied dc current. (*Reproduced from [23] with permission.* © *Plenum, New York.*)

commonly used for the magnetization of 9–12 m joints of oil field tubular materials. Figure 14-4 illustrates the apparent simplicity of the technique, in which an insulated rod is connected to a capacitor discharge (CD) bank. Modern CD systems consist of a charging circuit, a bank of capacitors, a silicon-controlled rectifier (SCR), and firing and safety circuits. The SCR ensures that the pulse is unipolar, while the safety circuit ensures that the SCR is closed after the pulse has been fired. The pulse current I_c is as high as practically possible because of low-resistance copper cables and an aluminum rod and connectors, although pulse duration is the more important parameter in CD design.

The pulse shape is governed by the physics of the *L-C-R* circuit [24]; L is the total inductance in the current circuit, which is dependent upon the electrical, magnetic, and dimensional parameters of the tube undergoing magnetization, while C and R are the total capacitance and resistance in the circuit. C within the unit is so large (2–8 F) that other distributed capacitances can be ignored,

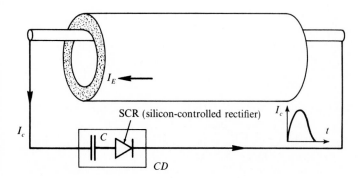

FIGURE 14-4

Schematic for capacitive discharge method for magnetization.

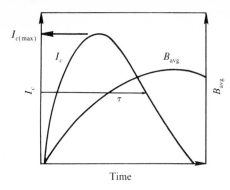

FIGURE 14-5
The form of the time variation of the central conductor current and the average induced flux density in the part. Note the time lag in the maxima due to I_E. (*From the Nondestructive Testing Handbook, Vol. 4, 2d ed. Reprinted with permission.*)

while R includes the resistance of the rod, cables, connections, and SCR. The voltages around this circuit are at any instant given by Ohm's Law, i.e.,

$$O = (d/dt)[\underset{\text{[inductor]}}{LI_c}] + \underset{\text{[resistor]}}{I_c R_c} + (1/C)\underset{\text{[capacitor]}}{\int I_c \, dt} \tag{14-2}$$

If it is assumed that L remains constant during the passage of I_c and the subsequent magnetization process, then Eq. (14-2) simplifies considerably because the time dependence of the inductance can be ignored.

Under this condition, the general time variation of I versus t yields three solutions to Eq. (14-2), which depend upon the relative values of R, C, and L. The best solution so far as magnetization of the part is concerned is to arrange for the pulse to have as long a time duration as possible, perhaps 200 ms. The reason for this is that the rapid rise of magnetizing current (I_c) induces an eddy current within the part, the effect of which is to shield the interior of the part from the full magnetizing effect of I_c. It is therefore essential that $I(t)$, the solution of Eq. (14-2), be relatively high while the average of the flux density within the material is rising. This high value of $I(t)$ causes a high field intensity, which magnetizes the material despite the demagnetizing effect of the eddy current. Figure 14-6 shows the direction of the inner and outer surface eddy currents (I_E) in relation to the magnetizing current (I_c) for a centered rod. I_c causes an opposing I_E on the inner surface of the tube, which effectively shields the midwall region of the tube from the field caused by I_c. The outer surface current provides the return loop.

The magnetization problem is seen by considering Ampere's Rule. The magnetic field intensity at radius r within the tube, is given by

$$\oint \overline{H} \cdot d\overline{l} = (I_c + I_E)$$

where I_E is the contribution due to the inner surface eddy current (Fig. 14-6). Since I_c and I_E are in opposite directions, their effects subtract, so that while I_E exists, the field intensity at radius r can be quite small. Obviously a high field level inside the material can only be achieved after the eddy current has decayed against the resistance of the tube material. The average flux density level B_{avg} is shown in Fig. 14-5 in comparison to the current pulse $I_c(t)$.

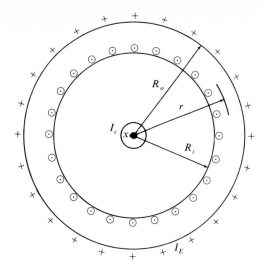

FIGURE 14-6
Configurations for I_c and I_e during internal conductor magnetization. The ID surface eddy currents shield the flux from penetration to the midwall regions of the tube. (*From the Nondestructive Testing Handbook, Vol. 4, 2d ed. Reprinted with permission.*)

14-3.3 Flux Density Within Tube Wall

Since inspection of tubes magnetized in this manner is generally performed in residual induction, the efficiency of the magnetization process need only be determined by the relative level of magnetization achieved by a current pulse in relation to the remanence (B_r) for the material. While many inspection specifications require only a sufficient field to give MPI indications from that surface, in the case of tubes, ID imperfections are often sought and found by their MFL at the OD. In order to obtain such MFL, it is essential that the material be at B_r. For a tube that is either unmagnetized or magnetized longitudinally, a fluxmeter method, employing a coil of perhaps one turn, which is arranged so as to surround an area that is perpendicular the final flux, can be used to measure the average flux density induced by each "shot" from the CD box.

A simpler, more direct method is to detect the current with a suitable inductive ammeter and monitor the pulse duration and amplitude. When these quantities stabilize, the material is saturated.

Figure 14-7 illustrates the penetration of the field (a) and flux density (b) from the ID and OD of a tube toward the central midwall region with the passage of time from the firing of the pulse. The field penetration Fig. 14-7(a) shows the relative slowness with which the midwall regions are exposed to high field levels, while the flux density Fig. 14-7(b), which is the effect of a typical B-H curve applied to Fig. 14-7(a), shows that the midwall region of the tube wall is the last to be magnetized. It is clear from Fig. 14-7 that a field strength that is sufficient to saturate the material must be felt by each and every part of the sample. If this is not so, then a flaw close to the ID will merely raise the midwall of the tube directly under it to saturation, but may not cause any MFL at the OD surface, where inspection is most likely to occur. At some midwall point, it is

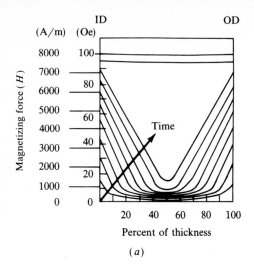

(a)

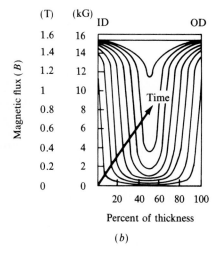

(b)

FIGURE 14-7
Computer simulation of the penetration of (a) magnetizing force and (b) magnetic flux density into the wall of a tube. As time increases, the magnetic flux penetrates the material, with the midwall region being magnetized last. (*From the Nondestructive Testing Handbook, Vol. 4, 2d ed. Reprinted with permission.*)

found that

$$H(r) = (I_c + I_E)/2\pi r > 3200 \text{ A/m} \tag{14-3}$$

is a sufficient condition for raising the induction to a value close to B_r.

14-3.4 Need for Multiple Pulsing

The worst case that an inspector must consider is that of the material initially magnetized to remanence in the opposite direction to that in which inspection will be performed. Normally, one expects the material to be initially at $(0,0)$ on the B-H curve, and if the magnetization current is both long enough and strong enough, then all parts of the material will be taken through the path Obc of Fig.

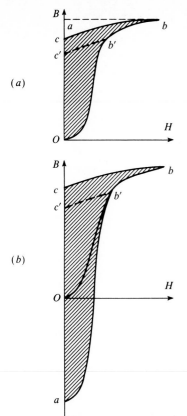

(a)

(b)

FIGURE 14-8
The shaded regions represent the energy densities for magnetization from (a) the unmagnetized state and (b) remanence in the opposite direction.

14-8. If the pulse is not strong enough to saturate the material, then it may follow a path such as $Ob'c'$. The worst possible case involves the path abc of Fig. 14-8(b). What may happen is that the material follows a path such as $ab'c'$ at the time of the first pulse.

Since the inductance of the tube is given by

$$L = (l/2\pi)(dB/dH)(\ln(R_o/R_i)) \qquad (14\text{-}4)$$

where l is the length of the tube, R_o and R_i are outer and inner radii, and dB/dH is the average differential permeability over the magnetization process. It is clear that a relatively high value for dB/dH, such as will be encountered during the passage of the first pulse, will lead to a large value of inductance in Eq. (14-2). High values of L tend to lower the maximum value of I_{max}, the peak current achieved by I_c, and elongate the pulse duration.

During the passage of a second pulse, the material is initially at c' and follows a path such as $c'b'bc$. The value of dB/dH is now much reduced below its original value, permitting higher values for I_{max} and shorter pulse durations.

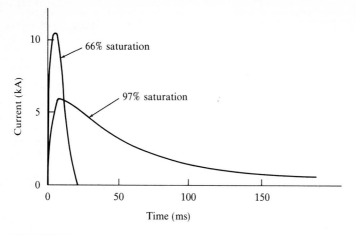

FIGURE 14-9

Long and short current pulses and their effect upon the same tube sample. (*Reproduced with permission from NDT International.*)

14-3.5 Effect of Pulse Duration

Pulse effectiveness is dependent upon how well they magnetize rather than how large a maximum current they can provide. Figure 14-9 illustrates the magnetizing ability of a 10,500 A pulse of short duration and a 6000 A pulse of relatively long duration. The fraction of the remanence value for the sample used, as determined with a fluxmeter, indicates that the long pulse is a more effective magnetizer than the short pulse. Clearly, the eddy current that is induced during the rapid rise of I_c still exists after the short pulse has died away and it limits the penetration of the field into the sample. On the other hand, the eddy current has died away before the longer pulse, so that the field intensity from the pulse is available to magnetize the midwall regions of the material.

What happens with short pulses is that the surface layer of the tube is magnetized, but the midwall region is not. This may provide sufficient surface magnetization for MPI, but can be inadequate for causing MFL at the outer surface of a tube from flaws that break the ID surface.

The effect of using relatively weak pulses to magnetize a sample of tube is shown in Fig. 14-10. Here a 60 ms long pulse raises the average flux density from zero to $0.72B_r$. Then, with the material at this level, a second pulse raises the average level to $0.90B_r$. The fraction of B_r that is achieved depends upon the amplitude and duration of the pulse used, the dimensions, and the magnetic and electrical properties of the sample [see Eq. (14-4)]. The size of the induced eddy current depends on the resistance of the sample material through its dimensions and electrical conductivity.

Close inspection of the average flux curves of Fig. 14-10 shows that while the peak currents are both at about 8 ms, the flux peaks at about 30 ms during the first pulse and 20 ms during the second. These values confirm in a qualitative

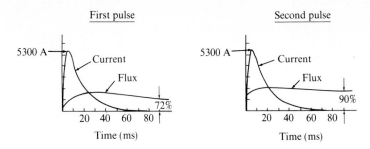

FIGURE 14-10
The waveform for a 60 ms current pulse and the magnetic flux that is induced in a sample during two successive pulses. Note that the flux peaks after the current. (*Reproduced with permission*, *ASNT*, *1983*.)

way the ability of the eddy current to retard the penetration of the field into the sample. The eddy current is also longer and stronger during the first pulse because its value also depends on the average value of dB/dH.

14-3.6 Typical Pulse Specifications

In view of the indirect connection between measurable current pulse parameters and induced flux density, two methods have evolved for specifying pulses. In the first method (Table 14-1), pulses from various types of pulsing units are classified according to duration [25] and tests with each type are performed so as to produce a peak current versus part weight table. The values in Table 14-1 indicate that pulses can be classed as long, moderately long, and short, depending upon their length, which is measured as the time from the start of the pulse to the instant during decay at which it reaches $0.5I_{max}$. For the long pulse, only the outer part diameter (D) is important, while for the shorter pulses, the mass per

TABLE 14-1
Magnetization specification for tubes by the CD method.
The CD systems are classed by pulse duration and the need for multiple pulsing is incorporated

Magnetization system	Decay time $\Delta t_{1/2}$ (ms)	Current requirement equation
Long pulse	> 100	$I = 300\,(D)$
Moderate pulse	40–100	$I = 110\,(W)$
Single short pulse	10–40	$I = 240\,(W)$
Double short pulse	10–40	$I = 180\,(W)$
Triple short pulse	10–40	$I = 145\,(W)$

TABLE 14-2
The number of current pulses that are required to completely magnetize variously sized oil field tubes. (CD system of 8.6 F operating at 50 V.)

Wall thickness (cm)	Outer diameter (cm)				
	19.4	21.9	24.4	27.3	34.0
0.64	1	1	1	1	1
1.27	1	1	1	1	1
1.91	1	2	2	2	2
2.54	2	2	3	3	3

Source: Courtesy American Society for Nondestructive Testing.

unit length (in this case, weight per foot W) is more important. Multiple pulsing is required depending upon the current values in the third column of the table.

The second method lists the number of pulses required to saturate specific diameters and wall thicknesses of tubes by a particular pulsing unit. A typical specification is provided in Table 14-2.

14-3.7   Pulse Rod Off-centering

Despite the complexity of the magnetization process, one practical advantage of the method over dc is that the conductor does not need to be centered within the bore of the part (Fig. 14-11). The rod field is stronger at the lower region of the tube, so the density of induced eddy currents is higher than at the top of the tube. This tends to equalize the actual magnetizing field at each point within the

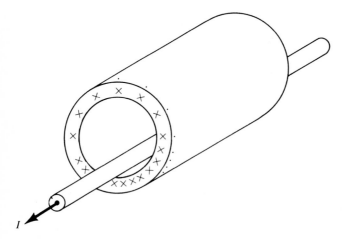

FIGURE 14-11
Off-centered magnetizing rod within a hollow part. For pulse magnetization, this is often not a problem.

Coil

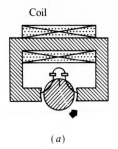

(*a*)

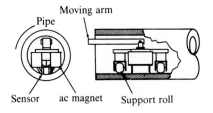

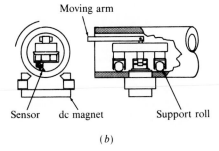

(*b*)

FIGURE 14-12
Applications of various yoke magnetizations. [(*b*) © 1977, *American Society for Nondestructive Testing*: *reproduced with permission of ASNT and the authors* [*27*]. (*c*) *From the Nondestructive Testing Handbook*, *Vol. 4, 2d ed. Reprinted with permission.*]

material. The phenomenon occurs also for weak pulses, and the uniformity of the resulting magnetization can be confirmed with a fluxmeter.

Other than the resistance it presents to the *L-C-R* circuit of the pulse, the material of the rod is unimportant.

14-4 YOKE OR PART ROTATION METHOD

Cylindrical parts (bars, tubes) are often inspected for longitudinal flaws by relative motion between an electromagnet yoke and the part. Figure 14-12 shows four examples of yoke magnetization, including an internally placed yoke [26–29]. Either all or a local part of the inspected part is magnetized to a high level of induction. Translational and rotational speeds are such as to provide in excess of 100% coverage by the sensor package. The test part forms part of the flux loop, with adjustable pole pieces being used to minimize the air gap for various diameters of the inspected part [30].

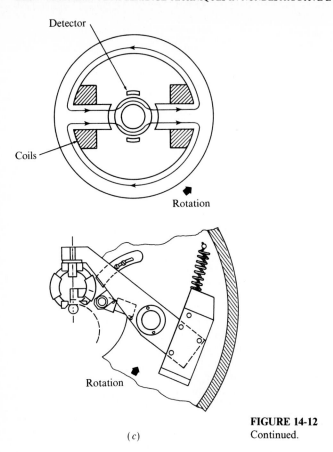

Detector

Coils

Rotation

Rotation

(c)

FIGURE 14-12
Continued.

14-4.1 Eddy Currents in Rotating Yoke Inspection

Relative motion between the yoke and the test piece induces an eddy current distribution within the part. In the case of a round bar, the resultant flux density distribution is a combination of the flux created by the yoke and a generated reaction field. This occurs as shown in Fig. 14-13. In Fig. 14-13(a), the static case, the flux lines obey the static boundary conditions (Chap. 12). However, upon rotation [Fig. 14-13(b)], a voltage is generated in the outer regions of the part that obeys

$$E = \int (\bar{v} \times \bar{B}) \cdot d\bar{l} \tag{14-5}$$

where the vector product of the velocity $\bar{v}$, with which the part cuts the applied flux density $\bar{B}$ is integrated along a contour p over the part. If the velocity is perpendicular to $\bar{B}$, then $(\bar{v} \times \bar{B})$ is a vector directed parallel to the length of the part and Eq. (14-5) reduces to

$$E = Blv \tag{14-6}$$

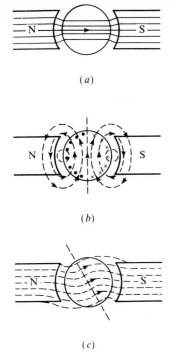

(a)

(b)

(c)

FIGURE 14-13
Distortion of the magnetic flux caused by the rotation of a cylindrical part in the field of a magnet. Rotation induces eddy current that follow as shown in (b) and that distort the field (c).

The eddy current distribution that is created by this EMF is shown in Fig. 14-13(b) and is dipolar. In common inspection situations (Fig. 14-12), it is clear that l is ill-defined, but the net effect is that the final eddy current distribution distorts the final flux density to that which is shown in Fig. 14-13(c). The neutral magnetic axis is rotated by some angle. Clearly, the faster the relative rotational motion and the stronger the field, the stronger the eddy current and the larger the distortion of the flux. The optimal point for placement of the MFL sensors moves around the circumference as these conditions are changed.

14-4.2 Tube Inspection

Good flux loop design consists of optimizing the reluctances of the air gap, the pole pieces, and the return loop so as to minimize flux leakage losses, since these represent wasted ampere turns on the magnetizing coils.

Rotation again creates an eddy current sheath, which has a different distribution from those created within the bar, because of the bore of the tube. Their net effect, however, is to shield any inner wall flaws from the full magnetic field, thereby reducing the sensitivity of MFL sensors at the OD to such flaws. This effect can be minimized by using as low a rotational speed as possible and velocity-independent sensors such as Hall elements or magnetodiodes (see Chap. 18).

CHAPTER
15

FIELD
LEVELS IN
MAGNETIC
NDE

15-1 INTRODUCTION

There are no universal rules regarding magnetization levels for MFL inspection. The criterion for good inspection is that the value of the leakage field from the smallest required flaw is relatively large in comparison to the noise that the magnetization process causes from surface and subsurface irregularities in the part, or other "nonrelevant" indications. The application of this criterion to the magnetic particle inspection of relatively smooth, shiny steel surfaces for tight cracks may require a different type and level of field intensity from that required for the detection of small pits on the inside surfaces of tubes, when the sensor is constrained to scan only the outer surface. While a low active ac field may be specified for the former, so as to hold the particles to relevant cracks while moving them away from nonrelevant surface indentations and extraneous flux leakage caused by local changes in material permeability, a high dc field is required for the latter, the value of which depends upon part wall thickness and surface MFL from the same causes. In effect, each case should be treated on its own merits and the inspector should beware of applying methods and formulas from one particular application to another.

In this chapter, the field levels that cause MFL are discussed, along with surface noise.

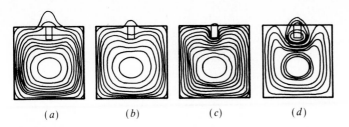

(a) (b) (c) (d)

FIGURE 15-1
Illustration of the leakage field reversal at low values of residual induction [31]: (a) high active excitation, (b) high residual induction following a high active excitation, (c) low active excitation, and (d) low residual induction. The flux lines shown are for direct current along a slotted bar. (*Reproduced from reference 31 with permission.*)

15-2 LEAKAGE FIELDS FROM A SLOT

In a series of tests with a slot in a rectangular bar, Heath was able to explain a remarkable field reversal effect that takes place in the MFL from flaws at low levels of residual induction [31].

By passing current along a bar and measuring the normal component of the MFL field just above the slot, the data shown in Fig. 15-2 were obtained, the explanation of which is presented in Fig. 15-1. At low active fields, the material in the vicinity of the slot is not highly magnetized. The slot presents a relatively high reluctance path to the flux, the majority of which seeks the relatively low reluctance path beneath the slot [Fig. 15-1(c)]. The high density of flux lines beneath the slot is equivalent to driving the material from 0 to C (Fig. 11-4). In this region, as the local value of H drives the value of B higher, the permeability also rises, so that the material is ready to accept even more flux. From C to D, however, the incremental permeability (dB/dH) has passed its peak (Fig. 11-17) at C, and falls rapidly. The permeability (B/H) levels out and begins to fall at a slightly higher local value of H, so that the reluctance of the material beneath the slot begins to rise. There are three paths the flux can take, beneath, through, and above the slot, and, as the reluctance beneath the slot falls, the air path becomes an increasingly more reasonable alternative.

At high active field levels, the situation shown in Fig. 15-1(a) is applicable, with the MFL field lines being continuous with those that magnetize the outer layers of the bar. Further, at high residual inductions, the situation is not much different, except that now most of the material is close to B_r, and that in the vicinity of the slot cannot have an induction that is higher than B_r, so the additional flux is forced out. Its value is much lower than it was for the active case. Figure 15-2 shows values of about 12, 22, and 37 Oe above the slot as the active field is raised, but after the field that caused the 37 Oe value is removed, the resulting residual flux leakage is only about 1 Oe.

By controlling the applied field carefully and scanning over the slot in residual induction, the MFL field reversal effect is seen. After relatively low applied fields the residual MFL reverses (Fig. 15-2), and a point is reached at

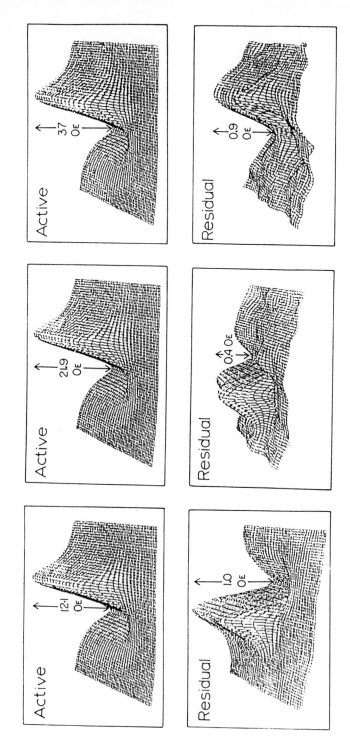

Residual Leakage Field Polarity Reversal
With Increasing Applied Field

FIGURE 15-2

Computer plots from a Hall element gaussmeter of the normal field (H_y) over a notch as the applied field is raised. Note the direction change in the residual induction leakage field [31]. (*Reproduced from reference 31 with permission.*)

which it might be zero. The variation of the residual MFL with increasing applied field is shown in Fig. 15-3 for slots of the same depth but differing widths [31]. The thinnest slot shows saturation in its MFL field [Fig. 15-3(a)] and requires less applied field to reach the zero MFL situation than the wider slots.

The explanation of the MFL field reversal is seen in Fig. 15-1(d). The collapse of the activating field, which only reached the steep part of the initial magnetization curve, creates a closed flux loop that contains the slot and some of the air above it. As it is made wider, the slot contributes more reluctance to this isolated magnetic circuit, with the result that the air field is lowered and for one point on the local magnetization curve, a field path is created for which the reluctance of the slot itself is equal to that of the surrounding material. At this point, there is no MFL.

This explanation is confirmed by the use of a finite element computer code that models the slot and its surroundings. The phenomenon has been observed with natural tight seam-like discontinuities.

15-3 SURFACE INSPECTION—ACTIVE FIELD

If parts are to be inspected in an active field, then Fig. 15-2 indicates that relatively large leakage fields can be obtained for quite low applied field strengths. Since many industrial steels have coercivities in the range of 5–20 Oe (400–1590 A/m) and the coercivity is the point at which dB/dH peaks (Fig. 11-17), application of two or three times the coercivity is generally considered to be adequate for sufficient surface magnetization to produce good magnetic particle indications from tight flaws. Some specifications state that 2390 A/m (30 Oe) are required, while others require 0.72 T in the part. An ac coil that produces this field intensity level at its center (at 50–60 Hz), so that the skin depth is roughly 1 mm, will provide good magnetization for MPI with either the wet or dry method.

15-4 SURFACE INSPECTION—RESIDUAL INDUCTION

Figure 15-3 indicates that peak-to-peak MFL fields of a few gauss are to be expected over surface-breaking flaws in residual induction and that tighter cracks

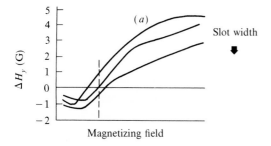

FIGURE 15-3
Variation of the peak-to-peak magnetic flux leakage residual field ΔH_y over slots of different widths but the same depth, measured with a Hall element gaussmeter following various levels of active magnetizing field H_a. Curve (a) shows saturation of the leakage field for the thinnest slot [31]. (*Reproduced from reference 31 with permission.*)

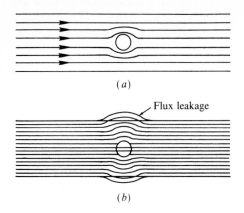

(a)

(b)

FIGURE 15-4
Effects of induction on flux lines in the presence of a discontinuity. (a) Compression of flux lines at low levels of induction around a discontinuity, so that no surface flux leakage occurs. (b) Lack of compression at high induction, showing some surface magnetic flux leakage. (*Courtesy American Society for Nondestructive Testing. From the Nondestructive Testing Handbook, Vol. 4, 2d ed. Reprinted with permission.*)

saturate for smaller applied fields than do wider flaws. Fields of roughly $3H_c$ will cause the MFL from surface-breaking cracks to saturate. Lower levels may be used so as to reduce noise in the form of false magnetic particle indications.

15-5 SUBSURFACE INSPECTION—ACTIVE FIELD

The field intensity in a material must be raised to such a level that there is adequate MFL from subsurface defects [32]. At low levels, as in Fig. 15-1(c), the diversion of flux merely raises the material around the flaw to a higher level of induction. This is seen in Fig. 15-4(a), with no resulting MFL. At higher applied field levels, as the permeability falls, the relative reluctance of the material above and below the flaw rises in comparison to that of the flaw and, eventually, MFL is produced [Fig. 15-4(b)]. Unless the material is at a value close to saturation, there may be little detectable MFL. The problem that is commonly encountered in this situation is that the depth of the flaw causes the surface MFL pattern to broaden, so that its width is similar to that of surface noise caused by local permeability variations, etc. Competing effects of obtaining adequate signal while minimizing noise therefore occur. In common inspection situations, the material is saturated and noise reduction sensors and electronics are used.

15-6 SUBSURFACE INSPECTION—RESIDUAL INDUCTION

It is clear from the foregoing discussion that the material should be relatively close to B_r, in order for there to be much MFL from subsurface flaws. Under this circumstance, the possibility arises for obscuring small flaw signals with surface noise.

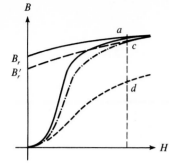

FIGURE 15-5
B-H curve for "good", locally work-hardened, and altered materials.

15-7   CAUSES OF SURFACE NOISE

Minor irregularities and small differences in permeability are major causes of surface noise. Sometimes, large changes in permeability occur, for example, if a welding torch has been applied to the surface of the part, and these can also be detected through a change in hardness. Figure 15-5 shows *B-H* curves for such material. Curve OaB_r represents the magnetization process of "normal" material for either active field (a) or residual field (B_r) inspection. Curve OcB_r' is that for material of slightly different permeability. Around c, the material may still be magnetized to a higher level for active field inspection and the flux remains in the material, but at B_r' the balance of the flux density $B_r - B_r'$ is ejected. An important point to consider is that MFL from flaws is an order of magnitude greater at a than at B_r, so better signal-to-noise ratios can be achieved when performing inspection in an active field.

Curve Od is that for a material that has been permanently altered. Here large active and residual inspection noise will be encountered.

15-7.1   Examples of Surface Noise

STRAIGHTENING. Figure 15-6 shows the normal component of the surface (H_y) around a 5 cm long section of a tube that had been straightened [29]. An area of high internal stress, caused by the straightening process, lowers B_r and causes the MFL at the 9–12 cm region. The signal is 3 cm wide and has a peak-to-peak amplitude of 18.4 G. Such wide MFL fields either do not give a magnetic particle indication or cause a very fuzzy one. They do, however, cause indications on automated MFL inspection units, unless filtered out.

Similar indications are caused when material is gripped, such as occurs when the ends of tubes are threaded or connections are bucked on.

JOINTS OF METALS. Figure 15-7 shows examples of externally and internally upset drill pipe [33]. The box (female) and pin (male) connections are inertially welded onto the pipe and the welded area is tempered in an attempt to obtain a uniform change of properties from the connections to the tube. In this area,

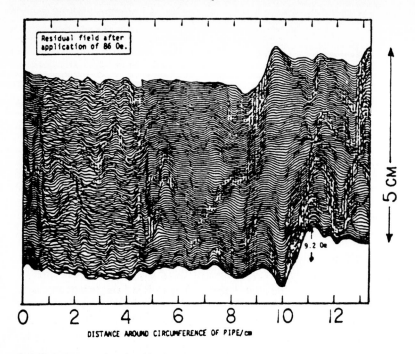

Residual field after
application of 86 Oe.

9.2 Oe

5 CM

0   2   4   6   8   10   12

DISTANCE AROUND CIRCUMFERENCE OF PIPE/cm

FIGURE 15-6
Field map of the normal component of the leakage field from a 13 × 5 cm section of L-80 tubing in a saturated residual circular field. The broad pattern at 9–12 cm is due to stress that is created during straightening. (*Courtesy Gordon and Breach, 1985.*)

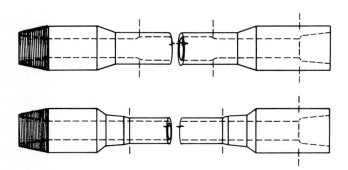

FIGURE 15-7
Externally and internally upset drill pipe.

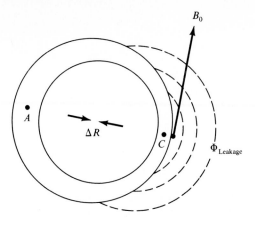

FIGURE 15-8
Broad magnetic flux leakage field from an eccentric circumferentially magnetized tube with points A and C at remanence B_r. (© 1987, American Society for Nondestructive Testing. Reproduced with permission.)

which is often inspected by wet and dry MPI and ultrasonics, large leakage fields sometimes exist when the saturation values of the two metals are not the same. In longitudinal magnetization, the flux from the connection, which has a larger cross-sectional area than the tube, also contributes to this leakage flux.

ECCENTRICITY IN TUBES. A common cause of surface noise from circumferentially magnetized seamless tubes is geometrical eccentricity [34]. As Fig. 15-8 shows, the conservation of flux when the tube is at remanence gives

$$B_A A_A = B_C A_C + \phi_{\text{leakage}}$$

by considering the flux at the points A and C. The total leakage flux is then

$$\phi_{\text{leakage}} = (A_A - A_C) B_r$$

i.e., proportional to the difference in the two cross-sectional areas, or in effect, to the wall thickness difference. Assuming that the leakage flux falls off in an inverse manner with distance from the tube surface, then the maximum leakage field (B_0) is given by

$$B_0 \simeq B_r (\Delta R / R) \tag{15-1}$$

where ΔR is the tube eccentricity and R is its outer radius. Equation (15-1) shows that stronger leakage fields are caused by higher values of B_r, ΔR (as expected), and smaller tube radii.

CHAPTER
16

MAGNETIC FLUX LEAKAGE FROM TIGHT FLAWS

16-1 INTRODUCTION

The previous six chapters have concentrated on magnetization and the methods that are used to achieve it. In this chapter, various classical (pole) approaches to the MFL from surface-breaking flaws are discussed. Although this approach often results in unwieldy mathematical relations, once they have been derived, it is relatively simple to use a computer to determine the approximate response of a sensor to such fields.

Many of the calculations that have been performed to give MFL field topographies are based on replacing a defect by a surface or volume charge density. At relatively large distances from the mouth of a defect, the use of constant values of surface charge density σ within the defect results in equations that are roughly consistent with experimental data, at least for defects that enter the surface normally. For angled defects, such as laps, this approximation breaks down.

The chapter begins with a discussion regarding the angle of the applied field (H_a) to the flaw and then the results of the classical models are developed.

16-2 ANGLE OF APPLIED FIELD TO FLAW

How the MFL field behaves across a slot that is magnetized at an angle to the opening of the slot is shown in Fig. 16-1 [35]. Using magnetic particles it is clear that the leakage field is perpendicular to the faces of the slot. This is seen clearly in the data for $\theta = 60$ and $30°$.

Using the coordinate system of Fig. 16-1(d), the tangential component of the MFL field can be divided into the components

1. $H_\perp = H_x(\theta = 0)\sin\theta$ for the MFL component perpendicular to the slot
2. $H_x(\theta) = H_\perp \sin\theta = H_x(\theta = 0)\sin^2\theta$ for the MFL component in the direction of the applied field H_a

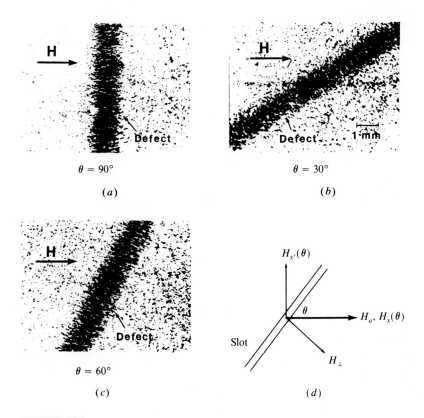

$\theta = 90°$

(a)

$\theta = 30°$

(b)

$\theta = 60°$

(c)

(d)

FIGURE 16-1
(a), (b), (c) Magnetic particle pictures showing the MFL field crossing a slot in a direction perpendicular to its faces. (d) The coordinate system used in analysis of the slot leakage field. (*Reproduced by permission of Ichizo Uetake, National Research Institute for Metals, Tokyo, Japan.*)

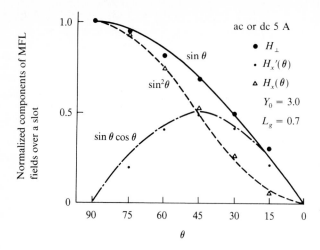

FIGURE 16-2
Comparison between theoretical predictions (full lines) and experimental data for the MFL components over a slot, as the exciting field direction (either ac or dc) is altered. (*Reproduced by permission of Ichizo Uetake, National Research Institute for Metals, Tokyo, Japan.*)

3. $H_y(\theta) = H_\perp \cos\theta = H_x(\theta = 0)\sin\theta\cos\theta$ for the MFL component perpendicular to the applied field

where

$$H_\perp^2 = H_x(\theta)^2 + H_y(\theta)^2 \qquad (16\text{-}1)$$

These theoretical relations and the experimental data (taken with a Hall element) are shown for both ac and dc excitation modes in Fig. 16-2. It is clear that the experimental data substantiate the theoretical predictions of Eq. (16-1), i.e.,

As the angle of the flaw to the exciting field falls, the MFL field component perpendicular to the slot falls off as $\sin\theta$ and that which is parallel to the exciting field falls off as $\sin^2\theta$.

As the angle of the flaw to the exciting field falls, the MFL component perpendicular to the exciting field rises as $\sin\theta\cos\theta$.

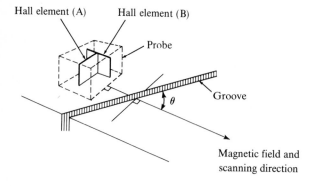

FIGURE 16-3
Hall element arrangement to retrieve the field $H_x(\theta = \theta)$. (*Reproduced by permission of Ichizo Uetake, National Research Institute for Metals, Tokyo, Japan.*)

One way to retrieve the $\theta = 0$ MFL field signal amplitude is to scan with perpendicular Hall elements, as shown in Fig. 16-3. Element A records $H_x(\theta)$ while element B records $H_x'(\theta)$, and the value of $H_x(x = 0)$ is found from

$$H_x(x = 0) = \frac{H_x(\theta)^2 + H_x'(\theta)^2}{H_x(\theta)^2} \qquad (16\text{-}2)$$

16-3 BASIC FÖRSTER THEORY FOR A THROUGH SLIT

In the following paragraphs, the magnetic circuit for a through slit in a ring and its resulting MFL field are discussed.

16-3.1 Magnetic Circuit for a Through Slit

Figure 16-4 shows the through slit geometry used in the determination of a relation between the field applied to the ring by the coil and that which occurs deep within the slit [36]. This slit causes a demagnetization field H_d that opposes the applied field H_a, which is assumed to be produced by a coil of MMF equal to NI ampere turns. The permeability of the ring is assumed constant, but since H_d is not constant around the ring, especially near to the slit faces, this is not a particularly good approximation. However, pursuit of this idea leads to some simple results.

The reluctance R of the ring is the sum of two terms, $L_m/\mu_0 u_r A_m$ for the material and $L_g/u_0 A_g$ for the gap. The flux is then

$$\phi = NI/\mu_0 \left(L_m \mu_r A_m + L_g/A_g \right) \qquad (16\text{-}3)$$

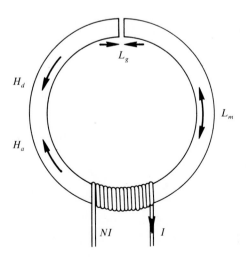

FIGURE 16-4
Slot ring sample geometry used to determine the relation between the applied field from a coil on a ring and the resulting gap field. (*From the Nondestructive Testing Handbook, Vol. 4, 2d ed. Reprinted with permission.*)

and the boundary condition on B yields

$$B_g = \mu_0 H_g = \phi/A_g$$
$$= B_m$$

Equation (16-3) then gives

$$H_g = \frac{NI/L_m}{L_g/L_m + A_g/\mu_r A_m} \tag{16-4}$$

Recalling that $H_a = NI/(L_m + L_g)$ allows removal of the MMF, so that

$$H_g = \left[\frac{(L_m + L_g)/L_m}{L_g/L_m + A_g/\mu_r A_m} \right] H_a \tag{16-5}$$

Equation (16-5) relates the field strength inside the gap to the applied magnetic field intensity and various geometrical parameters of the magnetic circuit. Further simplification is possible if it is assumed that

$$A_g = A_m$$

In this case, the gap field is given by

$$H_g = \left[\frac{\mu_r(1 + L_g/L_m)}{1 + \mu L_g/L_m} \right] H_a \tag{16-6}$$

Now, in many practical cases, the gap length L_g is very much less than the length within the material, so that Eq. (16-6) becomes

$$H_g = \left[\frac{\mu_r}{1 + \mu_r L_g/L_m} \right] H_a \tag{16-7}$$

Note that while it is possible to ignore L_g/L_m in the numerator, it is not possible to ignore it in the denominator because it is multiplied by μ_r, and so the $\mu_r L_g/L_m$ term may be comparable to unity. An estimate of the size of H_g can be obtained by putting typical values into Eq. (16-7).

Example 16-1. With $L_m = 400$ mm and $L_g = 2.5$ mm, the value of L_g/L_m is 6.4×10^{-3} and can be ignored in the numerator of Eq. (16-6). However, with $\mu_r = 200$, Eq. (16-7) gives $H_g = 88.8 H_a$, while Eq. (16-6) gives $H_g = 89.3 H_a$. Considering all of the other assumptions that have been made in deriving these formulas, this approximation is justified.

The expressions in brackets are factors by which H_a must be multiplied in order to give the field intensity within the gap. The preceding theory also shows that the demagnetization factor caused by the gap is

$$N_d = L_g/(L_g + L_m) \tag{16-8}$$

which becomes larger as the slit is made wider. Large slit widths lead to lower values of μ and negate the "infinite permeability" assumption made below.

16-3.2 Magnetic Leakage Field around the Through Slit

Using a conformal mapping technique [36], Förster showed that the MFL from a through slot should appear as in Fig. 16-5(a), which is drawn with equal fluxes between consecutive lines. It is evident that

1. the MFL field intensity at distances greater than about x(or y) = L_g shows an almost semicircular shape
2. the field intensity a short distance ($y \sim L_g/2$) into the gap is uniform
3. the boundary condition obeyed by the MFL field lines at the air–steel interfaces $(\mu \cot \theta)_{air} = (\mu \cot \theta)_{steel}$ indicates that the permeability of the material is infinitely high

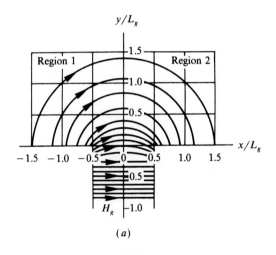

(a)

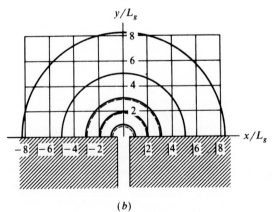

(b)

FIGURE 16-5
Schematic representation of Förster theory for (a) leakage field lines in the constant permeability approximation for a through slot and (b) the difference between the field lines and semicircles in this theory. (*Reproduced from [10] with permission.* © *Plenum, New York.*)

The great advantage of the Förster theory is that it leads to very simple equations for the MFL field components, i.e.,

$$H_x = \frac{H_g L_g}{\pi} \frac{y}{x^2 + y^2}$$

$$H_y = \frac{H_g L_g}{\pi} \frac{x}{x^2 + y^2}$$

(16-9)

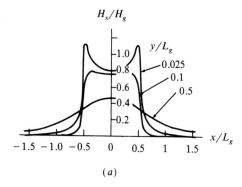

(a)

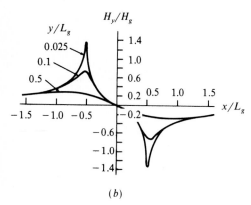

(b)

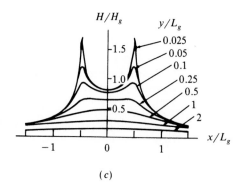

(c)

FIGURE 16-6
Values of (a) H_x/H_g, (b) H_y/H_g, and (c) H/H_g from Förster theory, plotted at various values of normalized lift-off. The coordinate axes are those shown in Fig. 16-5, with the sides of the slot at $x = \pm 0.5L_g$. (*Reproduced from [10] with permission.* © *Plenum, New York.*)

These relations express the MFL field in terms of slot width (L_g), the field deep within it (H_g), and geometrical parameters. Magnetic and geometrical terms are separate, and Eqs. (16-9) hold for x, $y > L_g$. It is useful for explaining observed phenomena at relatively large lift-off distances such as occur when steel parts are scanned with MFL sensors.

At the mouth of the slot, the theory predicts the curves shown in Fig. 16-6, which represent the tangential (H_x), normal (H_y), and total field values ($H_x^2 + H_y^2)^{1/2}$, normalized by the value of H_g. The abscissa are normalized by L_g, so that the slot faces occur at $x = \pm 0.5 L_g$. The tangential MFL field shows a double-peaked structure, which becomes less pronounced with lift-off. H_x is also always positive. The normal component H_y also shows a double-peaked structure, with equal positive and negative peaks that occur at $x = \pm L_g/2$ for small lift-off, but move farther apart at larger lift-off. The total field also shows two peaks that gradually merge as the lift-off is increased.

At the mouth of the slot, the theory predicts that $H_x(x = 0, \; y = 0) = 0.83 H_g$ and for about $y/L_g \sim 0.1$, the MFL field shows only one peak. Thus for tight cracks, the preceding approximation may be used if they are either through-wall or much smaller width than the crack opening.

Equations (16-6) and (16-9) may be combined to eliminate H_g.

16-4 KARLQVIST'S EQUATIONS FOR A THROUGH SLIT [38]

In an analysis of the leakage fields around tape recorder heads, Karlqvist was able to deduce the relations

$$H_x = \frac{H_g}{\pi}\left[\tan^{-1}\frac{x + L_g/2}{y} - \tan^{-1}\frac{x - L_g/2}{y}\right]$$

$$H_y = \left(\frac{H_g}{2\pi}\right)\ln\left[\frac{(x - L_g/2)^2 + y^2}{(x + L_g/2)^2 + y^2}\right]$$

(16-10)

The shapes of these MFL field components are similar to those shown in Fig. 16-6.

16-5 LEAKAGE FIELD THEORIES FOR A FINITE SLOT

16-5.1 Förster's Theory [36]

For a slot of depth Y_0, Förster has suggested that a good approximation can be obtained merely by filling in the lower part of the slot with metal. The contribution to the MFL field from this region is then subtracted from Eq. (16-9). The

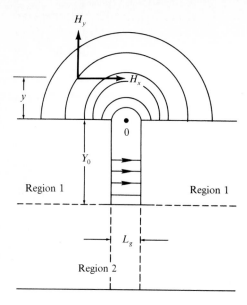

FIGURE 16-7
Filling-in the lower portion of a deep slot with ferromagnetic material to create a slot of finite depth. (*From the Nondestructive Testing Handbook, Vol. 4, 2d ed. Reprinted with permission.*)

results are

and

$$H_x = \frac{H_g L_g}{\pi} \left[\frac{y}{x^2 + y^2} - \frac{y + Y_0}{x^2 + (y + Y_0)^2} \right]$$

$$H_y = \frac{H_g L_g}{\pi} \left[\frac{x}{x^2 + y^2} - \frac{x}{x^2 + (y + Y_0)^2} \right]$$

$$(16\text{-}11)$$

Figure 16-7 shows the filling-in of the through slit, the second term in Eq. (16-11) coming from region 2. From this equation, it can be seen that (1) as Y_0 is made smaller, the second term subtracts less and less from the first term, and so the MFL field intensity collapses toward zero and (2) the magnitude of the defect

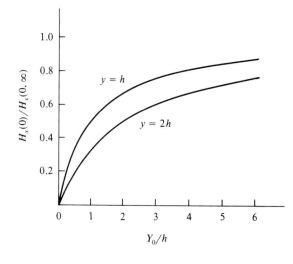

FIGURE 16-8
Behavior of the tangential component of the leakage field above a slot as the depth is increased. (*From the Nondestructive Testing Handbook, Vol. 4, 2d ed. Reprinted with permission.*)

field does not increase linearly with depth. The prediction of Eq. (16-11) for the tangential field strength at $x = 0$ and $y = h, 2h$ is shown in Fig. 16-8. The ordinate of the figure is the value of H_x divided by that for an infinitely deep slot [Eq. (16-9)]. The abscissa shows the slot depth, divided by a constant lift-off distance h. It is seen that, initially, the deeper the slot the more intense the MFL field is. Eventually, however, the slot becomes so deep that the MFL field becomes relatively insensitive to further increases in depth. In the light of Chap. 15, this theory is valid for relatively large active fields.

16-5.2 Theory of Zatsepin and Shcherbinin [39]

If the surfaces of the slot are replaced by opposing layers of surface charge, as shown in Fig. 16-9, then the total charge on each face is $Q = \sigma(\eta)\,dS$, where η varies between 0 and Y_0 and dS is the element of area. The element of charge is $dQ = \sigma(\eta)\,dS$ and the field strength at some point D is given by

$$d\overline{H}_1 = 2\sigma(\eta)\,d\eta\,\bar{r}/r^3$$

For the general case, the geometry of the slot shown in Fig. 16-9 yields

$$dH_{1x} = 2\sigma(\eta)\left[\frac{x + L_g/2}{\left(x + L_g/2\right)^2 + \left(y + \eta\right)^2}\right]d\eta$$

$$dH_{1y} = 2\sigma(\eta)\left[\frac{y + \eta}{\left(x + L_g/2\right)^2 + \left(y + \eta\right)^2}\right]d\eta$$

(16-12)

for the positively charged face. The contribution from the other face is found by replacing L_g with $-L_g$ and $\sigma(\eta)$ by $-\sigma(\eta)$.

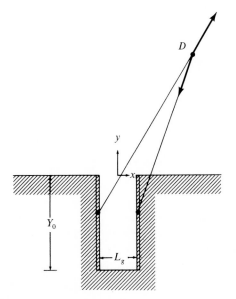

FIGURE 16-9
Replacement of the physical sides of a slot by two layers of surface charge in the Zatsepin and Shcherbinin model.

The surface charge density can be varied to match the MFL field, but the simplest case is that in which $\sigma(\eta)$ is held constant along the faces of the slot. In this approximation, $\sigma(\eta)$ is written as σ_s and the result for the MFL fields is

$$H_x = \frac{H_g}{\pi}\left[\tan^{-1}\frac{(x + L_g/2)Y_0}{(x + L_g/2)^2 + y(y + Y_0)}\right.$$
$$\left. - \tan^{-1}\frac{(x - L_g/2)Y_0}{(x + L_g/2)^2 + y(y + Y_0)}\right] \tag{16-13}$$

and

$$H_y = \frac{H_g}{2\pi}\ln\left[\frac{(x + L_g/2)^2 + (y + Y_0)^2}{(x - L_g/2)^2 + (y + Y_0)^2} \cdot \frac{(x - L_g/2)^2 + y^2}{(x + L_g/2)^2 + y^2}\right]$$

where σ_s has been replaced by $H_g/2\pi$, a result deduced by Förster. In the region well away from the mouth of the slot, Förster has shown that Eqs. (16-11) and (16-13) are equivalent. Small differences do occur, however, at the slot opening.

16-5.3 Depressed Dipole Model

Neither the Zatsepin–Shcherbinin nor the Förster model allow for the presence of volume charges (div M) within the metal around the slot. One simple way around this is to replace both surface charges by a linear dipole, as shown in Fig. 16-10. Setting $L_g/2 + \delta = p$, the MFL field equations become

$$H_x = 2m\left[\frac{x + p}{(x + p)^2 + (y + h)^2} - \frac{x - p}{(x - p)^2 + (y + h)^2}\right]$$
$$H_y = 2m\left[\frac{y + h}{(x + p)^2 + (y + h)^2} - \frac{y + h}{(x - p)^2 + (y + h)^2}\right] \tag{16-14}$$

This model was investigated by Novikova and Miroshin, who presented the data shown in Fig. 16-11, which are taken at a magnetizing field strength of 513 A/cm (645 Oe). Slot dimensions are given in the figure [14].

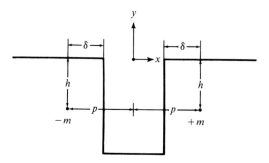

FIGURE 16-10
Depressed dipole model geometry.

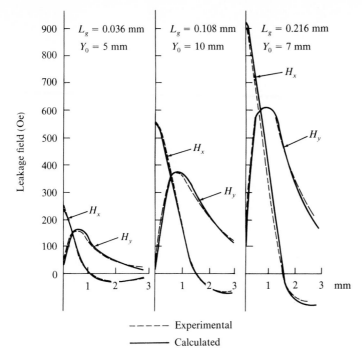

FIGURE 16-11

Measured and calculated H_x and H_y for rectangular slots. (*Reproduced from "Investigation of the Fields of Artificial Open Flaws in a Uniform Constant Magnetic Field," I. A. Novikova and N. V. Miroshin, Defektoskopiya 4, 1973, with permission. © Plenum, New York.*)

16-6 COMMENTS ON DIPOLE THEORIES FOR PERPENDICULAR SLOTS

The theories just discussed are relatively simple attempts to explain the MFL fields from machined slots that enter the surface perpendicularly. In reality, tight flaws do not look much like slots. Two basic types of tight flaws exist: (1) the heat check, or quench-crack-type, in which the metal tears in a direction that is roughly perpendicular to the metal surface and for which there are no changes in metallurgy at the faces of the crack and (2) the fatigue crack, which is due to local cyclic stresses that hold the material at its yield point for long periods of time, perhaps also in the presence of hydrogen. Such cracks may have a layer of material of differing *B-H* properties from the bulk of the material at their faces.

Relatively recently, the introduction of electrodischarge machined (EDM) notches and saw blades of thickness 0.08 mm has made it possible to study controlled tight simulated flaw MFL patterns. Computer programs have been written to simulate the effect of varying amounts of permeability change at crack faces and small sensors have been developed to measure the MFL fields.

The use of a constant charge density along the slot faces, which is equivalent to taking H_g constant within the gap, is an assumption that cannot be

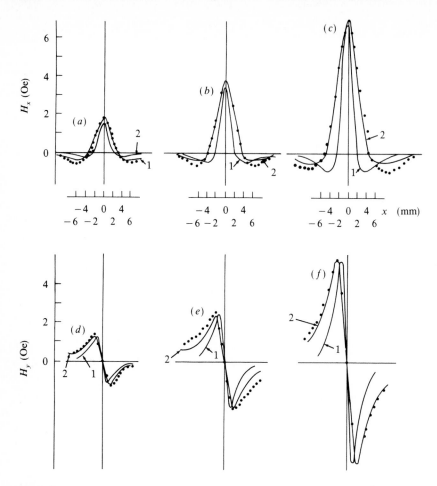

FIGURE 16-12
Experimental data (circles) and (curve 1) linear dipole model and (curve 2) Eqs. (16-13), for slots of width 1.2 mm and depths (a, d) 1.8 mm, (b, e) 3.6 mm, and (c, f) 7.2 mm. Residual induction. (*Reproduced from* [40] *with permission.* © *Plenum, New York.*)

justified, except perhaps at high levels of active field excitation. It works reasonably well for saturated residual induction also, as shown in Fig. 6-12.

16-7 COMPARISON WITH TEST DATA

For slots of width $L_g = 1.2$ mm and depths $Y_0 = 1.8$, 3.6, and 7.2 mm, Zatsepin and Shcherbinin [40] have recorded data at 2 mm lift-off with a small ferroprobe (Chap. 18) in residual induction. Figure 16-13 shows a comparison between experiment and two theories; curve 1 is the linear dipole model (see Problem 16-3), which is not a good fit, and the strip dipole model [Eq. (16-13)]. The latter

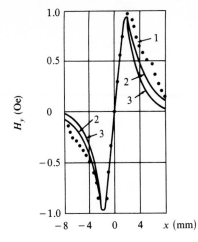

FIGURE 16-13

Normal component H_y above a tight crack (curve 1), in comparison with strip dipole theory [curve 2, Eqs. (16-13)], and line dipole theory (curve 3). Residual induction. (*Reproduced from [40] with permission.* © *Plenum, New York.*)

gives a better fit. The surface charge density was adjusted to provide the best fit at the peaks.

For slots of constant depth but differing widths, Fig. 15-3 shows that the MFL field amplitude falls as the slot width rises, indicating that the values of σ_s and $H_g L_g$ fall with rising slot width in residual induction.

For a tight perpendicular crack of depth 0.4 mm and width 0.01 mm, the same investigators found reasonably good agreement between their theory and experimental data at 2 mm lift-off for the central part of the normal field H_y. The data do, however, show some structure outside of the extrema, the peak-to-peak MFL field being about 2 Oe (160 A/m).

16-8   PREDICTIONS OF DIPOLE MODELS FOR PERPENDICULAR SURFACE FLAWS

The advantage of the models just discussed is that they can be easily manipulated, either algebraically, or by a small computer, so as to predict what might happen when parameters are varied. Three examples follow.

16-8.1   Variation of MFL Field Amplitude with Applied Field

One prediction of the preceding theories is that the magnetic and dimensional parts of the MFL field are separate, and so the shape of the MFL field does not change as the applied field strength changes. For the through slit case, H_g (the gap field) and H_a (the applied field) are related by Eq. (16-5); thus the leakage fields are given by

$$\begin{vmatrix} H_x \\ H_y \end{vmatrix} = \frac{\mu_r H_a L_g (1 + L_g/L_m)}{\pi (1 + \mu_r L_g/L_m)(x^2 + y^2)} \begin{vmatrix} y \\ x \end{vmatrix} \qquad (16\text{-}15)$$

for a ring sample. In this example, the geometrical parameters can be lumped into one constant K, so that Eq. (16-15) becomes

$$\left|\begin{matrix} H_x \\ H_y \end{matrix}\right| = \frac{K\mu_r H_a}{1 + \mu_r L_g/L_m} \left|\begin{matrix} y \\ x \end{matrix}\right|$$

$$K = \frac{L_g\left(1 + L_g/L_m\right)}{\pi\left(x^2 + y^2\right)}$$

(16-16)

which indicates that the MFL field components depend upon how the term $\mu_r/(1 + \mu_r L_g/L_m)$ varies with H_a. This variation is shown in Fig. 16-14, which has been constructed from the permeability data of Fig. 11-17 and typical slot dimensions. The model predicts that the MFL field follows a curve that is somewhat similar to an initial B-H curve, i.e., it is small for small H_a, rises most steeply where $\mu_r(H)$ rises steepest, and eventually rises at a lower rate as the applied field drives $\mu_r(H)$ to lower values as saturation is approached. The other curves in Fig. 16-14 are the measured MFL fields over a 0.19 × 0.32 mm sawcut slot and an angled lap. All the data are normalized to the field at $H_a = 60$ A/cm

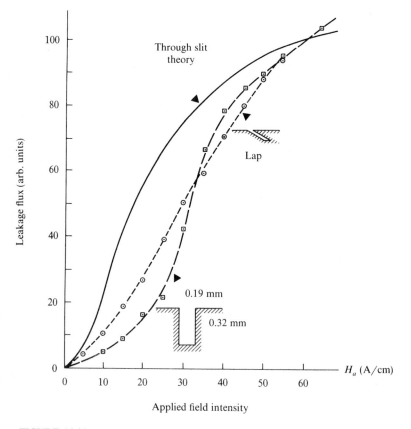

FIGURE 16-14
Comparisons of MFL field for a through slit, a sawcut slot, and a lap, with applied surface field intensity.

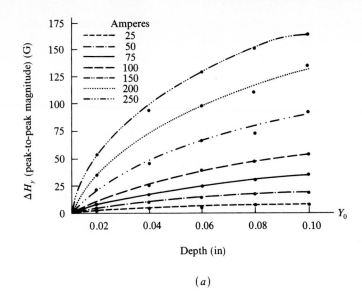

(a)

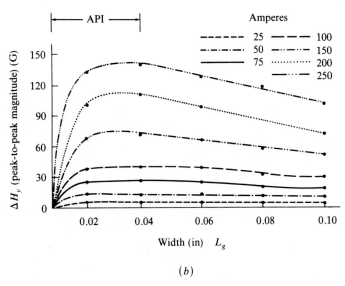

(b)

FIGURE 16-15

(a) Peak-to-peak active leakage field magnitude versus slot depth. (b) Peak-to-peak active leakage field magnitude versus slot width. "API" designates the American Petroleum Institute maximum test flaw width for MFL testing. (*Reproduced from reference 31 with permission.*)

(75 Oe). In general, the magnitude of the MFL field above a surface-breaking tight flaw is a monotonically increasing function of the applied field intensity H_a. As shown in Chap. 15, this is not the case with residual induction.

16-8.2 Predictions of MFL Field Trends with Flaw Dimensions

The preceding theories also predict how MFL field magnitudes change with flaw dimensions. These are left as exercises. Experimental data to test the validity of such models over a wide range of H_a are shown in Fig. 16-15 [31]. Figure 16-15(a) illustrates the peak-to-peak variation in H_y, i.e., ΔH_y, for increasing slot depth at constant slot width, while Fig. 16-15(b) shows the variation in ΔH_y for increasing slot width at constant slot depth.

As might be expected, ΔH_y does not continue to rise ad infinitum with slot depth, but rather saturates, as less and less effect is felt at the sensor from the additional poles at the bottom of the slot. The width variation is more complex, showing first an increase and then a decrease. At lower field levels, the MFL signal ΔH_y appears to be independent of slot width, although this occurs in a region that is wider than commonly occurring tight flaws.

16-9 CLASSICAL MFL THEORY FOR OBLIQUE SURFACE FLAWS

Zatsepin and Shcherbinin [39] have extended their theory to include angled lap-like flaws. For the positively charged face 1 of the oblique slot shown in Fig. 16-16, the MFL equations are given by Eqs. (16-17). From these equations, the MFL components H_{2x} and H_{2y} are found by putting $-\sigma_s$ for σ_1 and $-L_g$ for L_g. The full field components are then given by

$$H_x = H_{1x} + H_{2x}$$

and
$$H_y = H_{1y} + H_{2y}$$

The reader should verify that when the angle $\gamma = 0$, the result is Eqs. (16-13).

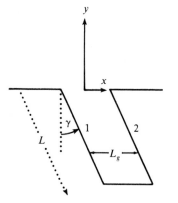

FIGURE 16-16
Angled slot parameters.

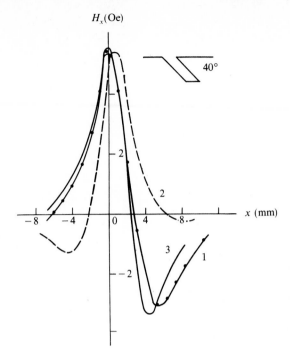

FIGURE 16-17
Tangential MFL fields in saturated residual induction over a 40° slot; curve 1 is experimental data at 2 mm lift-off, curve 2 is the uniform charge model, and curve 3 is a model with increased charge on the acute face of the slot. (*Reproduced from* [40] *with permission.* © *Plenum, New York.*)

A comparison between this theory and experimental results (curve 1) is given in Fig. 16-17. A uniform surface charge model ($\sigma_1 = \sigma_2$, curve 2) does not fit the data, but by adjusting the surface charge densities so that the upper part of the face within the acute angle has a higher pole density (curve 3), a much better fit is obtained. This comparison is performed for saturated residual induction:

$$H_x = 2\sigma_1 \left\{ \cos\gamma \cdot \tan^{-1} \frac{L\left[(x + L_g/2)\cos\gamma + y\sin\gamma\right]}{(x + L_g/2)^2 + y^2 + L\left[y\cos\gamma - (x + L_g/2)\sin\gamma\right]} \right.$$

$$\left. - \left(\tfrac{1}{2}\right)\sin\gamma \cdot \ln \frac{L^2 + (x + L_g/2)^2 + y^2 + 2L\left[y\cos\gamma - (x + L_g/2)\sin\gamma\right]}{(x + L_g/2)^2 + y^2} \right\}$$

$$H_{1y} = \sigma_1 \left\{ \cos\gamma \cdot \ln \frac{(x + L_g/2)^2 + y^2 + L^2 + 2L\left[y\cos\gamma - (x + L_g/2)\sin\gamma\right]}{(x + L_g/2)^2 + y^2} \right.$$

$$\left. + 2\sin\gamma \cdot \tan^{-1} \frac{L\left[(x + L_g/2)\cos\gamma + y\sin\gamma\right]}{(x + L_g/2)^2 + y^2 + L\left[y\cos\gamma - (x + L_g/2)\sin\gamma\right]} \right\} \quad (16\text{-}17)$$

CHAPTER
17

MAGNETIC FLUX LEAKAGE FROM SUBSURFACE FLAWS

17-1 INTRODUCTION

In many inspection situations, not all flaws are accessible at small lift-off and a layer of metal exists between the flaw and the sensor. Typical examples are:

1. external pitting or erosion in tubes that can only be inspected from the inside
2. internal pitting and cracking in tubes that can only be inspected from the outer surface
3. subsurfaces flaws such as laminations or porosity

In some situations, it might be possible to use very simple models for the MFL fields when active excitation methods are used. In this chapter simple theoretical models for two- and three-dimensional flaws are discussed.

294

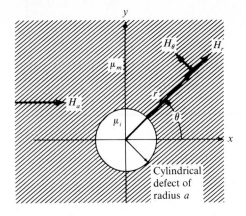

FIGURE 17-1
Coordinate system for a cylindrical hole (permeability μ_i) in a material (μ_m). (*Courtesy NTIAC, Southwest Research Institute, San Antonio, TX [41].*)

17-2 SUBSURFACE CYLINDRICAL HOLE [41]

The magnetic field from cylindrical subsurface holes drilled in a material of permeability μ_m and magnetized as shown in Fig. 17-1 is given by [41]

$$H_r = -m\cos\theta/r^2$$
$$H_\theta = -m\sin\theta/r^2 \tag{17-1}$$

where m is the dipole moment per unit length, which is given by the relation

$$m = \frac{\mu_m - \mu_i}{\mu_m + \mu_i} H_a a^2 \tag{17-2}$$

In this equation, a is the radius of the hole, μ_i is the permeability of its contents, and H_a is the applied field strength. The dipole moment of the hole is proportional to $H_a a^2$ and a permeability term that tends toward unity as the relative permeability of the material is raised.

When boundaries are introduced, the field intensity in the vicinity of the boundary is governed in the classical case of constant permeability by the method of images solution to a classical magnetic problem. In this method, the external field is constructed from a series of image dipoles that are located within the hole. When the depth of the hole (h) is greater than its diameter (a), then the image dipoles are located near to the center of the hole, as shown in Fig. 17-2, and the result for the dipole moment is

$$m = \frac{2\mu_m}{\mu_m + \mu_i}\left[1 - \left(\frac{\mu_m - \mu_i}{\mu_m + \mu_i}\right)^2\left(\frac{a}{2h}\right)^2\right]^{-1}\frac{\mu_m - \mu_i}{\mu_m + \mu_i}H_a a^2 \tag{17-3}$$

In order to see the difference between Eqs. (17-2) and (17-3), consider a case in which $\mu_m = 100$, $\mu_i = 1$, and $a = 0.25h$. The result is $m = 2.01 \times$ the value predicted by Eq. (17-2). Now, if μ_m is lowered to 50, the multiplier becomes 1.99. Clearly, since $\mu_m \gg \mu_i$, the effect of introducing the surface is to approximately double the dipolar field strength over the infinite medium case.

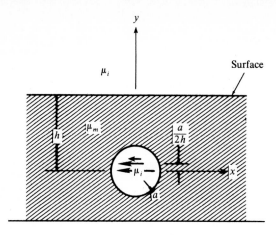

FIGURE 17-2
Image dipoles in the material surface, located inside a cylindrical hole. (*Courtesy NTIAC, Southwest Research Institute, San Antonio, TX [41].*)

Equation (17-3) may be written as

$$m_{eff} \sim 2 \frac{\mu_m - \mu_i}{\mu_m + \mu_i} H_a a^2 \tag{17-4}$$

and the permeability factor will be close to unity as long as the hole is not filled with material that has a permeability that is not close to 1.

Under these circumstances, the tangential (H_x) and normal (H_y) MFL fields are given by

$$\left| \begin{matrix} H_x \\ H_y \end{matrix} \right| = \frac{m}{r^4} \left| \begin{matrix} x^2 - y^2 \\ 2xy \end{matrix} \right| \sim -2 H_a a^2 \left| \begin{matrix} (x^2 - y^2)/r^4 \\ 2xy/r^4 \end{matrix} \right| \tag{17-5}$$

17-2.1 Predictions

Algebraic manipulation of Eq. (17-5) permits the following predictions to be made:

1. The magnitude of the MFL field, given by $H = (H_x^2 + H_y^2)^{1/2}$, is

$$H = 2 H_a a^2 / r^2 \tag{17-6}$$

2. The tangential and normal components of the MFL field take the form shown in Fig. 17-3. These curves have been calculated for $H_a = 2500$ A/m (31.5 Oe), $h = 3.6$ mm, and $a = 0.9$ mm. The MFL field components are seen to be quite small at this relatively low applied field strength.
3. The tangential component is zero when $y = \pm x$, i.e., $x = \pm h$. Thus the distance between the two positions at which $H_x = 0$ is a measure of the depth of the hole.
4. The tangential component has turning points at $x = 0$ and $\pm h\sqrt{3}$. These values could again be used to assess the depth h.

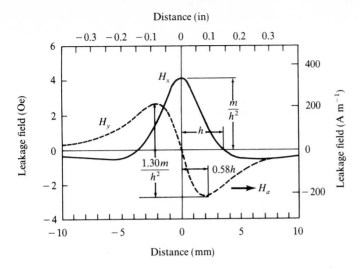

FIGURE 17-3

Calculated components of the magnetic leakage field at the surface from a subsurface cylindrical defect in a linear isotropic magnetic medium with high permeability and with the applied field in the positive x direction. (*Courtesy NTIAC, Southwest Research Institute, San Antonio, TX [41].*)

5. The tangential component has a peak amplitude of m/h^2 at $x = 0$ and a value of $-m/8h^2$ at its most negative point, which occurs at $x = \pm h\sqrt{3}$. The peak-to-peak variation in H_x (i.e., ΔH_x) is $9m/8h^2$, or in the simple approximation given, roughly $2.25H_a(a/h)^2$.

6. The normal component has turning points at $x = \pm 0.577h$.

7. The value of H_y at $x = \pm 0.577h$ is given by $H_y(\text{max}) = 3m\sqrt{3}/8h^2$, so the peak-to-peak variation in H_y is

$$\Delta H_y = 1.30\, m/h^2 \sim 2.6\, H_a\,(a/h)^2 \tag{17-7}$$

17-2.2 Comparison with Experimental Results

Leakage fields from holes of the same radius but differing depths (1.8–12.4 mm) have been made by Swartzendruber [41] using a Hall element with an active area of 0.8×1.5 mm at a lift-off in the region of 1.0 mm. Figure 17-4 illustrates typical data, recorded in residual induction after applying $H_a = 100$ A/cm (126 Oe). At this level of active field, the material around and above the hole is saturated.

By using the preceding relations, it is possible to assign equivalent dipole strengths and hole depths to each of eight holes. The data then indicate that

1. for shallow hole depths, the true and equivalent hole depths are almost the same

2. for deeper holes, the true and equivalent hole depths are quite different

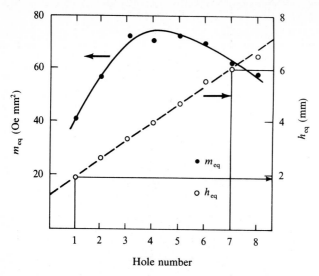

FIGURE 17-4
Equivalent hole depths (h_{eq}) and equivalent dipole strengths (m_{eq}) for the residual fields resulting from application of 100 A/cm. (*Courtesy NTIAC, Southwest Research Institute, San Antonio, TX* [*41*].)

3. the dipole strength is relatively constant at about 70 Oe mm^2 for all but the shallowest dipole. This may be caused by a variation in permeability in the metal around the hole.

Tests may also be performed to show that for a certain threshold applied field ($H_{a,\,min}$), the general shape of H_x and H_y are relatively independent of H_a, so that the constant permeability approximation does seem to give the MFL topography. It may, however, lead to inconsistencies for hole depths and dipole strengths.

17-3   SUBSURFACE SPHERICAL INCLUSIONS

Using a simple constant-permeability model, the diverted field around a spherical inclusion (Fig. 17-5) can be derived from the magnetic scalar potential outside of a sphere that is given by

$$\Omega(\bar{r}) = H_a\left[1 + \frac{\mu_m - \mu_i}{2\mu_m + \mu_i}\left(\frac{a}{r}\right)^3\right]x \qquad (17\text{-}8)$$

In Eq. (17-8), the first term is that for the unperturbed field (H_a) applied along the negative x direction and the second represents that due to the sphere of

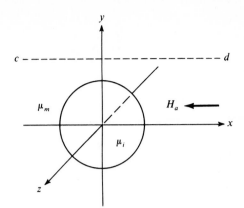

FIGURE 17-5
Model for a spherical inclusion in a uniform applied field H_a.

radius a, relative permeability μ_i, in a medium μ_m. Using $H = -\nabla\Omega$, the magnetic field components outside of the sphere are

$$\begin{vmatrix} H_x \\ H_y \\ H_z \end{vmatrix} = \begin{vmatrix} H_a \\ 0 \\ 0 \end{vmatrix} + \frac{\mu_m - \mu_i}{2\mu_m + \mu_i} H_a a^3 \begin{vmatrix} (3x^2 - r^2)/r^5 \\ 3xy/r^5 \\ 3xz/r^5 \end{vmatrix} \tag{17-9}$$

In this three-dimensional case, the variations on the y and z axes are the same.

17-3.1 Predictions

Figure 17-6 shows that the MFL field components H_x and H_y $(= H_z)$ vary at constant lift-off $y = h$. From Eq. (17-9), the following predictions can be made:

1. The extremes in H_y occur at $x_{1,2} = 1/2 \, (h^2 + z^2)^{1/2}$.
2. The maximum and minimum in H_y have a magnitude of

$$H_{y,m} = \frac{48h}{5^{3/2}(h^2 + z^2)^2} \frac{\mu_m - \mu_i}{2\mu_m + \mu_i} H_a a^3 \tag{17-10}$$

and so the peak-to-peak amplitude is given by

$$\Delta H_y = \frac{1.717}{h^3} \frac{\mu_m - \mu_i}{2\mu_m + \mu_i} H_a a^3 \tag{17-11}$$

3. The distance between the maximum and minimum in H_y is given by

$$x_1 - x_2 = h \tag{17-12}$$

4. The shape of the MFL field does not vary as the inclusion radius changes; only the magnitude is affected. The distance between the points e and f yields

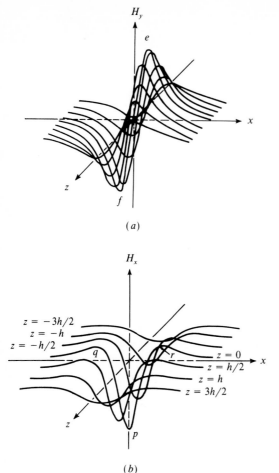

FIGURE 17-6
Prediction of Eq. (17-9) for the flux diverted by a sphere in an infinite medium. (a) H_y; (b) H_x.

information regarding the depth of the inclusion and the magnitude of the MFL field $\Delta H_{y,m}$ yields information regarding its size. This term, however, contains $(a/h)^3$ rather than just a^3, so it is not clear from the size of the leakage field alone whether one is scanning over a shallow smaller inclusion or a deeper larger one.

5. The sign of the MFL field is dictated by the permeability difference $\mu_m - \mu_i$. Often around subsurface flaws there is a region of stress in the material that can affect μ_m and cause additional structure in the MFL field.

6. The extreme of H_x occurs at $x_0 = 0$, and $x_{1,2} = \pm[1.5\ (h^2 + z^2)]^{1/2}$. At $x_0 = 0$, the MFL field has a magnitude of

$$H_x(0, h, z) = -\frac{1}{(h^2 + z^2)^{3/2}} \frac{\mu_m - \mu_i}{2\mu_m + \mu_i} H_a a^3 \qquad (17\text{-}13)$$

and the maximum value of this occurs at $z = 0$, at which point

$$H_x(0, h, 0) = -\frac{\mu_m - \mu_i}{2\mu_m + \mu_i} H_a \left(\frac{a}{h}\right)^3 \qquad (17\text{-}14)$$

This is represented by the point p in Fig. 17-6. At $x_{1,2}$, the tangential field has a magnitude of

$$H_x(x_{1,2}, h, z) = \frac{0.202}{(h^2 + z^2)^{3/2}} \frac{\mu_m - \mu_i}{2\mu_m + \mu_i} H_a a^3$$

and these maxima have their own maximum at $z = 0$, i.e.,

$$H_x(x_{1,2}, h, 0) = 0.202 \frac{\mu_m - \mu_i}{2\mu_m + \mu_i} H_a \left(\frac{a}{h}\right)^3 \qquad (17\text{-}15)$$

These points are denoted by q and r in Fig. 17-6 and occur at $x_{1,2} = \pm h\sqrt{1.5}$. The distance between the two maxima is $h\sqrt{6}$.

7. The tangential component of the MFL field does not change shape with the size of the inclusion. However, the distance $x_1 - x_2 = h\sqrt{6}$ gives an indication of the depth of the inclusion and the peak-to-peak value of H_x at $z = 0$, which is given by

$$\Delta H_{x, m} = H(x_{1,2}, h, 0) - H(0, h, 0)$$

$$= 1.202 \frac{\mu_m - \mu_i}{2\mu_m + \mu_i} H_a \left(\frac{a}{h}\right)^3 \qquad (17\text{-}16)$$

yields information regarding the size of the inclusion.

It should be noted that in deriving the preceding relations, a constant permeability assumption was made and the physical presence of an air–metal boundary was ignored. This will introduce image charges that will affect the dipole strength in a manner similar to Eq. (17-3).

17-3.2 Comparison with Experimental Results

The predictions of the simple model just discussed have been compared with experimental data that were recorded over inclusions in AISI 4340 steel using a small Hall element [14]. Figure 17-7 shows the peak-to-peak normal field amplitude ($\Delta H_{y, m}$), plotted against the diameter of five inclusions. The data fit the line

$$S = \Delta H_{y, m} = Ka^3$$

which is derived from Eq. (17-11). The fit is quite good, despite the fact that the

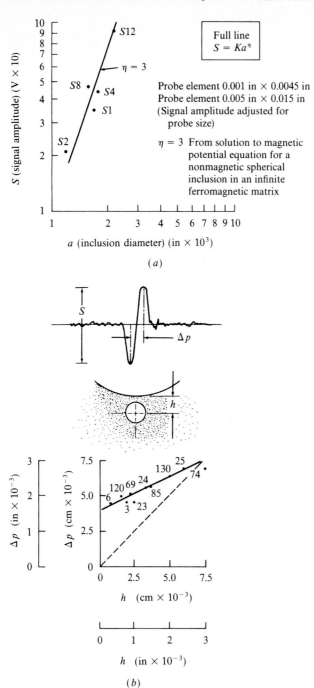

FIGURE 17-7
Predicted and experimental data for (a) peak-to-peak signal of the normal MFL field and (b) peak-to-peak width dependence for subsurface inclusions. (*Courtesy NTIAC, Southwest Research Institute, San Antonio, TX [14].*)

presence of the surface of the steel is not taken into account. Figure 17-7(b) shows data for the peak-to-peak width ($\Delta p = h = x_2 - x_1$) and inclusion depth h. These data predict a different depth dependence, which is probably due to the refractive effect of the metal surface.

In determining the shape of the MFL fields, the finite size of the Hall element was taken into account and corrected for mathematically.

CHAPTER

18

MAGNETIC FLUX LEAKAGE SENSORS

18-1 INTRODUCTION

As important as the need to understand the magnetization process and the nature of the MFL fields around flaws is the need for the system designer to select the sensor that will best detect the MFL. In most cases, the sensor is a package that consists of a detector or an array of detectors, the necessary electronics, and some form of presentation of the flaw data. Many sensor types can be used in magnetic NDE, some of which are selected as much for their reliability in hostile environments as for their ability to detect magnetic fields.

The more commonly used sensors are pickup coils, Hall elements, magnetodiodes, ferroprobes, and magnetic particles. Some of these sensors are passive (coils, magnetic particles), while the rest require some form of excitation. All except the magnetic particle require electronic signal processing. The magnetic particle can be mixed with a dye that fluoresces under ultraviolet light, emitting the absorbed photons at lower energy in the yellow-green part of the optical electromagnetic spectrum. Such sensors are discussed in this chapter.

18-2 HALL ELEMENT

Hall elements are crystals of specially grown semiconductor materials that, when excited by the passage of current, react to the presence of an external magnetic

304

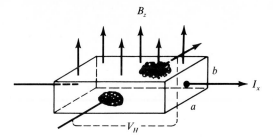

FIGURE 18-1
Relation between the Hall voltage V_H and the current and external magnetic field (I_x, B_z) that cause it. Measurement of V_H enables B_z to be measured.

field by developing a voltage across two parallel faces (Fig. 18-1). Such crystals can be made very small and can often be considered to be point detectors; if not, finite size corrections can be made.

18-2.1   Theory

Conducting condensed matter is almost transparent to the flow of conduction electrons, since crystal lattices do not deflect conduction electrons in the classical billiard-ball-like fashion. Rather, their "mean-free path" (λ) for scattering events is several orders of magnitude larger than the interionic distance, e.g., 10^{-6} m rather than 10^{-10} m. This is a result of the wave-like properties that are exhibited by conduction electrons and the Pauli Exclusion Principle. There is, then, a free (or almost free) electron gas within conductors, known as the Fermi gas. The effect of a magnetic field on the Fermi gas leads to the Hall effect, which is also found for holes in semiconducting materials. The Lorentz force $\bar{F}$ on a charged particle in a magnetic and electric field is given by [42]

$$\bar{F} = -e(\bar{E} + \bar{v} \times \bar{B}) \tag{18-1}$$

where $\bar{E}$ = electric field intensity
   $\bar{v}$ = particle velocity
   $\bar{B}$ = applied flux density

Under this force, the particle accelerates for a distance λ, collides, and then accelerates again for a time τ until another collision occurs. A simple equation that governs this kind of motion is

$$-e(\bar{E} + \delta\bar{v} \times \bar{B}) = m^*(d/dt + 1/\tau)\,\delta\bar{v} \tag{18-2}$$

where the right hand side takes into account the slowing of the particles by repeated collisions. The effective mass m^* of the particles may be very different from their rest mass in space, since it is determined by the particle-lattice interaction energy. Assuming a steady state such as occurs when direct current is passed through a crystal, Eq. (18-2) reduces to

$$m^*\,\delta\bar{v}/\tau = -e(\bar{E} + \delta\bar{v} \times \bar{B})$$

and arranging the flux density to be in the z direction, i.e., $\overline{B} = (0, 0, B_z)$,

$$\begin{vmatrix} \delta v_x \\ \delta v_y \\ \delta v_z \end{vmatrix} = \frac{e\tau}{m^*} \begin{vmatrix} E_x + \delta v_y B_z \\ E_y - \delta v_x B_z \\ E_z \end{vmatrix}$$

From this, it can be seen that

$$\left[1 + (e\tau B_z/m^*)^2\right] \delta v_y = -e\tau/m^* \left[E_y + e\tau E_x B_z/m^*\right]$$

Over a relatively long time, the drift velocity in the y direction is zero, so the result is

$$E_y = -e\tau E_x B_z/m^* \tag{18-3}$$

E_y is, of course, the electric field intensity in the y direction, which leads to the Hall voltage V_H as shown in Fig. 18-1. E_x is the electric field that is caused by the current density j_x, or current I_x, so replacing E_x by σj_x (where σ is the electrical conductivity of the material), Eq. (18-3) becomes

$$E_y = -e\sigma\tau j_x B_z/m^* \tag{18-4}$$

A similar argument shows that the electrical conductivity can be written as

$$\sigma = ne^2\tau/m^* \tag{18-5}$$

where n is the number of charge carriers per unit volume. Combining Eqs. (18-4) and (18-5) gives

$$E_y = -j_x B_z/ne \tag{18-6}$$

The quantity $R_H = E_y/j_x B_z$ is known as the Hall coefficient. In this simple example, its value is $-1/ne$, so that simple macroscopic measurements on E_y, j_x, and B_z yield the carrier concentration n.

The expression for the Hall coefficient becomes more complicated when both electrons and holes contribute to charge transport. For measuring magnetic fields, as large a Hall voltage as possible is required for a given field. The Hall voltage is, from Eq. (18-6),

$$V_H = aE_y = -aj_x B_z/ne = R_H j_x a B_z$$

$$= R_H I_x B_z/b \tag{18-7}$$

This relation is derived using $j_x = I_x/ab$, where a and b are the sample dimensions (Fig. 18-1). Thus the passage of current (I_x) through a suitable crystal of thickness b will yield a Hall voltage that is proportional to the applied field B_z.

Since conduction in the solid state can be either by electrons or holes, R_H can be either positive or negative and, depending upon the interactions that occur between the charge carriers and the crystal lattice, can be either large or small.

This effect was discovered in metals by Edwin Hall in 1897; it was later found that the Hall voltage in semiconductors can be 3 or 4 orders of magnitude larger than values found for metals. Typical values for semiconductors are 10–60 mV/kG with excitation currents around 100 mA ac [14].

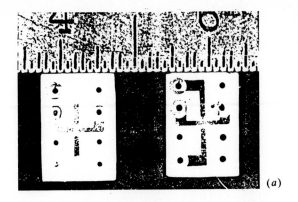

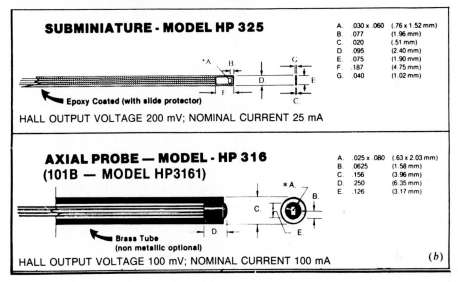

FIGURE 18-2
Subminiature Hall effect sensors. (*Reproduced by permission, LDJ Electronics, Inc., Troy, Michigan* (*a*) *Courtesy NTIAC, Southwest Research Institute, San Antonio, TX [14].*)

18-2.2 Manufacture

Hall effect generators are classified into two types, depending upon their method of manufacture. Bulk Hall crystals are bismuth-doped semiconductor material, e.g., InSb, that are cut into rectangular plates and bonded onto a substrate before the I_x and V_H leads are attached. Finally they are encapsulated. Thin film Hall generators are made by material deposition using *IC* manufacturing technology. Figure 18-2 shows examples of Hall sensors made for (*a*) general use [43] and (*b*) bearing race inspection [14]. The latter, which were made by evaporating bismuth onto an alumina substrate, have active areas as small as 0.1 × 0.025 mm and a sensitivity of 0.5 V/A T (0.05 V/A kG).

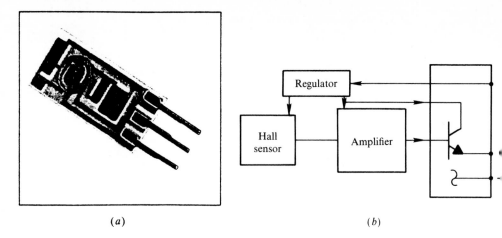

(a)   (b)

FIGURE 18-3
Photograph (a) and schematic (b) of an integrated Hall element. (*Courtesy of MICROSWITCH, a Honeywell Division.*)

A recent advance has been to combine the Hall sensors with a regulated power supply and amplifier on the same chip. Figure 18-3 illustrates such a device while Fig. 18-4 shows both the output and temperature characteristics of the device [44].

The major advantages of such devices over sensing coils are that their active areas are very small, so that they approximate to point sensors, and that their output does not depend upon the relative velocity with which the sensor passes through the magnetic field.

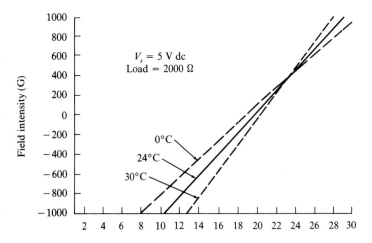

FIGURE 18-4
Output and temperature characteristics of an integrated Hall sensor. (*Courtesy of MICROSWITCH, a Honeywell Division.*)

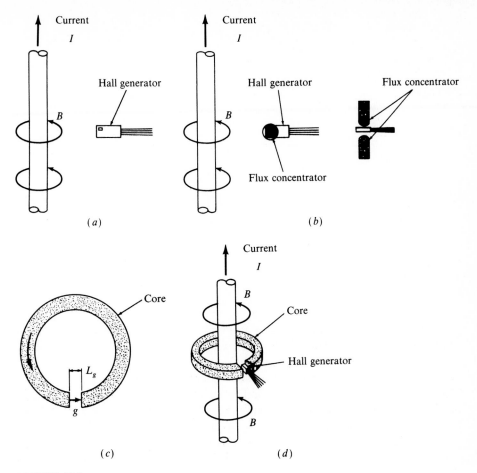

FIGURE 18-5
(*a*) Use of a Hall element as a current sensor. (*b*, *c*, *d*). Small currents may require a flux concentrator. (*b*) Shows that flux concentrators placed perpendicular to a Hall generator increase the sensitivity of the current sensor. (*Courtesy Electronic Products Magazine.*)

18-2.3 Typical Applications

CURRENT SENSOR. Hall elements can be used with or without flux concentrators for the measurement of current. Figure 18-5 illustrates typical situations. The current in the conductor can be found from the relation

$$I = 2\pi aB/\mu_0 \qquad (18\text{-}8)$$

where a is the distance between the Hall element and the center of the conductor. Ferrite is used as a flux concentrator and some form of calibration is required.

FLUX LEAKAGE SENSOR. The Hall sensor is the most useful laboratory device for the sensing of MFL. It may be used to detect both active and residual leakage

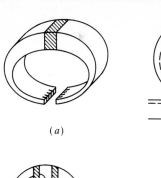

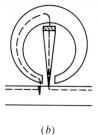

(a)

(b)

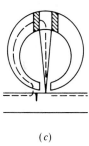

(c)

FIGURE 18-6

Flux leakage pickup sensors made from ferrite C cores and Hall elements (shaded). (a) The sensor gives a unipolar output as it passes over a discontinuity. (b) The sensor gives a bipolar output. (c) The Hall elements are connected in opposition so that the output is also bipolar. [*From U.S. Patent 3,529,236 (1970). Reprinted with permission.*]

fields, although in the case of active fields, it may be necessary to remove the value of the ambient active field from the signal. This can be done by connecting two similar sensors back-to-back in opposition.

An application involving ferrite flux concentrators is shown in Fig. 18-6. This is used in pipeline pigging [45]. The ferrite near the flaw is elongated and thin, while that near the Hall sensor is less elongated and broader. Unipolar [Fig. 18-6(a)] or bipolar [Fig. 18-6(b), (c)] outputs can occur depending upon the position of the Hall sensor(s). Such sensors are driven by high audio-frequency ac and, in the case [Fig. 18-6(c)], a differential amplifier would be needed prior to frequency discrimination.

Eddy currents at the pole tips are reduced by cutting slots in the ferrite as shown. These slots also serve to reduce the warping of the MFL field by the sensor.

A severe problem with rifled gun tubes is the detection of the presence of cracking in the rifling, as shown in Fig. 18-7 [46]. The riflings themselves contribute to the MFL but, as shown, in a rather regular manner in comparison to the signal from a crack. With the aid of filtering, so as to allow only very few frequencies to pass, the MFL signals from the rifling can be greatly reduced. In the application shown, the tangential component of the MFL field is detected.

18-2.4 Advantages and Disadvantages

Hall sensors can be used to measure normal and tangential fields, and can be connected so as to take differences in such fields. This can aid in suppression of material surface noise. Sensors can be made extremely small, so that they approximate point detectors, but then arrays of them are required for 100%

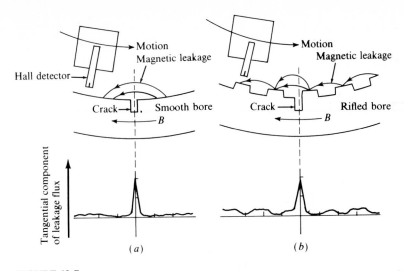

FIGURE 18-7
Electronic detection of magnetic leakage associated with cracks in cannon tube: Tangential component of magnetic flux leakage (*a*) for smooth bore tube and (*b*) for rifled tube. (*From the Nondestructive Testing Handbook, Vol. 4, 2d ed. Reprinted with permission.*)

surface coverage. A disadvantage is that no two elements have exactly the same sensitivity, so that some balancing of the sensor array is required. Further, ac activation is required and, since the active elements are not very robust, they require encapsulation. This adds to the lift-off distance between the element and the inspected surface.

18-3 MAGNETODIODE
18-3.1 Theory

The magnetodiode is a solid state device, the resistance of which changes with applied field strength. It is fabricated to consist of p and n zones in a semiconductor, with a central intrinsic zone (i) on one side of which is a recombination zone (r), as shown in Fig. 18-8 [47]. In this r zone, electrons have a relatively long lifetime in comparison to their lifetime in the i zone. In this situation, if electrons are deflected by an external B, more recombination of electron-hole pairs can occur, with the result that the diode resistance increases.

Obviously, this magnetoresistance is dependent upon ambient temperature, but this dependence can be reduced by connecting two such diodes in series electrically, but in opposition magnetically as shown in Fig. 18-8(*b*). Typical temperature characteristics [Fig. 18-9(*c*)] show that the center potential (V_m) is roughly independent of temperature for small changes in the ambient temperature.

The sensitivity of the magnetoresistance of the chip, which is the basis of field detection by this device, is quite high. For a 1000 Oe field, V_m is typically 1.2 V for an applied voltage of 6 V [Fig. 18-9(*a*)]. The active volume is quite small, typically $3.0 \times 0.6 \times 0.4$ mm, and the response is linear at low fields and above

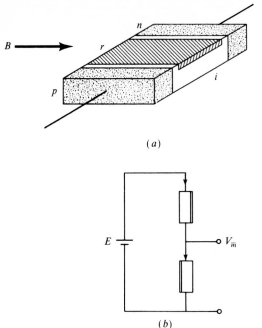

(a)

(b)

FIGURE 18-8
(a) Schematic of the magnetodiode showing p and n zones in semiconductor material, along with the intrinsic (i) and recombination (r) zones. (b) Simple temperature compensation scheme with two magnetodiodes. (*Copyright Sony Corp., reprinted with permission.*)

500 Oe [Fig. 18-9(a)]. Further, their frequency response is flat [Fig. 18-9(b)] from 0 to 100 kHz, which makes them very useful for ac MFL applications.

18-3.2 Typical Application

Sumitomo Metal Industries [48] have reported the use of magnetodiodes in their SAM (Sumitomo Automated Magnetic) inspection system for hot rolled bars and steel tubes. Figure 18-10 illustrates the differential connection of the sensors [Fig. 18-10(a)] and electrical block diagram [Fig. 18-10(b)]. Alternating current excitation is used for surface-breaking flaw detection with three sets of six pairs of diodes, the tube being rotated as it passes slowly through the inspection system. Differential signals are amplified, demodulated, and unified so that the largest of simultaneously occurring signals pass to a notch filter and recorder.

Surface cracks as shallow as 0.3 mm can be detected with this approach. The length of the flaw can also be measured and their total number counted and recorded. In a recent development, all of the analogue electronics have been replaced by a microprocessor [49].

18-4 MAGNETIC RECORDING TAPE

In the testing of steel billets, it is often necessary to detect surface cracks and to distinguish between such flaws and subsurface ones. The former can be removed

MD-230A ΔV-H
Typical characteristic curve

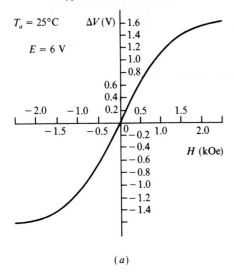

$T_a = 25°C$ ΔV (V)

$E = 6$ V

H (kOe)

(a)

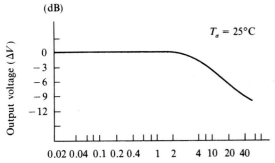

(dB)

$T_a = 25°C$

Output voltage (ΔV)

0
-3
-6
-9
-12

0.02 0.04 0.1 0.2 0.4   1 2   4 10 20 40

Magnetic field frequency (kHz)

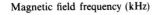

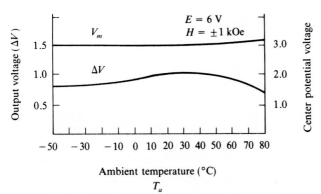

Output voltage (ΔV)

$E = 6$ V
$H = \pm 1$ kOe

V_m

1.5

ΔV

1.0

0.5

Center potential voltage

3.0

2.0

1.0

-50 -30 -10 0 10 20 30 40 50 60 70 80

Ambient temperature (°C)
T_a

FIGURE 18-9
(a) Characteristic curve of the magnetodiode. (b) Frequency response. (c) Temperature. (*Copyright Sony Corp., reprinted with permission.*)

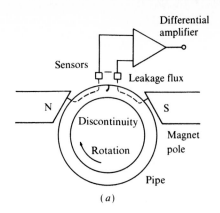

(a)

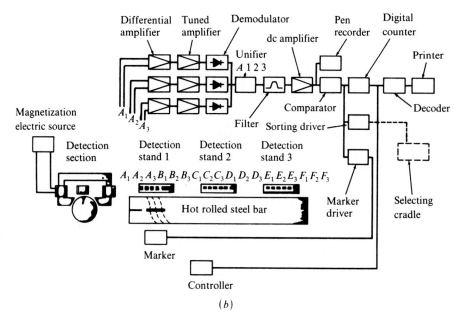

(b)

FIGURE 18-10
SAM electromagnetic testing system: (a) The principle of the ac magnetizing method and (b) electrical block diagram. (*From the Nondestructive Testing Handbook, Vol. 4, 2d ed. Reprinted with permissions from ASNT and Sumitomo Metal Industries.*)

by grinding, while the latter are subsequently worked into the final product, perhaps as laminations. Magnetic tapes have been developed for such applications.

18-4.1   THEORY

The magnetic tape is made from finely divided ferrite embedded in wear-resistant neoprene. The particles become magnetized by flaw MFL fields, as shown in Fig.

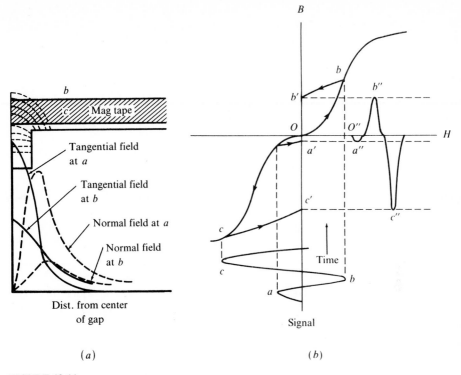

FIGURE 18-11
(a) Flaw leakage fields as they affect magnetic tape and (b) signal transfer to tape.

18-11(a), the leakage fields taking localized regions of the tape through parts of its *B-H* characteristics. The incoming relatively small signal [a of Fig. 18-11(b)] takes the tape through the lower part of the *B-H* curve, after which it falls back to its residual induction a' [50]. Then, a larger peak b in the signal b magnetizes another portion of the tape along *Obb'* and the peak at c takes a third portion of the tape along *Occ'*. The resulting output for this impressed signal is *O''a''b''c''*. Magnetization of the ferrite particles is low because of the large demagnetization factors associated with finely ground materials.

18-4.2 Typical Applications

DRILL PIPE END AREAS. A particularly difficult region to inspect is that for about 1.5 m inside the end of a drill pipe. In this region corrosion and erosion occur, and the magnetic properties of the steels are often different in progressing from the tool joint to the tube. The area is also prone to fatigue cracking. A balloon is inflated with compressed air and a high coil field is passed over the OD of the tube. The balloon is reinflated inside a very thin nonmagnetic tube and scanned with either magnetic particles or other MFL-sensitive devices. The

FIGURE 18-12
Scheme for inspecting a particularly difficult region of drill pipe using an expandable magnetic rubber balloon.

problem of distinguishing between MFL from the edges of pits and cracks within the pits is sometimes difficult. Fig. 18-12 illustrates the method.

BILLET INSPECTION. Figure 18-13 shows a typical billet inspection system, built by Inst. Dr. Förster (West Germany) [32]. The billet is fed with current that causes MFL from longitudinal flaws and passed under a continuous band. Flaw signals are "read" from the band by Förster microprobes (see later) and then erased [51]. Four such systems are required for scanning a billet. Unfortunately, the method of magnetization leads to a nonuniform induction across each face of the billet, and so the pickup signals must be amplified more at each corner. Figure 18-14 shows that selective amplification can compensate for this geometrical ambient induction loss, which is performed in the "field strength equalizer" section of the electronics. Once this is done, if the cracks are all tight, then their amplitudes correspond roughly to the depths of the flaws, so long as they are perpendicular to the billet surface.

The method requires that the billets be bright metal, or pickled, so as to provide low lift-off between the billet and the tape. Surface roughness should not exceed about 100 μm and there should be no burrs or flakes. Under these conditions, flaws as shallow as 0.3 mm can be found.

Electronic filtering discriminates against subsurface flaws, which present a broader MFL pattern to the band, which is then filtered out [52].

ERW LINES. Another serious problem in the inspection of tubes is that of the inspection of the weld in electric resistance welded (ERW) materials. Such tubes are manufactured from flat plate by belling the edges, forming the plate into a cylinder, welding along the join, and cropping the excess metal in the region of the join, both on the ID and OD of the tube. When magnetized circumferentially, signals from longitudinal flaws in and on the edge of the weld may be obscured by MFL from the weld itself, which may be caused by differences in metallurgy across the weld or excessive undercut (Fig. 18-22).

By rolling a resilient wheel of magnetic band along the weld, such signals can be detected and the nonrelevant signals can be discriminated against with a high-pass filter [53].

18-5 PICKUP COIL

While the coil may be the easiest sensor to design and use, a complete discussion of its capabilities requires not only a knowledge of the MFL field through which

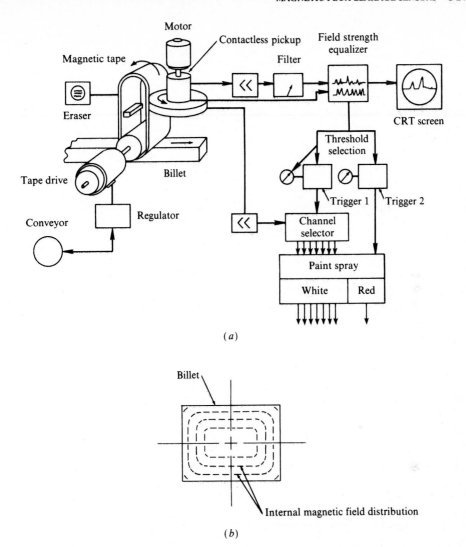

(a)

(b)

FIGURE 18-13
Rectangular billet tested with magnetic flux leakage. (*a*) A circuit for collection of data using magnetic tape. (*b*) Circular magnetization in a rectangular billet. (*After Förster and Muller. Reprinted with permission.*)

it is passed, but also its dimensions and orientation to the flaw field. This treatment begins with the voltage induced in a straight wire and proceeds to rectangular coils, the signal from which is fed to a very high resistance amplifier, so that no current flows in the coil to set up currents that themselves create fields that can affect the field being detected.

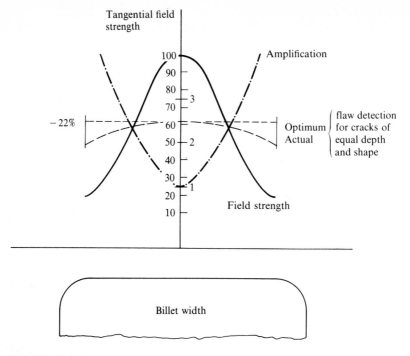

FIGURE 18-14
Electronic amplification required to compensate for loss of surface field toward the edges of the billet.
(*After Förster and Muller.*)

18-5.1   Straight Wire

When a straight-wire conductor is passed through a magnetic field, Faraday's Law of Induction indicates that a voltage is induced between the ends of the conductor according to the relation $E = -d\phi/dt$ where $d\phi/dt$ is the rate of change of magnetic flux (in webers per second). The negative sign is a mathematical statement of Lenz's Law because when a voltage is generated by the flux-cutting process, if that voltage is allowed to cause a current, then the current will create its own magnetic field and this field must act to oppose the field that created it.

For a wire traveling at speed v perpendicular to a flaw opening [Fig. 18-15(a)] that is larger than the length of the wire, then, in moving a distance dx in the x direction, the flux cut is given by $\phi = B_\perp (L\,dx)$, where $B_\perp$ is the component of the field that is perpendicular to the direction of travel and L is the length of the wire. The voltage between its ends is then

$$E = -d\phi/dt = -B_\perp L\,dx/dt$$
$$= -B_\perp Lv \tag{18-9}$$

Equation (18-9) indicates that the induced voltage is related to $B_\perp$, the sensor length, and the relative velocity. This is also true of coils. If $B_\perp$ is

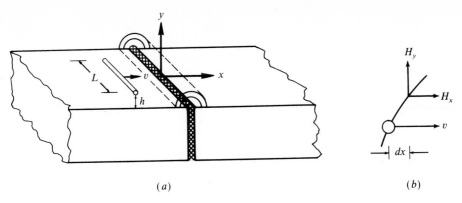

FIGURE 18-15
Coordinate system and dimensions used in deriving the voltage generated between the ends of a straight wire that scans through an MFL field. (a) Schematic and (b) v perpendicular to H_y.

constant along the length of the coil, then Eq. (18-9) is valid; if not, then an integral of $B_\perp$ along the sensor direction must be performed.

As an example, the voltage developed when such a wire passes through the MFL field given by the simple Förster equation is

$$E = -\frac{\mu_0 H_g L_g}{\pi} \frac{x}{x^2 + y^2} Lv \tag{18-10}$$

and since the lift-off distance is generally held constant ($y = h$), the result is

$$E_{(y=h)} = -\frac{\mu_0 H_g L_g}{\pi} \frac{x}{x^2 + h^2} Lv \tag{18-10a}$$

The general shape of the signal is shown in Fig. 18-16. It is easy to show that the turning points in E occur at $x = \pm h$ and thus at these particular points $E(x = \pm h)$ is proportional to $1/2h$. From this follows the result that the peak-to-peak signal is given by

$$\Delta E = \mu_0 H_g L_g Lv / \pi h \tag{18-11}$$

That is, the signal that would normally be measured decreases inversely with the sensor lift-off.

The simplest MFL sensors that have evolved from the straight wire are the perpendicular and parallel coils.

18-5.2 Perpendicular Coil

If a coil is constructed so that one set of wires travels at height h from the inspected surface and the return wires travel at some distance ($2b$) higher, as shown in Fig. 18-17, then the EMF will be the difference between the signals developed in the lower and upper branches. In the simple Förster approximation and redefining the lift-off so that it is measured from the center of the coil (i.e., putting $h = H - b$ and $h + 2b = H + b$), then the EMF developed between the

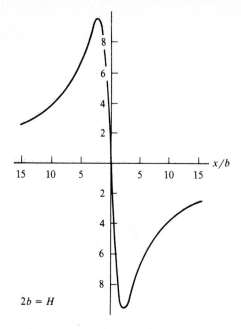

FIGURE 18-16
Shape of output EMF from single wire and perpendicular coil sensors.

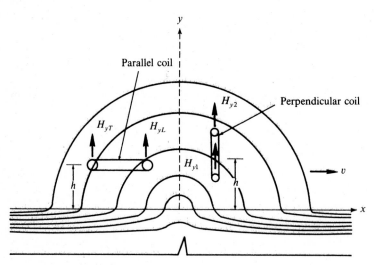

$$E_{parallel} \propto (H_{yL} - H_{yT}) Lv$$
$$E_{perpendicular} \propto (H_{y1} - H_{y2}) Lv$$
$$L = \text{coil length into page}$$

FIGURE 18-17
Parallel and perpendicular coils cutting magnetic flux leakage fields from a discontinuity at a speed v. For discontinuity fields longer than the coil, the output of the coils is given by the formulas shown. (*From the Nondestructive Testing Handbook, Vol. 4, 2d ed. Reprinted with permission.*)

ends of the coil is given by

$$E = -\frac{\mu_0 H_g L_g A_c \cdot 2vH}{\pi} \times \frac{x}{\left\{x^2 + (H - b)^2\right\}\left\{x^2 + (H + b)^2\right\}} \quad (18\text{-}12)$$

where A_c, the area of the coil, is given by $2Lb$. The general shape of this function is plotted for x/b and $2b = H$ in Fig. 18-16. The turning points in the signal, found from $dE/dx = 0$, occur at

$$x_0^2 = \left[2\sqrt{H^4 - (bH)^2 + b^4} - (H^2 + b^2)\right]\Big/3 \quad (18\text{-}13)$$

and so the distance between the turning points is given by

$$2x_0 = \left(\frac{2}{\sqrt{3}}\right)\left[2\sqrt{H^4 - (bH)^2 + b^4} - (H^2 + b^2)\right]^{1/2} \quad (18\text{-}14)$$

Upon numerical solution, the turning points can be found and used to determine the maximum sensitivity of such a coil to tight crack fields.

18-5.3 Applications of Perpendicular Coils

In Chap. 13 the magnetic circuit was discussed. In the application to installed oil well casing, perpendicular coils may be used. Figure 18-18 shows a pad system in which such MFL coils scan the ID of such tubes [54]. Rings of overlapping coils are pressed against the ID as the tool is withdrawn at constant speed from the bottom of a well. They scan through MFL from internal and external surface pits. Also included within the sensor package is a high frequency eddy current coil, the purpose of which is to discriminate whether the MFL signal originates from the ID or OD of the tube. Since the eddy currents are confined to the inner surface, the lack of an eddy current signal indicates that the MFL originates from the OD of the tube.

One problem with such coils is that the active excitation field passes through the coil. So long as its value remains constant, it does not contribute the induced EMF. The problem does not occur with residual induction and, since the MFL fields are now smaller, the coil can be wound on a ferrite core. If the core is short, its permeability is much below that of the ring sample and linear with the field. Ferrite-cored coils find application where the ambient temperature is relatively constant.

18-5.4 Parallel Coil

If the coil is oriented so that one set of wires follows the other as shown in Fig. 18-17, then the resultant signal is the difference between the EMFs induced in the leading and trailing edges [55]. If the components of the MFL field that are perpendicular to the direction of motion of the coil and $B_{\perp L}$ and $B_{\perp T}$ at the leading and trailing edges, then the signal is given by

$$E = -[B_{\perp L} - B_{\perp T}]Lv$$

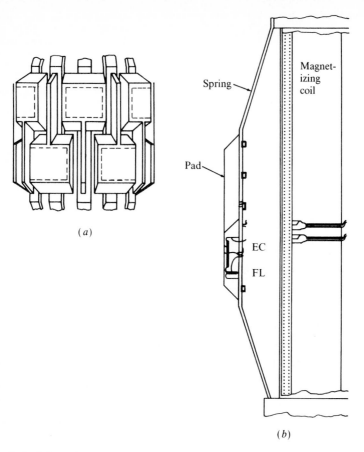

FIGURE 18-18
Example of the use of coils that scan the tangential component of the flux leakage from a flaw.
(*a*) Sensor pad configuration in the inspection head for the inspection of tubes from the inside
surface. (*b*) The coil configuration within a pad. In this particular system, a high frequency eddy
current coil (EC) is included so as to provide discrimination regarding the position of the flaw. (An
ID flaw will give an eddy current signal, but an OD flaw, at the frequencies employed, will not give a
signal.)

If the simple Förster equation [Eq. (16-9)] is used and we substitute
$x_T = x + b$ and $x_L = x - b$, the preceding expression yields

$$E = \frac{H_g L_g Lv}{\pi} 2b \frac{h^2 + b^2 - x^2}{\left[(x + b)^2 + h^2\right]\left[(x - b)^2 + h^2\right]} \qquad (18\text{-}15)$$

The shape of the voltage signal that is predicted by Eq. (18-15) is shown in
Fig. 18-19. The dashed lines are the voltages from the leading and trailing edges,
while the full line is the difference between these two curves. The signal consists
of one major peak centered around $x = 0$, and two smaller side peaks. Manipula-

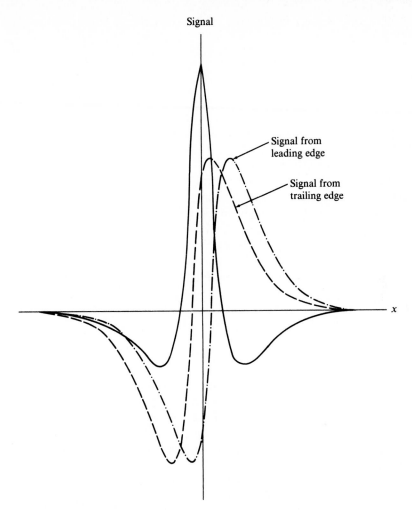

FIGURE 18-19
Parallel coil signal. (*Reproduced from [55] with permission.*)

tion of Eq. (18-15) indicates that the roots of the equation occur at

$$x_0 = \pm(h^2 + b^2)^{1/2}$$

so that the distance between the $E = 0$ points is given by

$$2x_0 = 2(h^2 + b^2)^{1/2} \tag{18-16}$$

Differentiation of Eq. (18-15) reveals the turning points and is left as an exercise. Some similarities can be seen between Eq. (18-15) and the first derivative of Eq. (18-10), especially at small values of b, so the parallel coil may be thought of as taking the first derivative of the normal component of the MFL field.

One of the differences between the operation of the parallel and perpendicular coils is the fact that the parallel coil is somewhat tuned dimensionally to the MFL field that it senses. This can be seen as follows. The maximum value of the coil signal [E(max)] occurs at $x = 0$. Differentiation of the form of E(max) with respect to b shows that the maximum value attained by the central peak of the signal occurs when $b = h$. In practical terms, once the lift-off has been set by mechanical considerations, the signal can be maximized by making the coil spacing ($2b$) equal to twice the lift-off.

The advantage of the use of this orientation of coil is that when designed for tight flaws, they provide some noise reduction for longer range surface noise. This can also be accomplished with electronic filtering. Some high-noise situations do occur, however, when scanning materials in residual induction, for which the noise level is so high that the coil cannot effectively filter it. This occurs when the strength of the signal [effectively the $H_g L_g$ term of Eq. (18-15)] is much larger than that of the flaw signal, so that the differentiating effect of the coil is swamped.

18-5.5   Applications of Parallel Coils

Parallel coils are used extensively in the inspection of tubes. As shown in Fig. 18-20, when searching for transversely oriented or three-dimensional flaws on the ID or OD of a tube, or cracks in rods, the material is passed through a longitudinal magnetizing system and scanned by overlapping rings of sensors. Constant relative velocity is needed when coil sensors are used, typically 0.3–2.2 m/s.

Often, in order to increase coverage, coil arrays are used, the elements of which may be scanned digitally, or connected in some way so as to assist in noise reduction. Figure 18-21 is an example of overlapping angled coils, the angle of which has been preset for the detection of a specific tight flaw, e.g., a seam in a tube.

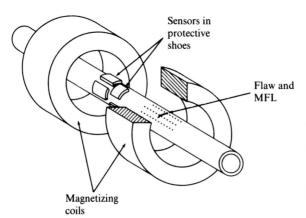

Sensors in protective shoes

Flaw and MFL

Magnetizing coils

FIGURE 18-20
Application of parallel sense coils to inspection of cylindrical material for flaws that have a component that is perpendicular to the long axis.

FIGURE 18-21
Angled coil array.

A commonly encountered inspection situation is that of rotating sensors through the MFL field of an electric resistance weld in a tube. A relatively broad MFL field may occur in the weld region due to either poor internal weld cropping procedure or lack of constancy of *B-H* properties across the weld. Such fields may obscure the field from a tight flaw within the weld region. One method that suppresses the weld signal is the use of oppositely connected coils (figure 8) in conjunction with a high-pass filter. It is expected that one of the pair will scan

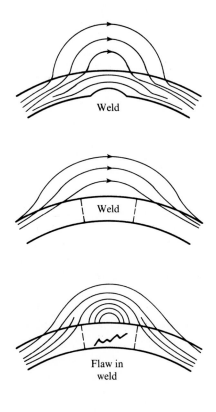

Weld

Weld

Flaw in
weld

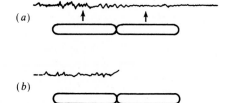

(*a*)

(*b*)

FIGURE 18-22
Board MFL configurations for circularly magnetize welded tubes. The weld flaw signal is expected to have a tighter pattern. A figure 8 coil will respond to the deepest part of or the end of the flaw.

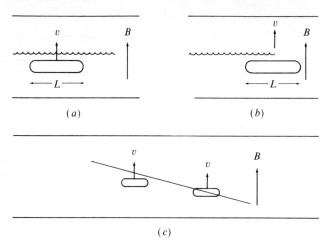

FIGURE 18-23
The size and angle at which discontinuity leakage fields are scanned by finite-sized coil sensors affects the signal generated: (*a*) coil smaller than discontinuity; (*b*) coil scans only the end of the discontinuity (the same situation occurs if the discontinuity is smaller than the coil); and (*c*) coil scans discontinuity at an angle (some signal cancellation can occur in this situation). (*From the Nondestructive Testing Handbook, Vol. 4, 2d ed. Reprinted with permission.*)

through the flaw MFL field. Figure 18-22 illustrates the problem of scanning such welds, and the use of figure 8 coils to suppress the weld signal.

18-5.6 Advantages and Disadvantages of Coils

The advantages of coils are that they are robust, easy to manufacture, can be shaped to suit the inspection task, and do not require excitation. They require encapsulation in nonconducting media, so as to avoid eddy current effects. Finite size and angle effects [32] are shown in Figs. 18-23 through 18-25. As Fig. 18-23(*a*) and (*b*) shows, the coil voltage is an integral over the flaw field and the coil dimensions, so even if it is smaller than the flaw [Fig. 18-23(*a*)], any depth information in the MFL signal will be averaged. If the coil scans only the end of the flaw [Fig. 18-23(*b*)], the signal will not be related to any flaw parameters. As shown in Fig. 18-23(*c*), when a coil scans an angled flaw, the edges of the coil can simultaneously sense the opposing leakage fields on either side of the flaw, and so some cancellation of the signal occurs. Figure 18-24 shows a comparison between ultrasonic and magnetodiode sensors, while Fig. 18-25 compares the magnetodiode with the coil (length unknown).

Finally, the velocity dependence of the coil is a severe problem when the material first begins to move and when it slows down prior to stopping.

SCANNING THROUGH MFL FIELDS. The leakage field models of the previous chapters may be used to simulate the response of various types of sensor. In Fig. 18-26, the model of Zatsepin and Shcherbinin has been used to create the signal

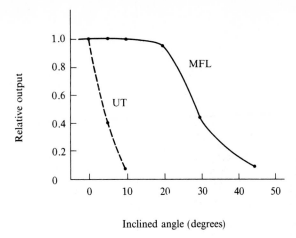

FIGURE 18-24
Relation between inclined angle and output of magnetic flux leakage and ultrasonic tests. (*From the Nondestructive Testing Handbook, Vol. 4, 2d ed. Reprinted with permission.*)

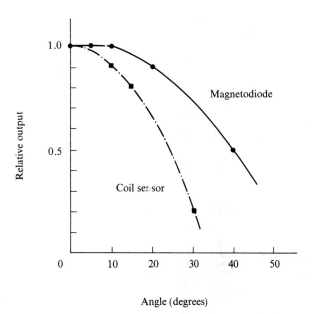

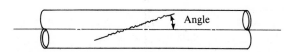

FIGURE 18-25
An inclined discontinuity cannot be detected by a large coil sensor as easily as by the small magnetodiode. (*Courtesy American Society for Nondestructive Testing. From the Nondestructive Testing Handbook, Vol. 4, 2d ed. Reprinted with permission.*)

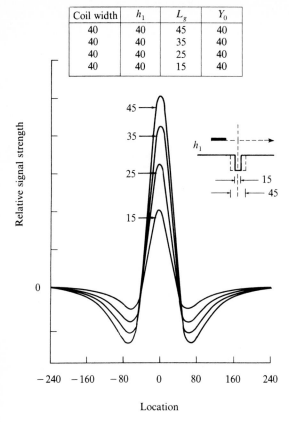

Coil width	h_1	L_g	Y_0
40	40	45	40
40	40	35	40
40	40	25	40
40	40	15	40

FIGURE 18-26
Computer simulation of coils stepped through a MFL field (Zatsepin and Shcherbinin model).

expected from either a parallel coil or a pair of Hall elements connected in opposition. Relative magnitudes of the sensor lift-off (h_1), slot width (L_g), and depth (Y_0) are shown in the inset table. The model shows that (1) perceptible signals are seen for about 200 units on either side of the slot, (2) the peak signal rises as the slot is made wider but the relative increase falls off as L_g gets larger, and (3) the zero crossings at about ± 40 units do not appreciably change as L_g is increased, but the two minima move farther apart.

18-6 FÖRSTER MICROPROBE

This sensor consists of a ferrite core with one or more coils wrapped tightly around it. The development of this probe for NDE originates with Förster. Its principle of operation makes use of the nonlinear B-H characteristics of ferrite. As shown in Fig. 18-27, the ferrite is excited at some relatively high frequency, e.g., 140 kHz, and the signal at twice this frequency is taken from this or other coils. Typical dimensions for the core are length 2 mm and diameter 0.1 mm [51, 56, 57].

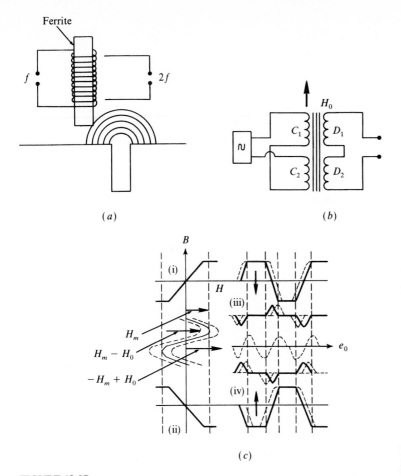

FIGURE 18-27
(a) Second harmonic probe. (b) More complex version. (c) Operation of type (b) showing how output voltage e_0 is generated.

18-6.1 Theory

Ideal ferrite core materials have initial values of dB/dH of 10–12, which reduces to 2–3 as the applied field strength is raised and show little or no hysteresis. They also saturate at relatively low fields.

A complex form of such a detector is shown in Fig. 18-27(b) and its mode of operation in Fig. 18-27(c). In this configuration, a sinusoidal current is passed through opposed coils (C_1 and C_2) on a ferrite core, so that voltages of equal magnitude but opposite direction are induced in pickups D_1 and D_2. The voltages are not purely sinusoidal but rather contain large amounts of odd harmonics. With no external field present, the output (e_0) is $D_1 - D_2 = 0$, but with an external field H_0, applied in a direction parallel to the core, its magnetization varies as shown by (i) and (ii) in Fig. 18-27(c). The voltages in D_1 and D_2 are then phase shifted in opposite directions [(iii) and (iv) of Fig. 18-27(c)], so that the output (e_0) contains only even harmonics of the excitation frequency. The second harmonic output has a voltage given by

$$E_2(\text{rms}) = (44 \times 10^{-9})(Nafh\mu_r H_0) \tag{18-17}$$

where N = the number of turns
   a = the cross-sectional area (cm^2)
   f = the excitation frequency
   h = a factor that depends upon the ratio of the magnetizing field strength (H_m) and the field (H_s) at which the core saturates; in this particular situation, it is given by $h = (4/\pi)\,[\sin(2H_s/H_m)]$

18-6.2 Examples of Microprobe Use

Differentially connected probes with the parameters length = 2 mm, diameter = 0.1 mm, excited coil = 350 turns, pickup coil = 250 turns, excited frequency = 140 kHz, and the second harmonic detected have been used to scan residual MFL fields. The range of MFL field strengths that can be measured can be increased by raising the amplitude of the excitation field H_e. The region of linearity between e_0 and H_e increases, but the sensitivity falls. In order to obtain a higher sensitivity to MFL, the effective permeability (μ_{app}) of the core must be reduced in such a way that $H_e = H_{e,0}$, its optimal value. This optimal field is determined from

$$\mu_{\text{app}} H_{e,0} = \sqrt{2}\, B_s \tag{18-18}$$

Calibrated curves for such a probe are shown in Fig. 18-28. In Fig. 18-28(a), curve 1 shows the situation for $H_{e,0} = 55$ A/cm (69 Oe), while curve 2 shows what happens when the ferroprobe is overexcited at $H_e = 165$ A/cm (207 Oe). By changing the core diameter to 0.25 mm and keeping H_e at 165 A/cm, curve 3 is obtained and by underexcitation of this probe at $H_e = 100$ A/cm (126 Oe), curve 4 is obtained. It can be seen from these curves that:

1. The sensitivity for curve 1 is high up to 25 A/cm only.
2. The sensitivity of the probe for curve 2 is about 30 mV cm/A up to $H_e = 100$ A/cm.
3. The sensitivity of the probe for curve 3 is about 60 mV cm/A up to 100 A/cm. This permits detection of fields of the order of 1–2 A/cm (1.3–2.5 Oe) in a background field of $H_a = 50$–100 A/cm (63–126 Oe).
4. The sensitivity in curve 4 is as good as in curve 3 as long as H_e is between 60–150 A/cm (75–188 Oe), but is 0 for $H_e < 50$ A/cm (63 Oe). This threshold can be used to reduce or eliminate the effects of the background field H_a and the threshold itself can be raised by reducing the effective permeability of the core.

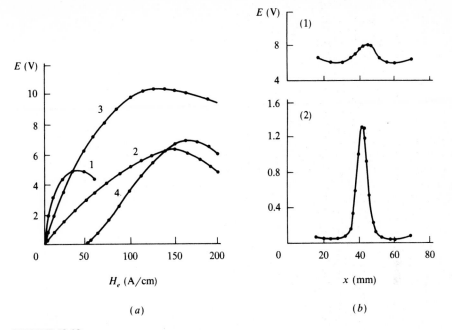

FIGURE 18-28

(*a*) Ferroprobe calibration curve and (*b*) signals recorded over defects by a ferroprobe in the optimum (1) and underexcited (2) modes. (*Reproduced from* [*56*] *with permission.* © *Plenum, New York.*)

The sensitivity of the underexcited probe of low effective permeability to flaws is shown in Fig. 18-28(*b*). Curve 1 of this figure corresponds to excitation according to curve 3 of Fig. 18-28(*a*), while curve 2 corresponds to curve 4 of the same figure.

Förster has reported that 1 mm diameter probes show a wide frequency range up to about 1 MHz when driven at 10–30 MHz. The signal per oersted of MFL field is reported as being roughly 10^4 times that of a conventional Hall element. Such probes are used in billet inspection (Fig. 18-13).

A second example [57] outlines the use of elongated probes with cores of dimensions $5 \times 3 \times 1$ mm, made by vacuum deposition of Fe–Ni alloy (Fig. 18-29). Such probes, the calibration curves for which are shown in Fig. 18-29(*b*), are useful for the measurement of MFL fields of up to 10–15 A/cm (12.5–19 Oe), which is within the range of many of the MFL fields experienced in residual induction inspection.

Some idea of the sensitivity of such elongated probes to small flaws can be seen by integrating the MFL field from a point dipole over the long dimension of the sensor core. In this somewhat artificial situation, the normal component of the MFL field H_y is given by

$$H_y = Qy/\left(x^2 + y^2 + z^2\right)^{3/2}$$

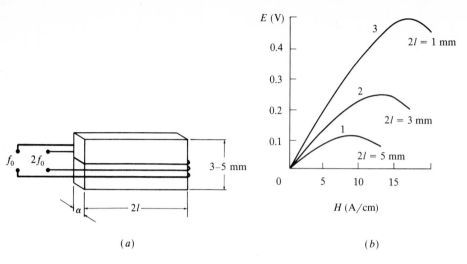

(a)   (b)

FIGURE 18-29
(a) Simple elongated second harmonic MFL probe and (b) calibration curves. (*Reproduced from* [57] *with permission.* © *Plenum, New York.*)

where Q represents the dipole strength, which is related to the size of the flaw. The average value along the length $2l$ of the sensor is given by

$$\overline{H}_y = \frac{1}{2l} \int_{z-l}^{z+l} H_y \, dz$$

$$= \frac{Qy}{2l(x^2 + y^2)} \left[\frac{z+l}{\sqrt{x^2 + y^2 + (z+l)^2}} - \frac{z-l}{\sqrt{x^2 + y^2 + (z-l)^2}} \right]$$

At the point $(0, y, 0)$, i.e., directly over the flaw, this average reduces to

$$\overline{H}_y(0, y, 0) = Q/y(y^2 + l^2)^{1/2} \tag{18-19}$$

At a lift-off of $y = h$, and assuming that the sensor length $(2l)$ is very much greater than the lift-off, Eq. (18-19) reduces to

$$H_y(0, y, 0) = Q/hl \tag{18-19a}$$

Thus for small flaws, the average value of the normal component of the MFL field over the length of the sensor falls off as $1/l$.

18-7   MAGNETIC PARTICLE

Magnetic particle inspection (MPI) is a surface or near surface flaw detection method that employs small pieces of soft iron or ferrite as the sensor [58]. These differ from the sensors previously described in two ways: it is possible to place them closer to the flaw and the force that holds them to the flaw field is quite complex. Their major advantage is that they can be applied to irregularly shaped parts [59].

18-7.1   Current Requirements for MPI

Since many of the parts inspected with MPI have little or no symmetry, and numerous magnetization methods are available, it is essential to select a method that provides adequate magnetic field intensity around any suspected flaws and remains within code restrictions. Commonly used methods will now be discussed.

CURRENT ALONG PART [53]. Here, current is passed through copper mesh connections into the part to establish circumferential magnetization according to $\oint \overline{H} \cdot d\overline{l} = I$. A gaussmeter can be used to determine the level of surface field strength for good MPI indications for the type of current and particles used. If the part has a variety of diameters, so that different currents are required to produce the same value of H at the surface, then the thinner sections should be inspected first. This avoids excessive false indications when the wider sections are magnetized. In some older equipment, the resistance of the part, if it is long, can limit the maximum current available from the power supply. Some inspection specifications do not permit this form of magnetization because of the possibility of arc burn damage.

CURRENT THROUGH PART [58]. In this method, two prods are attached to the part and a current is passed. A gaussmeter midway between the prods should read roughly 32 A/cm (40 Oe) while the current is passing. The prods are held 15–20 cm apart, but this distance depends upon the power supply. Banding of the particles is an indication of excessive field intensity. This method is useful in the inspection of welds.

INTERNAL CONDUCTOR. To establish circumferential magnetization, the part is threaded onto a conductor bar and magnetized by one or more dc pulses [60]. If the conductor is centered (central conductor method), simple ampere per unit diameter specifications can easily be found by the use of test flaws or a gaussmeter. In some applications for the magnetization of relatively short pieces, this is known as a *head shot*.

COIL METHODS. The current used in a coil to establish longitudinal magnetization depends on such factors as:

Part centering. If the part is centered in the coil and the coil is much larger than the part, MP indications will be produced for surface fields of 25–40 A/cm. If the part is hollow, higher OD surface fields may be needed to produce clear ID indications.

Wrapping turns. 4 × 0 welding cables are often used to wrap several turns around a part. Under these circumstances, the depth to penetration of the field is dependent upon the current excitation (i.e., long or short dc pulse), part dimensions (diameter, wall thickness), and the number of wraps used. With pulses, more than one are generally needed, so that the flux penetrates the material despite surface eddy currents caused in the part by the pulse.

Many specifications exist for the ampere turn product that is to be used prior to inspecting certain products. These depend upon factors such as part surface roughness, part prior magnetization, contrast between part and particles to be used, and use of wet or dry particles. The surface field can always be tested with portable flaws, which are strips of high permeability material that contain flaws of differing depths and widths [60, 61].

18-7.2 Forces on Particles

When a small piece of ferromagnetic material is passed through a MFL field, it is acted upon by the following forces [62]:

> leakage field forces
> excitation field forces, if present
> magnetostatic image forces
> interparticle interaction forces
> gravitational forces
> viscous and surface tension forces

Leakage field forces were discussed in Chaps. 16 and 17 and are much larger for active field magnetization than for residual for the same flaws. Once such fields are modeled, then it is relatively easy to find the force that it exerts on an object within its influence. Additionally, as the particle approaches the inspected surface, it is attracted to its image in the material under inspection. Such image-attraction forces are well known and their strength can be calculated. This is beyond the scope of this text.

When they are under the influence of such forces, the particles themselves become small magnets and so attract and repel one another. Such interparticle interactions are also well known and become more important as the density of the particles within the surrounding medium increases, since they tend to be short-ranged. They lead to particles sticking together in chains, which in itself raises the L/D ratio from that of one typical particle to that of the chain, which is lower, and thus the chain is easier to magnetize to a high B level. This, in turn, suggests that if a high B is to be achieved within the particles for a relatively low H (the MFL field), then they must be made from high permeability material.

Gravity provides the force $mg \cos \theta$ that acts to remove a particle from an MFL field in a situation as shown in Fig. 18-30. Other forces acting are those that

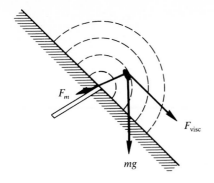

FIGURE 18-30
Magnetic, viscous, and gravitational forces on a magnetic particle in a flux leakage field.

are due to air currents, or the surface tension of the containing fluid, if the wet method is used. These are virtually impossible to assess. In this simple treatment, only the magnetic and gravitational forces are discussed.

In a MFL field, the particle becomes a magnetic dipole, with a magnetic moment $\bar{p}$ that is governed by its shape. The energy of this dipole in the external field is

$$E = -\mu_0 \bar{p} \cdot \overline{H}$$

and the force on the dipole is given by $\vec{F}_m = -\nabla E$. If the dipole moment is constant, then

$$\overline{F}_m = \mu_0 (\bar{p} \cdot \nabla) \overline{H}$$

but if the dipole moment is proportional to the ambient field, then the magnetic force is given by

$$F_m = \mu_0 \alpha V (\overline{H} \cdot \nabla) \overline{H} \tag{18-20}$$

where V is the volume of the particle and α depends on its shape through the demagnetization factor N_d. In Chap. 13, it was found that for spheres the magnetization is given roughly by $M \sim H_a/N_d$, and $N_d = 1/3$, so that $\alpha = 3$. For a particle with an L/D ratio of 2, α rises to 5.8. This shows the advantage of using elongated particles.

Equation (18-20) is as far as the discussion can be taken without assuming a field model for H. In the following paragraphs, the simple Förster model will be used for the MFL field of a tight crack and the linear dipole model will be used to represent the MFL field from a drilled hole in a test ring.

ISOLATED PARTICLE ABOVE CRACK. For a two-dimensional leakage field, Eq. (18-20) expands to

$$F_m = \alpha \mu_0 V \left[\left(\hat{i} H_x + \hat{j} H_y \right) \cdot \nabla \right] \left(\hat{i} H_x + \hat{j} H_y \right)$$

and, introducing the components of the vector operator,

$$\bar{F}_m = \alpha\mu_0 V\left[\left(\hat{i}H_x + \hat{j}H_y\right) \cdot \left(\hat{i}\,\partial/\partial x + \hat{j}\,\partial/\partial y\right)\right]\left(\hat{i}H_x + \hat{j}H_y\right)$$

$$= \alpha\mu_0 V\left[\hat{i}\left(H_x\,\partial H_x/\partial x + H_y\,\partial H_x/\partial y\right)\right.$$

$$\left. + \hat{j}\left(H_x\,\partial H_y/\partial x + H_y\,\partial H_y/\partial y\right)\right] \tag{18-21}$$

Use of the simple Förster relations

$$\left|\begin{matrix} H_x \\ H_y \end{matrix}\right| = \frac{H_g L_g}{\pi\left(x^2 + y^2\right)}\left|\begin{matrix} y \\ x \end{matrix}\right|$$

for the field close to, but not right in the mouth of, a deep crack, Eq. (18-21) becomes

$$\bar{F}_m = \alpha\mu_0 V\left(\frac{H_g L_g}{\pi}\right)^2\left[\hat{i}\frac{x^3 - 3xy^2}{\left(x^2 + y^2\right)^3} + \hat{j}\frac{y^3 - 3x^2 y}{\left(x^2 + y^2\right)^3}\right]$$

The magnitude of the force, found from $|\bar{F}| = (F_x^2 + F_y^2)^{1/2}$, is given by

$$|\bar{F}| = \frac{\alpha\mu_0 V}{r^3}\left(\frac{H_g L_g}{\pi}\right)^2 \tag{18-22}$$

where $r = (x^2 + y^2)^{1/2}$. That is, the force on an isolated particle in this field model is proportional to:

1. The square of the magnetic field intensity deep within the flaw H_g. This is governed by applied field and flaw dimensions, in the case of active field excitation, and the flux density within the sample and flaw dimensions, in the case of residual field excitation. In the latter case, the material should be above $0.7B_r$ so as to maximize H_g.
2. The square of the width of the flaw. This is true for small flaw widths. It should be pointed out, however, that for real flaws in residual induction, the product $(H_g L_g)$ diminishes with increasing flaw width (Fig. 15-3), so that the powder-holding ability of flaws also diminishes with increasing flaw width. This explains why tight cracks hold powder in residual induction, while wider test flaws do not, except perhaps at their corners.
3. The inverse cube of the distance from the mouth of the flaw. This is because both the field strength and its gradient diminish with distance. Such rapid fall off with distance is the reason why particles concentrate over flaws to give a sharp indication, as shown in Fig. 18-31.

LINEAR DIPOLE MODEL. A standard technique for testing the ability of a combination of the magnetization level and particle quality to provide indications from subsurface flaws is to require that a minimum number of cylindrical holes in a test ring give clear MP indications [62]. Test rings such as that shown in Fig. 18-32 are sometimes used; this particular ring is made from AISI 01 Ketos steel,

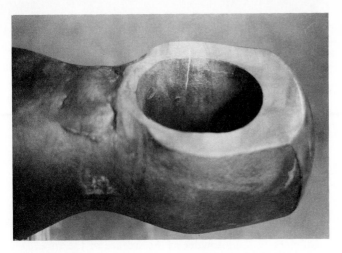

FIGURE 18-31
Crack in a railroad coupler seen in circumferential residual magnetization with dry particles.

with Rockwell B hardness between 90 and 95. It has a diameter of 127 mm, central hole diameter of 31.8 mm, thickness of 22 mm, and hole spacing of 19 mm. The holes are 1.78 mm diameter and located at $1.78N$ mm from the OD surface (N is the hole number). The detection of the minimum number of holes for specific central conductor currents and types of particles in accordance with military standard MIL-E-6868E is shown in Table 18-1. The contents of the table

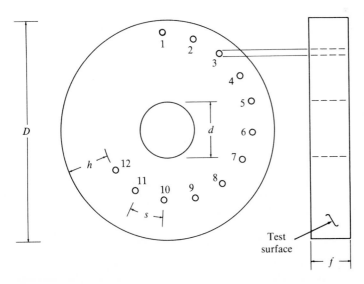

FIGURE 18-32
MIL-E-6868E test ring. (*Courtesy NTIAC, Southwest Research Institute, San Antonio, TX* [*41*]).

TABLE 18-1
Performance requirements for MIL-E-6868E

Type of particle	Central conductor current (dc A)	Minimum number of holes giving indications
Wet	1400	3
	2500	5
	3400	6
Dry	1400	4
	2500	6
	3400	7

indicate that at a certain magnetizing field level ($H_a \propto I$), the MFL field from a cylindrical hole must be sufficient to hold powder at the surface of the ring.

The tangential and normal MFL fields in air due to the presence of such holes are given by Eq. (17-5) as

$$\begin{vmatrix} H_x \\ H_y \end{vmatrix} = -\frac{m}{r^4} \begin{vmatrix} x^2 - y^2 \\ 2xy \end{vmatrix} \qquad (17\text{-}5)$$

where m is the dipole moment of the hole. As noted in Chap. 17, the effect of the surface is to effectively double the dipole moment over that of the infinite material case. Using the results of Chap. 17 and setting $(\mu_m - \mu_i)/(\mu_m + \mu_i) = 1$, the field equations become

$$\begin{vmatrix} H_x \\ H_y \end{vmatrix} = -\frac{2H_a a^2}{r^4} \begin{vmatrix} x^2 - y^2 \\ 2xy \end{vmatrix}$$

The quantities that are relevant to the computation of the force on an isolated particle at the surface of the ring are

$$\partial H_x/\partial x = -4H_a a^2 x(3y^2 - x^2)/r^6$$

$$\partial H_x/\partial y = -4H_a a^2 y(-3y^2 - y^2)/r^6$$

$$\partial H_y/\partial x = -2H_a a^2 y(-3x^2 - y^2)/r^6$$

$$\partial H_y/\partial y = -2H_a a^2 x(-3x^2 - y^2)/r^6$$

The x component of the force is proportional to

$$\left(H_x \, \partial H_x/\partial x + H_y \, \partial H_x/\partial y\right) = 8H_a^2 a^4 x(x^2 y^2 - x^4 - 2y^4)/r^{10}$$

and the y component of the force is proportional to

$$\left(H_x \, \partial H_y / Mx + H_y \, \partial H_x / \partial y \right) = 4H_a^2 a^4 y \left(2x^2 - y^2 \right) \left(-3x^3 + y^2 \right) / r^{10} \quad (18\text{-}23)$$

The point on the surface of the ring immediately over the hole has coordinates $(0, h)$, so it can be seen that

$$F_x = 0$$
$$F_y \propto 4H_a^2 a^4 / h^5 \quad (18\text{-}24)$$

From these relations, it can be seen that the force that binds magnetic particles to the leakage fields of elongated subsurface flaws is:

1. Proportional to the square of the applied field. In effect, the detectability of such subsurface flaws goes as the square of the applied field, or the current that causes it, in the active field. This is also true of the Förster approximation. (It should be noted, however, that if inspection is to be performed in residual induction following application of an active field, H_a need not be increased beyond that value that causes the material to saturate.)

2. Proportional to the fourth power of the radius of the flaw. This is a combination of effects caused by flaw size and the ability of particles to adhere in MFL fields that are not rapidly spatially changing,

3. Inversely proportional to the fifth power of the depth of the flaw. The indication here is that at relatively large depths, magnetic particles are not the best method to use for the detection of such flaws. Again, this is caused by a combination of the relatively spatially constant leakage fields and the ability of particles to be attracted to them. Obviously this is the result of the vector operator in Eq. (18-20).

18-7.3 Particle Size and Testing

Dry powder contains particles of all sizes up to that of a 100 mesh (ASTM) screen. Typically 75 wt% should be finer than 120 mesh and 15 wt% should be less than 325 mesh. Fillers used for contrast should be nontoxic and nonirritant.

Bulk permeability can be measured as shown in Fig. 11-16 or 13-10. A typical requirement is that a packed test tube of dimension 15×2.5 mm, when subject to an alternating field of strength 11,100 A/m (140 Oe) generates a peak-to-peak voltage of at least 2.5 V across the leads of a 50-turn coil wrapped around the tube.

Wet particles are generally ferrites and range in size from 1–600×10^{-4} mm. They do not have the jagged proturberances that dry particles have, so their L/D ratios are generally lower. Their concentration in a low viscosity fluid, e.g., kerosene, varsol or other light oil, or water that has been treated with anticorrosion, antifoaming, and wetting agents, should be 1–2 ml/100 ml for nonfluorescent or 0.1–0.4 ml/100 ml for fluorescent particles. Less noise, i.e., particles sticking to small MFL fields that are due to surface roughness rather than flaws,

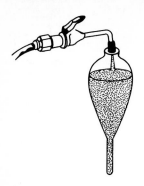

Demagnetization

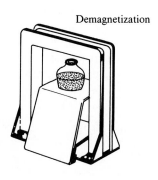

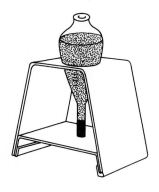

FIGURE 18-33
Wet particle concentration and demagnetization check.
(*Reprinted with permission from 'Nondestructive Testing, Magnetic Particle: Classroom Training Handbook' by The American Society for Nondestructive Testing, Inc., Columbus, Ohio; copyright 1977, General Dynamics, Convair Division, San Diego, California.*)

occurs at the lower concentrations. The transport fluid should be nontoxic and nonvolatile, and agitation is generally required [58, 59].

A centrifuge tube (Fig. 18-33) is used to check particle concentration. Since the particles will not settle into the lower graduated part of the tube if they are magnetized, it is often necessary to perform ac demagnetization; 60 Hz can be used since the skin depth at this frequency is larger than the particle size.

18-7.4 Use of Ultraviolet Light

One of the most startling changes one sees is the difference in clarity of objects in sunlight and under sodium light. Under sodium lamps, which emit light in a narrow wave band in the yellow region of the visible spectrum, the edges of objects appear to be much sharper than they appear in daylight, which is a mixture of wavelengths. The same principle is used in MPI: It is easier to see small indications that hold a small number of particles if the ambient light conditions are adjusted to give the best response as far as the eye is concerned.

With fluorescent MPI, a proprietary dye is added to the particles that absorb ultraviolet light and remit some of the energy in the yellow-green part of the visible spectrum, where the eye is most sensitive. The UV is provided by a mercury lamp and filter. The mercury spectrum consists of lines at 312 (UV), 365 (UV), 405 (violet), 435 (violet), 546 (green), 577 (yellow), and 626 and 670 (red) nm. The filter passes only the UV and some of the 405 nm violet, so that there is sufficient visible light present to permit movement. The mercury lamp spectrum and filter passband are shown in Fig. 18-34(a). Most of the energy falling on the inspected surface is at 365 nm, the near ultraviolet (or black) light region. It is not considered harmful.

The intensity at the inspected surface should be at least 800 $\mu W/cm^2$ and is easily measured. Since the output of mercury lamps decreases with age, the inspected surface intensity should be checked each day. Lamps are often left on because they wear out faster when continually turned on and off.

The emission spectrum of a typical dye is shown in Fig. 18-34(a). Such emissions can easily be photographed.

18-7.5 Specialized Applications

Magnetic particles can be provided in rubber solution for the inspection of relatively inaccessible regions, e.g., bolt holes. Because of the high viscosity of the solution, sufficient time must be allowed for the particles to migrate to the leakage field before the solution is cured.

Wet particle inspection can also be performed with the solution in a reusable sachet. In this application, the thickness of the containing sachet contributes a known lift-off.

18-7.6 Magnetic Particle Surface Field Indicators

Since the gaussmeter is a relatively recent invention, it is not widely used in magnetic NDE outside of the research laboratory. The presence of adequate surface field is sometimes tested by using test flaws of known dimensions. Figure 18-35 shows schematically three types of devices, all made from high permeability steel and operating on the principle that the relatively uniform surface field can be transformed to a crack leakage field that will hold powder. The devices are (1) a raised cross [Fig. 18-35(a)] in which high μ quadrants are separated by a

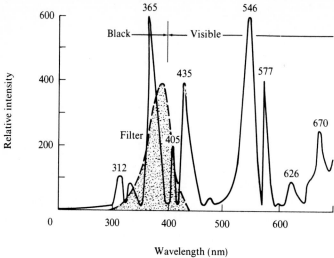

(*a*)

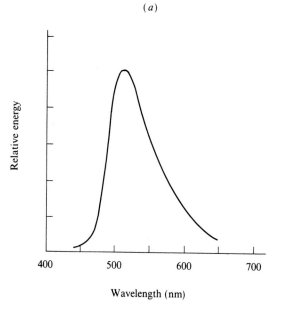

(*b*)

FIGURE 18-34
(*a*) Typical mercury lamp spectrum and filter passband. (*b*) Emission band of a wet fluorescent particle dye. (*Reproduced with permission of Magnaflux Corp.*)

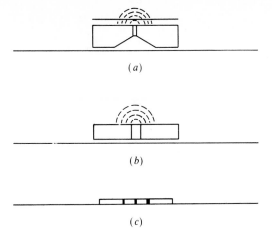

(a)

(b)

(c)

FIGURE 18-35
Portable field indicators. (a) Raised-cross type with nonmagnetic lid. (b) Pie-gauge type. (c) Burmah–Castrol strip. The widths of the respective slots (made from nonmagnetic materials) differ between the types. (© *1987, American Society for Nondestructive Testing. Reproduced with permission.*)

gap of 0.13 mm and capped with a nonmagnetic shield (which represents additional lift-off), (2) a flat pie gauge [Fig. 18-35(b)], which consists of six sectors that are separated by a gap of 0.75 mm, and (3) an encapsuled strip that contains slots of the same depth but differing widths.

In the active field, boundary conditions dictate that the field strength inside and on the outside surface of a part are equal, so that if an indication can be seen with the field indicator, then the same size flaw in the part also will yield a similar indication. In residual induction, the device is expected to share flux with the part. The closer the device to the part and the better the contact with the part, the more likely an indication will occur. Further, since the powder-holding ability of these "portable" flaws is proportional to $(H_g L_g)^2$, then the gap width L_g is a serious factor in determining the ability of the device to hold powder. While such devices may be used to indicate the general direction of surface fields, preferred devices are those that can place the simulated flaw as close as possible to the inspected surface. The Burmah–Castrol strip [Fig. 18-35(c)] and a similar device produced in Japan seem to meet this criterion well.

CHAPTER
19

DEMAGNETIZATION

19-1 INTRODUCTION

Demagnetization is the act of reducing the flux density level in an object to a level at which its external fields do not cause problems in subsequent use. It is required (1) if the part is to become a moving component of an assembly and might pick-up particles that would cause wear, (2) if the MFL field might interfere with cleaning of the part or affect nearby instruments, (3) if the part is to be magnetized to a lower level in another direction, and (4) if the MFL field would interfere with processes such as welding or machining [58]. Demagnetization to a specific external field strength level is often specified by procedural codes and can easily be measured with a gaussmeter. Typically no more than 8 A/cm (10 Oe) might be specified. In this chapter, commonly used forms of de- and remagnetization are outlined.

19-2 CIRCUMFERENTIAL REMAGNETIZATION

In some cases, after longitudinal magnetization for transverse flaw detection has been performed, it may only be necessary to remagnetize in the circular direction so that the resultant magnetization causes little or no external field. For tubes, this is easy to accomplish with the internal conductor method (Chap. 14). A gaussmeter held so as to check for longitudinal external field is all that is needed to check that the induction has been rotated.

For tubes that have been belled at their ends (i.e., upset) prior to the cutting of threads or the addition of tool joints, it is sometimes difficult to remove the longitudinal induction with one shot. Several shots are often needed.

344

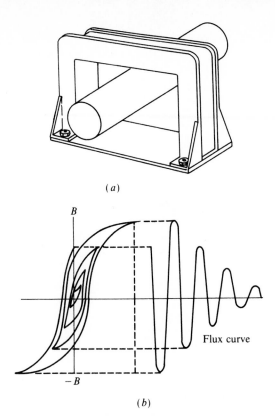

(a)

B

Flux curve

− *B*

(b)

FIGURE 19-1
(a) Alternating current coil demagnetization of an elongated part. (b) The *B-H*
loop for the outer region of the part.

19-3 ALTERNATING CURRENT DEMAGNETIZATION

For axially elongated products that have been magnetized longitudinally, passing
them through an ac coil, as shown in Fig. 19-1(a) will take the outer regions
through successively smaller *B-H* loops at the frequency of the coil [58]. The field
strength from the coil at the surface of the part must be strong enough to take the
material within the coil through an outer *B-H* curve. This material then experiences lower oscillating values of *H* as it is withdrawn and so the flux density falls
as shown in Fig. 19-1(b). Unfortunately, because of the skin effect, the ac field
causes eddy current shielding of the inner portion of the part (recall that one skin
depth for steel at 60 Hz is about 1 mm). The material emerges from the coil with
a demagnetized outer surface and, depending upon how thick it is, a high level of
longitudinal magnetization in its midregion.

19-4 DIRECT CURRENT COIL DEMAGNETIZATION

Partial demagnetization of elongated products can be accomplished by passing
them axially through a coil in which the field intensity opposes the direction of

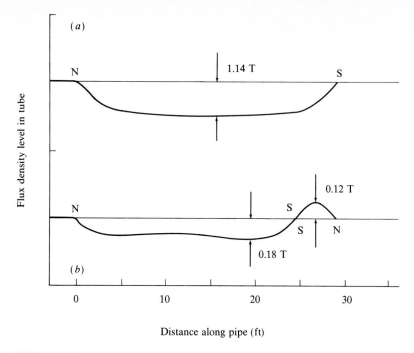

FIGURE 19-2
(*a*) Single dipole flux density field of an axially magnetized tube. (*b*) Double opposed dipole configuration of the internal field after dc demagnetization by balancing the external fields. (*Courtesy Gordon and Breach, 1985.*)

the residual induction [63]. Direct current demagnetization currents are often obtained by trial-and-error, since traditionally only the external fields at the ends of the part are available to the inspector. Presently accepted practice is to adjust the demagnetizing field so as to provide minimum air fields at either end of the part, but, as shown in Fig. 19-2(*b*), the part is left in an opposed double dipole configuration. Running a field indicator along the outside surface of the material will indicate the existence of two like poles close to either end and the double pole as shown. The example shown is that of a 10 m tube of material of remanence 1.14 T. Optimal dc demagnetization lowers the values of axial flux density in the material to 0.18 and 0.12 T in the two dipoles. Different values will be obtained with different lengths and thicknesses of parts.

The effect shown in Fig. 19-2(*b*) occurs because the *B-H* curve exhibited by the ends of the material (where the demagnetization coefficient is high) is different from that which is exhibited by the central region (where the demagnetizing field is lower and, in these cases, almost zero). In this situation a combination of dc followed by ac demagnetization would produce a low final induction in the material.

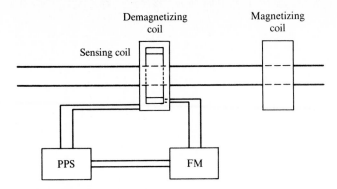

FIGURE 19-3
Scheme for optimizing the dc demagnetization of an elongated part. The voltage developed in a sensing coil as a magnetized material passes through it is fed to a fluxmeter (FM). The signal from the fluxmeter is proportional to the flux density within the part and is used to control the programmable power supply (PPS) of the demagnetizing coil.

19-5   FLUX-SENSED DEMAGNETIZATION

It is clear from the preceding text that traditional ac and dc coil methods do not totally demagnetize an object. For elongated products, Fig. 19-3 illustrates a method for sensing the flux density within a part with a fluxmeter (FM) and using the signal to control a programmable power supply (PPS) that attempts to keep the fluxmeter reading as low as possible. The combination of sense coil and fluxmeter is physically identical with Fig. 11-16.

19-6   DEMAGNETIZATION OF IRREGULARLY SHAPED PARTS

Once an irregularly shaped part has been magnetized, it is virtually impossible to demagnetize it. For industrial steels that have coercive forces that are much higher than the earth's field, it has been suggested that the bulk flux density may be lowered by suspending the material in the East–West direction and attempting to beat the object. This does not work. Surface demagnetization may produce external fields that meet code, but the part contains a high bulk induction that may reappear at the surface when the part is transported or placed close to other similar parts. Such localized demagnetization may be possible with coil wraps and be effective during subsequent working. The ends of pipe are often demagnetized in this manner prior to welding.

PROBLEMS FOR PART III

Chapter 10

10-1. What is the force between two poles each of strength 0.05 A-m separated by 10 cm? Is it attraction or repulsion? Assume $\mu_0 = 4\pi \times 10^{-7}$ H/m.

10-2. What is the force between a N pole of strength 0.05 A-m and a S pole of strength 0.02 A-m separated by 0.05 m?

10-3. Find the magnetic field intensity 10 cm from a pole of strength 0.03 A-m.

10-4. A pole of strength 0.03 A-m is located at $(-2, 0)$ and a pole of strength -0.04 A-m is located at $(2, 0)$. Calculate the magnetic field intensity (strength and direction) at $(0, 2)$, $(2, 2)$, and $(-2, 2)$.

10-5. Setting $y = 2d$ in Eq. (10-7), plot F_{1x}/μ_0 and F_{1y}/μ_0 for the interval $-5d < x < 5d$. Repeat at $y = 3d$ and $4d$. The shapes of the fields are those that a Hall element gaussmeter would see when scanning over the magnetic field of a bar magnet.

10-6. With some imagination, it can be seen that when a horizontal coil is passed through the field of a magnet at some constant lift-off $y = h$, the signal generated

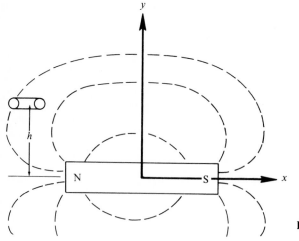

FIGURE P10-6

in it is roughly proportional to $[dF_{1y}/dx]\mu_0$. Compute and plot this quantity. What is the peak-to-peak separation of this signal? What is the peak-to-peak amplitude?

10-7. From Eq. (10-18), find the magnetic field components of the dipole in rectangular cartesian coordinates. What is the value of dH_y/dx at constant $y = h$?

10-8. Compute the magnetic field intensity 2 cm from a long straight wire that carries a current of 10 A. Give the result in amperes per meter, amperes per centimeter, and oersteds.

10-9. Compute the magnetic field intensity at the center of a one-turn square loop of side length L that carries a current of I A.

10-10. What is the magnetic field intensity at the center of a 300-turn coil of radius 30 cm that carries a current of 10 A?

10-11. Calculate and plot the magnetic field intensity along the axis of the coil of Problem 10-10.

10-12. Compute dH_P/dz and d^2H_P/dz^2 for a Helmholtz coil pair and show that at $z = b$ (Fig. 10-7) the first derivative vanishes. Show also that the second derivative vanishes when $a = 2b$.

10-13. Deduce Eq. (10-32).

10-14. Compute the field intensity at the center of a solenoid of length = 20 cm and diameter = 3 cm. What is the field at either end?

10-15. Show that the magnetic field intensity on the axis at P for a coil of n_1 turns per meter and n_2 layers per meter is given by

$$H_P = n_1 n_2 Il \left[\sinh^{-1}(y_2/l) - \sinh^{-1}(y_1/l) \right]$$

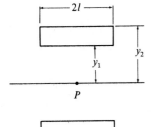

FIGURE P10-15

Chapter 11

11-1. A 1000-turn coil of cross-sectional area 0.5 m² is placed on a horizontal bench top. If the earth's vertical field component is 24 A/m, what is the vertical flux through the coil? The coil is turned over in 0.5 s. What is the flux change through the coil due to the earth's vertical field component? What is the EMF induced between the terminals of the coil?

11-2. A 100-turn coil has a cross-sectional area of 2 cm². A magnetic field of 100 Oe, which is perpendicular to the plane of the coil, decays to zero.
(a) What is the magnetic flux in the coil before decay?
(b) What is the magnetic flux linkage with the coil before decay?
(c) What is the EMF induced in the coil if the flux decays in 0.01 s?

11-3. The flux density through a 4 cm² 200-turn coil, which is wound on a ferrite ring, changes from 4000 G in one direction to 4000 G in the other direction.
(*a*) What is the flux change through the coil?
(*b*) What is the change in flux linkage ($\Delta N \Phi$) with the coil?
(*c*) What is the EMF induced in the coil if the flux change occurs in $1/50$ s?
The permeability of ferrite is 3000 at these values of *B*.

11-4. You are given a 5 in OD ring that is cut to be 1 in wide and has a wall thickness of 0.362 in. The ring is cut from a sample of pipe and you wish to measure the *B-H* properties of the material of the ring. Given an integrator that has a constant of $K = 1$ and a power supply that gives 0.1 Hz ac current up to 5 A (peak), design such an experiment. How many turns of wire do you need on the *H* coil to give ± 150 Oe in the ring? How many turns do you need on the *B* coil? Discuss.

Chapter 12

12-1. A coaxial cable consists of an inner conductor of radius R_1 and an outer conductor of radius R_2 as shown in Fig. P12-1. Assuming these conductors carry equal and opposite currents *I*, find:
(*a*) *B* and *H* at points between R_1 and R_2
(*b*) *B* and *H* at points outside the cable

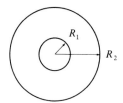

FIGURE P12-1

12-2. A scientist wishes to measure the *B-H* properties of a piece of ferromagnetic steel tubing. He cuts a ring sample from the tube and has it machined so that the inner and outer radii are R_1 and R_2, respectively, as shown in Fig. P12-2. He winds a pickup coil of N' turns on the ring and then a magnetizing coil of *N* turns. Compute *H* due to these windings:
(*a*) just inside the metal at $R = R_1$
(*b*) just inside the metal at $R = R_2$
(*c*) at any point in the metal with $R_1 < R < R_2$
[*Hint:* Use Ampere's Law.]
What is the average value of *H* in the metal in terms of R_1 and R_2? What can you

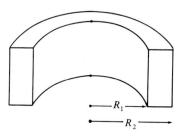

FIGURE P12-2

say about the thickness of the ring, in relation to the average radius, in order to ensure that the values of H at the inner and outer surfaces are within 2% of each other?

12-3. The Rowland ring experiment is generally performed with a toroid rather than with a "rectangular" ring (see Problem 12-2). Repeat Problem 12-2 for the field strengths at any point with the magnetizing coil winding. Is there any difference between the two cases?

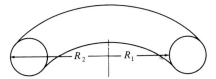

FIGURE P12-3

12-4. A permanent magnet is a right circular cylinder of length L. If the magnetization is uniform along the axis of the cylinder, what are the magnetization current densities J_M inside and on the surfaces, respectively?

12-5. A thin ferrite magnet of cross section A extends along the x axis from 0 to $x = L$. The magnetization is given by $\overline{M} = (M_x, 0, 0) = ax^2 + b$, where a and b are constants. Find the volume pole density $\rho_m(\bar{r})$ and the surface pole density $\sigma_m(\bar{r})$.

12-6. A cylindrical magnet (length L, radius R) of uniform magnetization lies on the z axis, with the coordinate origin at the center of the magnet. Determine the magnetic scalar potential $\Omega(\bar{r})$ on the z axis both inside and outside the magnet. Use the results to determine B_z on the axis.

12-7. A sphere of magnetic material (radius R) is placed at the origin of coordinates. The magnetization is given by $\overline{M} = (ax^2 + b)\bar{i}$. Determine all pole densities and magnetization currents.

12-8. Digitize the B-H curve of Fig. 11-17(a) and use a standard differentiation routine to produce the variation of dB/dH versus H that is shown in Fig. 11-17(b). Repeat the procedure for the remanent magnetism. How do the two curves compare?

12-9. What is the magnetic field intensity (H) at the surface of a copper wire of 1 mm diameter carrying a direct current of 10 A?

12-10. What are the magnetic field strength and flux density at the surface of a 1 mm diameter iron wire?

12-11. What dc current is required to produce a field strength of 30 Oe at the surface of a 7.625 in OD tube? Discuss how the length of the tube and its wall thickness might affect the power supply?

12-12. The formula for the field intensity, measured from the surface of a cylindrical bar carrying alternating current, is given by

$$H = H_0 \exp\left(-d\sqrt{\pi f \mu \sigma}\right)\sin\left(2\pi f t - d\sqrt{\pi f \mu \sigma}\right)$$

where d = the depth
f = the ac frequency
μ = the magnetic permeability
σ = the electrical conductivity

For a 2 cm diameter steel bar of average relative permeability 500, with a 30 Oe surface field strength, compute the variation of H with depth into the bar. Using the B-H curve of Fig. 11-17(a), plot the variation of B in the bar ($f = 60$ Hz; obtain σ from tables).

Chapter 13

13-1. An ellipsoid with principal axes of lengths ($2a$, $2b$) is magnetized uniformly in a direction parallel to the $2b$ axis. The magnetization is M_0. Find all pole densities.

13-2. A slit soft iron ring consists of 20 cm of iron, of cross-sectional area 4 cm^2, and a 1 cm air gap. A coil is wrapped around the iron to provide 2000 ampere turns.
(a) Calculate the field intensity in the air gap.
(b) Calculate B and H in the iron. What is the demagnetizing field strength? ($\mu = 3000\mu_0$.)

13-3. In Fig. P13-3 all cross-sectional area sections are 6 cm^2 and the permeability is $5000\mu_0$. Find the flux and its density through the central and outer legs. (*Adapted from ref.* [5], *with permission.*)

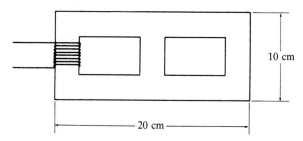

     10 cm

     — 20 cm —

FIGURE P13-3

13-4. The circuit of Fig. 13-3 has a permanent magnet 8 cm long, a soft iron path of 16 cm, and a cross-sectional area of 4 cm^2; the air gap is 0.8 cm long and has a cross-sectional area of 3 cm^2. With $\mu = 5000\mu_0$ for the soft iron:
(a) Find B in the gap for the Alnico 5 magnet steel of Fig. 11-19(b).
(b) Repeat the calculation for $L_g = 0.8$ mm.

13-5. Figure P13-5 shows a permanent magnet yoke system for providing a relatively uniform field in the gap(g). The yoke (y) is soft iron and there is some leakage (l). Considering the permanent magnet (PM) (p) and yoke (y) to be in series and the gap (g) and leakage (l) to be in parallel, compute the reluctance R. Show that $\phi_g \approx [R_l/(R_l + R_g)]\phi_p$ if R_y and R_m are small in comparison to R_l and R_g. Show also that

$$H_g B_g = \frac{R_l}{R_l + R_g} \frac{l_p A_p}{l_g A_g} H_p B_p$$

if there is negligible flux spreading out of the gap [8]. (*Adapted with permission of McGraw-Hill from ref.* [8], © *1938.*)

13-6. According to Johnson [16], the flux loop for a down-hole casing inspection system is given by Eq. (13-19). For 4.5 in OD casing, determine the flux loop for four casing wall thicknesses, 0.204, 0.225, 0.250, and 0.290 in. Keep the pole piece diameter constant, taking $t_g = 1$ cm at 0.25 in wall thickness. Putting $L_p = L_c = 20$ cm.
(a) Find the flux loop reluctances for 6000 G in the casing wall.
(b) What factors affect the flux loop the most?

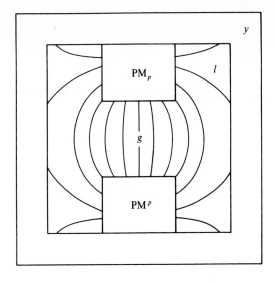

FIGURE P13-5

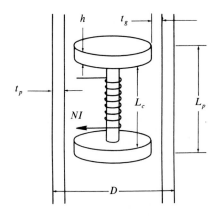

FIGURE P13-6

(c) Is the air gap term of the same order of magnitude as the steel terms?

(d) What is the effect of the L/D ratio of the core and the casing on the values of μ that must be used?

Johnson recommends using a core of low-μ material so as to minimize the effects of the variations in t_g and t_p. Do the results of your calculations substantiate this?

13-7. Drill pipe and tubing sometimes get stuck down hole. One way to determine the stuck point is to lower and clamp to the inside of the tube, a tool that contains two C cores. Tension on the pipe will then give an inductance change, which is due to a variation in the flux loop of the C cores with air gap, only if the tool is clamped above the stuck point. If each C core has length L_c, cross-sectional area A, and relative permeability μ_c:

(a) What is the equation of the flux loop?

(b) Using this equation, calculate the inductance of the system from $L = N\phi/I$.

(c) How does the inductance change for small variations in air gap L_g, i.e., find

dL/L in terms of dL_g/L_g and $d\mu_c/\mu_c$. (Remember that the permeability of the cores also varies with L_g.)

The inductance and its changes are measured with an LC tank circuit that oscillates at $\omega_0^2 = (1/LC)$, with C fixed.

(d) What is the relation between small changes in frequency ω_0 and changes in L, i.e., express $d\omega_0/\omega_0$ in terms of dL/L.

(e) Using the results of (c), how does ω_0 change as a function of dL_g and $d\mu_c$?

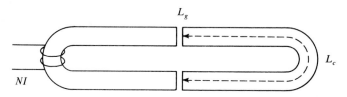

L_g

L_c

NI

FIGURE P13-7

13-8. ^{13}Cr steel heat exchanger tubing is difficult to inspect with conventional eddy current techniques because it is ferromagnetic. Design a permanent magnet flux loop for the longitudinal magnetization of tubing that has 2.54 cm OD and 0.21 cm wall thickness. Aim to place 1 T into the tube wall. How big are the pole pieces? What is the diameter of the central core? How big is the air gap? What kind of magnet material did you select and why?

13-9. A 1 mm gap is cut into a steel ring of diameter 127 mm. The magnetic field in the gap is measured as 1.5 T. What is the surface pole density on either face?

13-10. The ring of Problem 13-9 has applied to it an MMF of 500 ampere turns. What is the field strength deep inside the gap, if the permeability is (a) 200 μ_0; (b) 400μ_0? What are the values of the demagnetizing field?

13-11. Common wall thicknesses for 127 mm diameter well casing are 6.43, 7.65, 9.20, and 11.10 mm. Design a flux loop that will place a flux density of 1 T in the thickness of these materials, taking the permeability of all of the steel in the circuit as 50μ_0 and the distance between the pole pieces as about 40 cm. What flux density does this system induce in the thinnest walled material? Would it be a good idea to add ring-shaped shims to the pole pieces to cover this range of wall thicknesses?

13-12. A magnetizing force of $H_a = 200$ Oe (160 A/cm) is applied to cylinders for which the demagnetizing factors are listed in Table P13-12. If the materials have a permeability of 300μ_0, find the apparent permeabilities for the L/D ratios listed.

Chapter 14

14-1. For a hollow tube of inner and outer radii R_i and R_2, show that the magnetic field intensity created at points within the tube material due to flow of direct current along the tube is given by

$$H = \frac{I}{2\pi r} \frac{R^2 - R_i^2}{R_0^2 - R_i^2} \text{ A/m}$$

TABLE 13-12
Demagnetizing factors

Dimensional ratio L/D	$10^6 N$	
	Ellipsoid of revolution	Cylinder
5	55818	54107
10	20283	20290
15	10742	11140
20	6747	7146
25	4671	4997
30	3437	3660
40	2117	2180
50	1440	1456
60	1050	1042
70	804	788
80	637	621
90	527	502
100	430	414
150	207	198
200	127	119
300	64	64

where r = the radial distance from the center of the tube
R_i = the inner wall radius
R_0 = the outer wall radius

14-2. Produce a table of values for common pipe outer diameters, which ensures that 4000 A/m are caused at the outer surface by both the current along the part and central conductor methods, carrying direct current.

14-3. The values in Table P14-3 have been given for the magnetization of either solid bars, or tubulars, by either a centralized conductor carrying a direct current or current along the part. Investigate the values given in the table for the two general cases.

(*a*) For solid parts:
(i) Expand the table so as to show the range of field intensities at the surface.
(ii) At what distance below the surface is the field intensity equal to 3200 A/m?

(*b*) For the central conductor case:
(i) What ranges of field strength occur at the surfaces of the tubes for pure dc, full wave rectified ac, and half-wave rectified ac?
(ii) Discuss what differences may occur for the three excitation modes.

14-4. A direct current of 5000 A is applied to a length of tube stock of outer diameter 30.5 cm and wall thickness 2.0 cm, as shown in Fig. 14-1. Use the *B-H* curve of Fig. 11-17 to show the flux density variation within the tube wall.

14-5. For a current pulse system, assuming constant inductance, deduce the values of the maximum current ($I_{c,max}$) and the time (T_{max}) at which this occurs for the sin, sinh, and critically damped solutions to Eq. (14-2).

TABLE 14-3
Magnetizing current for circular
magnetization of solid and tubular articles

Greatest width or diameter centimeters	Magnetizing current (approx.) (A)
1.0	280–480
1.3	350–600
1.9	525–900
2.0	560–960
2.5	700–1200
3.0	840–1440
3.8	1050–1800
4.0	1120–1920
5.0	1400–2400
6.0	1680–2880
6.3	1750–3000
7.0	1960–3360
7.6	2100–3600
8.0	2240–3840
8.9	2450–4200
9.0	2520–4320
10.0	2800–4800

14-6. For the magnetizing system of Problem 14-5, derive a transcendental relation connecting T_{max} with L, C, and R. For the critically damped case, how do $T_{c, max}$ and T_{max} vary with the values of L, C, and R in the circuit?

14-7. What are the pulse widths (determined by the time from the start of the pulse to the time at which it falls to $0.5 I_{max}$) for the sin, sinh, and critically damped solutions to Eq. (14-2)?

14-8. A pulse magnetization system consists of C farads and resistance R, which is made up to 100 ft of 0000 welding cable in series with a 50 ft Al rod of diameter = 1.5 in, wall thickness = 0.25 in, and SCR of internal resistance 0.01 Ω, and contacts of the same resistance. If the rod and cables are laid on the floor, what inductance do they present? If the cables are coiled into a 2 ft diameter loop, what additional inductance appears in the circuit? What are the values of $I_{c, max}$ for these values of inductance for $C = 1$, 2, and 8 F and charging voltages of 50, 75, and 100 V? What are the values of T_{max} for these values?

14-9. Write a computer program that will allow you to vary C, R, and L by 100%, and so enable you to see the variation of T_{max} and I_{max}.

14-10. Compute the self-inductances of 10 m samples of tubes as follows:
(*a*) OD = 60 mm, wall thickness = 4.9 mm, average $dB/dH = 600\mu_0$.
(*b*) OD = 60 mm, wall thickness = 4.9 mm, average $dB/dH = 120\mu_0$.
(*c*) OD = 273 mm, wall thickness = 12.7 mm, average $dB/dH = 150\mu_0$.
Use these values to compute the wave shapes for CD systems with $C = 4$ F and $R = 4$ mΩ. What are $I_{c, max}$ and the pulse duration in each case?

14-11. A sleeve that is 10 cm long and has a wall thickness of 1 cm is threaded onto a CD rod. 10 turns of wire are wrapped around it, as shown in the figure, and connected to a fluxmeter. After the pulse is fired the fluxmeter finally reads 5 mWb. If the remanence value of the material is 1.2 T (12 kG), what can you conclude about the magnetization level of the sleeve?

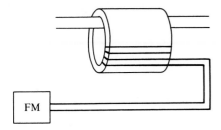

FM

FIGURE P14-11

14-12. 5 m long samples of cylindrical construction steel are to be inspected for longitudinal flaws prior to being welded into the superstructure of an offshore oil rig. The samples are 46 cm in diameter and have a wall thickness of 1.5 cm.
(*a*) What methods of inspection might be available to you?
(*b*) Which method would you choose and why?
(*c*) If, as a level III, you selected a magnetic method, how would you write the inspection procedure?
Be prepared to discuss your answers in class.

Chapter 15

15-1. The normal component of the flux leakage field of Fig. 15-6 can be represented by $B = B_n \times \exp(-a|x|)$. Determine roughly B_n and a from the figure. A detector passes through this field at 15 cm per second. By performing a Fourier transform of the flux leakage signal, determine the largest spectral component. (In this way, a filter in the sensor circuit will discriminate such broad-based noise from tight crack signals.)

15-2. For a 5 in OD tube of material of remanence 1 T (10 kG) and eccentricity 0.005 in, what is the maximum external tangential leakage field intensity.

Chapter 16

16-1. Show that the asymptotic expressions to $B_x = (B_0/\pi)[\tan^{-1}(N + x)/y + \tan^{-1}(N - x)/y]$ and $B_y = (B_0/2\pi)\ln[(N - x)^2 + y^2]/[(N + x)^2 + y^2]$, i.e., the Karlqvist equations in their original form, are

$$B_x = -B_0\left(2Nb/\pi x^2\right)\left[1 - (b/x)^2(1 - N^2/b^2) + \cdots\right] \rightarrow -2bNB_0/\pi x^2 + \cdots$$

$$B_y = B_0\left(2N/\pi x\right)\left[1 - (b/x)^2(1 - N^2/3b^2) + \cdots\right] \rightarrow 2NB_0/\pi x + \cdots$$

where $2N = L_g$.

16-2. The field intensity in the vicinity of a dipole of charge $\pm m$, separated by distance

$2b$, is

$$H_x = m\left[\frac{x + b}{\left[(x + b)^2 + y^2\right]^{3/2}} - \frac{x - b}{\left[(x - b)^2 + y^2\right]^{3/2}}\right]$$

$$H_y = m\left[\frac{y}{\left[(x + b)^2 + y^2\right]^{3/2}} - \frac{y}{\left[(x - b)^2 + y^2\right]^{3/2}}\right]$$

Plot these fields for various values of y. Show that:

(a) the maximum value of H_x is given by $H_{x,\max} = (2mb)/(y^2 + b^2)^{3/2}$

(b) if the point of observation is fixed and $2b$ is varied, the maximum value of $H_{x,\max}$ is $4m/3\sqrt{3}\,y^2$

(c) at large values of y, so that $y \gg b$, $h_{x,\max} = (2mb)/y^3$

(d) the values of x for which $H_x = 0$ are given by $x^6 + p_1 x^4 + p_2 x^2 + p_3 = 0$, where $p_1 = -(1/2)\,(2b^2 - 3y^2)$, $p_2 = (-b^4 + 3b^2 y^2)$, and $p_3 = -(1/2)(y^6 - 2b^6 - 3b^4 y^2)$

(e) for $y \gg b$, the distance between the points of intersection of H_x with the x axis is $2y/\sqrt{2}$

(f) from the equation $\partial H_y/\partial x = 0$, the ordinates of the points at which H_y reaches extreme are given by $(x + b)^2[(x - b) + y^2]^5 = (x - b)^2[(x + b)^2 + y^2]^5$

16-3. A positively charged, infinitely long filament creates a magnetic field of $H = 2\sigma_1 r/r^3$, where σ_1 is the linear charge density. Two such parallel filaments, separated by a distance $2b$, constitute a linear dipole, and the magnetic field intensity is given by

$$H_x = -4\sigma_1 b(x^2 - y^2 - b^2)\big/\left[(x + b)^2 + y^2\right]\left[(x - b)^2 + y^2\right]$$

$$H_y = -8\sigma_1 bxy\big/\left[(x + b)^2 + y^2\right]\left[(x - b)^2 + y^2\right]$$

Plot these field components for various values of y. Show that

(a) $H_{x,\max} = 4\sigma_1 b/(y^2 + b^2)$

(b) the maximum value of $H_{x,\max}$ is $2\sigma_1/y$ at $y = b$

(c) $H_x = 0$ at $x_{1,2} = \pm(y^2 + b^2)^{1/2}$

(d) H_x has extrema (i.e., $\partial H_x/\partial x = 0$) at $x_{3,4} = \pm[y^2 + b^2 + 2y(y^2 + b^2)^{1/2}]^{1/2}$. Plot how x_3 or x_4 vary with increased lift-off

(e) H_y has extrema at $x_{5,6} = \pm[b^2 - y^2 + 2(b^4 + b^2 y^2 + y^4)^{1/2}]^{1/2}$

16-4. A special case of Problem 16-3 arises when $y \gg b$. Show that, under this condition,

(a) $H_{x,\max} = 4b\sigma_1/y^2$

(b) $x_{1,2} = \pm y$

(c) $x_{3,4} = \pm y\sqrt{3}$

(d) $x_{5,6} = \pm y\sqrt{3}$

(e) H_y at $x_{5,6} = \pm(3\sqrt{3}/2)(\sigma_1 b/y^2)$

16-5. For a pipe of average diameter D, with a through slot of width L_g parallel to its long axis, show that the charge on both sides of the slot, for circumferential magnetization, is given by

$$\sigma = \pm \frac{H_a}{2\pi} \left[\frac{1 + L_g/(\pi D - L_g)}{1/\mu + L_g/(\pi D - L_g)} \right]$$

where μ = the permeability of the pipe material
H_a = the applied magnetic field intensity

16-6. For a slot of depth Y_0 and width L_g in a steel plate magnetized as shown in Fig. P16-6, Shiraiwa and Hiroshima have shown that

$$H_x = \sigma \left[\tan^{-1} \frac{y + Y_0}{x - L_g/2} - \tan^{-1} \frac{y}{x - L_g/2} \right.$$

$$\left. - \tan^{-1} \frac{y + Y_0}{x + L_g/2} + \tan^{-1} \frac{y}{x + L_g/2} \right]$$

for the x component of the MFL field above the slot. H_y is as given in Eq. (16-13). Compare this form for H_x with that given in Eq. (16-13) by selecting several values of Y_0 and y, and plotting the leakage field by computer.

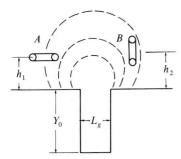

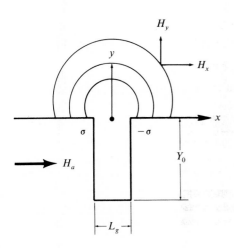

FIGURE P16-6

16-7. A small flat coil rides at a lift-off of h above a magnetized plate. If the coil is very narrow, its output can be considered to be proportional to the first derivative of the normal component of the MFL field. Using Förster's equation for H_y, find $\partial H_y / \partial x$ and graph the resulting function. The peak of this signal occurs at $x = 0$.

FIGURE P16-7

What is the lift-off characteristic for the peak of the signal, i.e., how does the coil signal vary with h? Repeat the calculation for a coil for which the distance between the leading and trailing edges is $2b$. (The arithmetic will be more complicated than above and you may wish to set this up on a computer.)

16-8. A small vertical coil rides at an average lift-off of h_2 over the same slot as in Problem 16-7. If the coil is very narrow, its output can be considered to be proportional to the first derivative of the tangential component H_x of the MFL field. Using Förster's equation, find $\partial H_x / \partial x$ and graph the resulting function. Where are the peaks of this function? What is the peak-to-peak amplitude? What is the interpeak distance along the x axis? What is the lift-off characteristic? Repeat for a coil of width $2b$.

16-9. Write out the equations for H_{2x} and H_{2y} for the negatively charged face 2 of the oblique slot (Fig. 16-16). Show that when $\gamma = 0$, the equations that are derived from Eq. (16-17) reduce to Eq. (16-13).

16-10. Use the Förster through slit theory to show that the peak-to-peak change ΔH_y in the MFL field is given by $H_g L_g / \pi h$, where h is the lift-off.

16-11. What does the Förster finite-depth slot theory predict for the peak-to-peak change in the normal component of the MFL field? (*Hint:* Solve $\partial H_y / \partial x = 0$ so as to obtain the x values at which H_y has its turning points, then program a computer to solve this equation for constant lift-off but variable slot depth, and return the result to the equation for H_y.)

16-12. For a deep crack, show that, in the Zatsepin–Shcherbinin model, if $x \gg \frac{1}{2} L_g$, then
(a) $H_x = 4\sigma_s b / y$ for $x \ll y$
(b) $H_x = 4\sigma_s by / x^2$ for $x \gg y$
(c) $H_y = 4\sigma_s bx / (x^2 + y^2)$
Further, show that H_y has extreme at $x = \pm (y^2 + b^2)^{1/2}$.

Chapter 17

17-1. With a Hall element, measure the peak-to-peak MFL fields over the holes in a Ketos steel test ring at various levels of magnetic field intensity from a central conductor.

17-2. Using Eq. (17-5), determine the EMF developed in a straight wire that cuts the normal MFL field component at height h parallel to the x direction. The equation for the EMF is given by $E = \mu_0 \int H_y v \, dx$.

17-3. Using Eq. (17-9) and the equation $E = \mu_0 \int \int H_y v \, dx \, dz$, calculate the EMF ($E$) induced in a straight wire traveling at speed v parallel to Ox through the leakage field from a subsurface inclusion.

17-4. From Problem 17-3, determine the EMF induced in a flat coil of N turns and spacing $2b$. Sketch the shape of the signal. What is the maximum EMF developed in the coil?

Chapter 19

19-1. A railway engine travels northward at 25 m/s. If the vertical component of the earth's magnetic field is 0.3 Oe, what is the EMF induced between the ends of an axle of length 4 ft 8.5 in? (Convert to SI units.)

19-2. A coil with its axis set along the horizontal component H_0 of the earth's field carries a current that is adjusted so that the resultant flux density at its center is zero. Carrying the same current, the coil is rotated by 60°. What is the resultant magnetic field intensity at its center? (*Source:* Cambridge University Overseas Higher School Certificate, 1969.)

19-3. The core of a 1000-turn coil has a cross-sectional area of 2 cm^2 and the magnetic induction in it is 0.5 Wb/m^2 (5000 G). Calculate the EMF induced in the coil when the flux is reversed in 0.01 s. (*Source:* Joint Matriculation Board GCE, 1969.)

19-4. A current of 10 A passes through a plane circular coil (radius = 10 cm; 100 turns). This current is reduced steadily to zero in 2.0 s. Find the EMF induced in a coplanar circular coil (16 turns; radius = 0.25 cm) at the center of the larger coil. What is its direction in relation to the current in the larger coil? (*Source:* Cambridge University East African Examination Council Advanced Level Certificate of Education, 1970.)

19-5. In a uniform field of flux density 10^{-3} Wb/m^2:

  (*a*) Find the flux linked with a coil of area 0.02 m^2 with 100 turns, set at right angles to the flux.
  (*b*) Find the flux linked with a coil of area 0.01 m^2 (100 turns), set with its plane at 60° to the flux. What is the field intensity in amperes per meter and oersted?

19-6. A coil has an area of 0.005 m^2 and 20 turns. Find the change in flux linkage when

  (*a*) the flux threading it changes from 5×10^{-5} to 3×10^{-5} Wb
  (*b*) the flux density at right angles to it is changed from 10^{-4} to 5×10^{-4} Wb/m^2
  (*c*) it is moved completely from a field of density 10^{-4} Wb/m^2 perpendicular to its plane
  (*d*) it is turned through 180° from its original position of (*c*)
  (*e*) it is turned through 60° from its original position of (*c*)

19-7. What is the distance between the minima of Eq. (9-15)? (Find $dE/dx = 0$.)

19-8. How does the peak value of the EMF of Eq. (9-15) vary with (*a*) b and (*b*) h? What is the optimal value of b in terms of h so as to maximize the signal?

19-9. One way to determine the degradation of wire rope is to integrate the positive part of a signal such as that shown in Fig. 18-19. Show that the area above the $E = 0$ axis from the MFL field from a broken strand is proportional to $Lv \ln[(\sqrt{1 + (h/b)^2} + 1)/(\sqrt{1 + (h/b)^2} - 1)]$.

19-10. Show that the output of an integrator that is connected to a toroidal air-cored coil of average radius R and winding radius a is given by $E_0 = K\mu_0 NI[R - \sqrt{R^2 - a^2}]$, where K is a constant for the integrator and N is the total number of turns on the coil.

19-11. A very small parallel coil with a turns configuration of $(N, -2N, N)$ rides at a lift-off of h from a steel surface that contains a crack that is magnetized perpendicular to its opening. Using the Förster and Zatsepin–Shcherbinin models, write a computer programmer that will show the form of the coil output as it passes through these fields.

19-12. A straight wire is passed through the leakage field from a subsurface inclusion. Show that, if the wire moves at speed v, the EMF developed between the ends of the wire is given by

$$E = \mu_0\left[(\mu_m - \mu_i)/(2\mu_m + \mu_i)\right] H_a a^3 v \left[4xh/(x^2 + h^2)^2\right]$$

19-13. Write a computer programme that will enable you to determine the signal developed in a flat coil of width $b = h$ as it passes through the MFL field of Problem 19-12.

REFERENCES
FOR
PART III

1. Betz, C., *Principles of Magnetic Particles Testing*, Magnaflux Corp., Chicago, IL, 1967.
2. Kaempffer, F. A., *The Elements of Physics. A New Approach*, Blaisdell, Waltham, MA, 1967.
3. Reitz, J. R., and Milford, F. J., *Foundations of Electromagnetic Theory*, 2nd ed., Addison-Wesley, Reading, MA, 1967.
4. Starling, S. G., and Woodall, A. J., *Physics*, 3d ed., English Language Book Society, UK, 1963.
5. Cullity, B. D., *Introduction to Magnetic Materials*, Addison-Wesley, Reading, MA, 1972.
6. Janicke, J. M., *Fluxmeter Handbook*, RFL Industries, Inc. Technical Literature, Boonton, NJ, 1978.
7. Bozorth, R. M., *Ferromagnetism*, Van Nostrand, New York, 1951.
8. Harnwell, G. P., *Principles of Electricity and Magnetism*, 2nd ed., McGraw-Hill, New York.
9. MIT Staff, *Magnetic Circuits and Transformers*, Wiley, New York, 1943.
10. Förster, F., "Nondestructive Inspection by the Method of Magnetic Leakage Fields. Theoretical and Experimental Foundations of the Detection of Surface Cracks of Finite and Infinite Depth." *Defektoskopiya*, vol. 11, pp. 3–25, 1982.
11. McClurg, G. O., "Theory and Application of Coil Magnetization," *Nondestructive Testing*, pp. 23–25, 1955.
12. Bergander, M. J., "Principles of Magnetic Defectoscopy of Steel Ropes," *Wire Journal*, vol. 11, no. 5, pp. 62–67, 1978.
13. Walters, W. T., and Nagel, D. D., US Patent 3,543,144, AMF, Inc., November 1970.
14. Beissner, R. E., Matzkanin, G. A., and Teller, C. M., "NDE Applications of Magnetic Leakage Field Methods, a State-of-the-Art Survey" NTIAC Report 80-1, Southwest Research Institute, January 1980.
15. Rogers, W. M., "New Methods for In-Place Inspection of Pipelines," Proceedings of the Conference on Mechanical Working and Steel Processing, Dolton, IL, 1974.
16. Johnson, W. M. Jr., US Patent 3,940,689, Schlumberger, February 1976.
17. Walters, W. T., and Reema, M. L., US Patent 3,209,243, AMF, Inc., September 1965.
18. Moyer, R. C., and Hintermaier, R. W., "Specifications for the Nondestructive Evaluation of API Oil Country Tubular Goods," specification, Exxon Co. USA, 1983.
19. Ryan, S. R., "Theory of Magnetic Reluctance Thickness Measurements in High Conductivity Ferromagnetic Materials," *NDT International*, June 1981.
20. "Flaw Detection in Rails," Report FRA/ORD/77-10, US Department of Transportation (Federal Railroad Administration) 1978.
21. Stanley, R. K., "The Production and Use of Residual Magnetic Field in Oilfield Tubular Inspection," *Proceedings of the ASNT National Conference*, 1982, pp. 340–344.
22. Moake, G. L., and Stanley, R. K., "Inspecting Oil Country Tubular Goods Using Capacitive Discharge Systems," *Materials Evaluation*, vol. 41, no. 7, pp. 779–782, 1983.

23. Zatsepin, N. N., Shcherbinin, V. E., and Pashagin, A. I., "Investigation of the Magnetic Field of a Defect on the Inner Surface of a Ferromagnetic Tube. 1. Basic Regularities and Mechanism of Formation of the Field of the Defect," *Defektoskopiya*, no. 2, 1968, pp. 8–16.

24. Moake, G. L., and Stanley, R. K., "Capacitor Discharge Magnetization of Oil Country Tubular Goods," in W. Lord (ed.), *Electromagnetic Methods of NDT*, Gordon and Breach, 1985, pp. 151–160.

25. Moyer, R. C., and Hintermaier, R. W., "Specification for the Nondestructive Evaluation of API Oilfield Tubular Goods," Exxon Co. USA Specification, 1st rev., 1984.

26. Kanbayashi, H. "Improved Flaw Detection Apparatus Using Specially Located Hall Elements," US Patent 3,579,099, May 1971.

27. Shiraiwa, T., Hiroshima, T., Hirota, T., and Sakamoto, T., "SAM Inspection Systems for Oil Country Tubular Goods," *Materials Evaluation*, vol. 35, no. 8, pp. 52–56, 1977.

28. See, for example, US Patent 3,582,771, AMF, Inc., June 1971 and US Patent 3,612,403, Förster Instruments, September 1978.

29. Stanley, R. K., "Basic Principles of Magnetic Flux Leakage Inspection Systems for the Evaluation of Oil Country Tubular Goods," in W. Lord (ed.), *Electromagnetic Methods of NDT*, Gordon and Breach, 1985, pp. 97–149.

30. Spierer, E. D., "Flux Density Measurement in Ferromagnetic Tubular Product," in W. Lord (ed.), *Electromagnetic Methods of NDT*, Gordon and Breach, 1985, pp. 161–173.

31. Heath, S., M.S. Thesis, Colorado State University, 1983, unpublished.

32. McIntire, P. and Mester, M. L. (eds.), *Nondestructive Testing Handbook*, vol. IV, *Electromagnetic Testing*, 2nd ed., 1986, sec. 21.

33. American Petroleum Institute Recommended Practice RP 7G.

34. Stanley, R. K., "Circumferential Magnetization of Tubes and the Measurement of Flux Density in Such Materials," *Materials Evaluation*, vol. 44, no. 7, pp. 966–970, 1986.

35. Uetake, I., private communication, 1987.

36. Förster, F., "Nondestructive Inspection by the Method of Magnetic Leakage Fields. Theoretical and Experimental Foundations of the Detection of Surface Cracks of Finite and Infinite Depth," *Soviet Journal of Nondestructive Testing*, vol. 3, no. 11, pp. 841–859, 1982.

37. Hwang, J. H., "Defect Characterization by Magnetic Leakage Fields," Ph.D. Dissertation, Colorado State University, 1975.

38. Karlqvist, O., "Calculation of the Magnetic Field in the Ferromagnetic Layer of a Magnetic Drum," *Transactions of the Royal Institute of Technology, Stockholm*, *vol. 86, no. 1*, 1954.

39. Zatsepin, N. N., and Shcherbinin, V. E., "Calculation of the Magnetostatic Field of Surface Defects. I. Field Topography of Defect Models," *Defektoskopiya*, no. 5, pp. 50–59, 1966.

40. Shcherbinin, V. E., and Zatsepin, N. N., "Calculation of the Magnetostatic Field of Surface Defects. II. Experimental Verification of the Principal Theoretical Relationships," *Defektoskopiya*, no. 5, pp. 59–65, 1966.

41. Swartzendruber, L. J., "Magnetic Leakage and Force Fields for Artificial Defects in Magnetic Particle Test Rings," *Proceedings of the XIIth Symposium on NDE*, San Antonio, Texas, April 1979.

42. Kittel, C., *Introduction to Solid State Physics*, 3d ed., Wiley, New York, 1968.

43. Hall element literature, F. W. Bell Company, Orlando, FL.

44. Microswitch literature, Honeywell Corp.

45. Proctor, N. B., "Magnetic Flux Sensors Having Core Structures of Generally Closed Configurations for Use in Nondestructive Testing," US Patent 3,529,236, AMF, Inc., September 1970.

46. Hatch, H. H., and Rosato, N., "The Magnetic Recording Borescope—Instrumentation for the Inspection of Cannon Tubes," Report AMMRC PTR 71-4, US Army Materials and Mechanics Research Center, 1974.

47. "What is the Sony Magnetodiode?" Sony Corp. publication.

48. Shiraiwa, T., Hiroshima, T., and Morishima, S., "An Automatic Magnetic Inspection Method Using Magnetoresistive Elements, and its Application," *Materials Evaluation*, vol. 31, pp. 90–96, May 1973.

49. Hiroshima, T., Hirota, T., Katayama, Y., and Furukawa, Y., "Real-Time MLFT System with Microprocessors," *NDT Journal Japan*, vol. 2, no. 1, pp. 19–22, 1984.
50. Lion, K. S., *Instrumentation for Scientific Research*, McGraw-Hill, New York, 1959.
51. Treppschuh, H., Stricker, F., and Fullung, D., "The Automatic Detection of Cracks in the Surfaces of Steel Billets by Means of Magnetographic Equipment," *Huttental-Geiswald, Stahl und Eisen*, vol. 90, no. 6, pp. 285–292, 1970.
52. Muller, P., "Experiences and Results in the Application of Magnetography in Industry," *Canadian Symposium on NDT*, 1970, pp. 174–187.
53. Feshchenko, Y. B., and Frolova, N. K., "Development of Metallographic Metal Inspection," *Soviet Journal of Nondestructive Testing*, vol. 9, no. 4, pp. 442–447, 1973.
54. Walters, W., et al., "Inspection Apparatus for Well Pipe Utilizing Detector Shoes with Outriggers, and Magnetic Latching Means," US Patent 3,543,144, AMF, Inc.
55. Stanley, R. K., "Basic Principles of Magnetic Field Flux Leakage Inspection of Oilfield Tubulars," paper summaries, ASNT Fall Conference, Houston, Texas, October 1980, pp. 157–166.
56. Zatsepin, N. N., Khalileev, P. A., and Shcherbinin, V. E., "Inspection of Articles for Discontinuities in Accordance with the Tangential Component of the Defect Field using a Ferroprobe Field Meter," *Soviet Journal of Nondestructive Testing*, no. 5, pp. 629–635, 1969.
57. Shcherbinin, V. E., Zatsepin, N. N., Vedenev, M. A., and Drozhzhina, V. I., "Film Ferroprobe Defektoskopic Transducer," *Soviet Journal of Nondestructive Testing*, no. 4, pp. 503–505, 1969.
58. *Nondestructive Testing—Magnetic Particle*, 2nd ed., CT-6-3, General Dynamics, Convair Division, 1977.
59. Betz, C., *Principles of Magnetic Particle Testing*, 2nd ed., Magnaflux Corp., Chicago, IL, June 1973.
60. Stanley, R. K., "Circumferential Magnetization of Tubes and the Measurement of Flux Density in Such Materials," *Materials Evaluation*, vol. 44, no. 7, pp. 966–970, 1986.
61. Stanley, R. K., "Streamline Magnetic Inspection of OCTGs," *Drilling*, vol. 48, pp. 10–13, January 1987.
62. Swartzendruber, L. J., "Magnetic Leakage and Force Fields for Artificial Defects in Magnetic Particle Test Rings," Proceedings of the XII Symposium on NDE, Southwest Research Institute, San Antonio, Texas, April 1979.
63. Lord, W. (ed.), *Electromagnetic Methods of Nondestructive Testing*, Gordon and Breach, New York, 1985, pp. 97–160.

PART IV

RADIOGRAPHIC TECHNIQUES IN NONDESTRUCTIVE EVALUATION

CHAPTER
20

PRINCIPLES OF
RADIATION

20-1 INTRODUCTION

Radiographic techniques for nondestructive evaluation certainly are historically important since they provided the first opportunity for inspecting items for internal defects. From the very basic beginnings, radiography has been expanded to where it now encompasses a variety of unique, versatile, and very useful methods that are available to the NDE engineer in the quest to assure safe and efficient performance of operating systems.

Generally, radiography is most satisfactory for finding internal, nonplanar defects such as porosity and voids. With proper orientation, however, planar defects may be located with radiography. It is also suitable for detecting changes in material composition, for thickness measurement, and for locating unwanted or defective components hidden from view in assembled parts. Usually, the depth of an object in the material cannot be easily determined based on a single radiographic inspection. Nonetheless, there are a large number of situations where radiography is either the best technique or, in some cases, the only technique that can be used to perform a particular inspection. The material in this part is generally based on Refs. 1–7 and other more specialized references and is intended only to provide an overview of the basic principles of radiography and the various radiographic techniques.

The reader may notice a complete absence of comments and references that relate to the various industry codes covering NDE. This is intentional since the emphasis here is on general information and most of the codes are rather specifically oriented toward a particular industry.

20-2   BASIC PRINCIPLES OF RADIOGRAPHIC INSPECTION

The mechanism of radiographic inspection is the propagation of energy from a source through an object and the evaluation of the energy pattern received on the opposite side. The radiation source emits energy that travels in straight lines and penetrates the object under investigation. Once the radiation energy has passed through the item, an image as received on a recording plane opposite to the source is used to evaluate the condition of the part being inspected. Although film is most frequently used as the recording medium for the image plane, there are a variety of other techniques that are often used such as fluoroscopic screens and digitized systems coupled with video monitors. Since the basic geometry and principles for inspection are quite similar for most of the display techniques, the present discussion will assume film as the recording medium.

A general layout for a radiographic inspection is shown in Fig. 20-1. The specimen to be inspected is arranged so that it is between the source of energy and the recording plane, i.e., the film. The radiographic energy, then, projects an image of any suitable abnormality onto the recording plane. Where film is used as the medium, it must be chemically developed in order to obtain the image. Other methods, however, may provide instant viewing of the image of the inspected object.

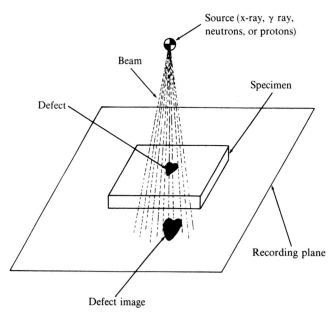

FIGURE 20-1
Layout for radiographic inspection.

From the arrangement seen in Fig. 20-1, it is clear that an image of a defect will occur on the film, provided that there is a sufficient difference in the radiation intensities received by the film under the defect as compared to that received through the remainder of the material. This dose differential to the film is a result of differences in the rate of absorption for rays passing through the defect and through the remaining material. Generally, where the defect is an absence of material, such as a void or porosity, the image will appear darker than the surroundings. If, on the other hand, the material in the defect is more absorptive than the surrounding material, the image will be lighter in appearance. This will be discussed in more detail in the material to follow.

20-3   PHYSICS OF RADIOGRAPHIC NDE

Radiographic energy is available in a variety of forms, each possessing characteristics that make it most suitable for particular inspection tasks. The most familiar form of radiographic energy is electromagnetic in nature, consisting of small packets of energy, the size of which is given by

$$E = h\nu = hc/\lambda \tag{20-1}$$

where h = Plank's constant
ν = frequency of the radiation (Hz)
c = speed (i.e., speed of light)
λ = wavelength (mm)

The term photon is generally used to identify this packet of energy.

Electromagnetic (EM) radiation is emitted by naturally occurring sources as well as artificially produced radioisotopes and by the process of accelerating electron beams into a target material, which results in the release of radiation. Electromagnetic radiation generally exists in the photon energy range from 10^{-3} to 10 MeV, where MeV designates mega electron volt (10^6 eV). According to Eq. (20-1), the wavelength in this energy range is from ~ 1 to 1 × 10^{-6} nm (1 to 10^{-5} Å). While the radiation energy is constant for any particular isotope, it is generally a function of the voltage across the tube plates for radiation excited by x-ray tubes.

While the basic structure of the atom is well known, it is suitable that a brief review be given at this time. Atomic and nuclear structures play an important role in radiographic NDE both in characterizing the source as well as the reaction with the material being inspected.

The atom consists of a positively charged nucleus surrounded by negatively charged electrons orbiting the nucleus in discrete shells determined by their energy level. The lower-energy electrons orbit closer to the nucleus and are thus more tightly bound than the higher-energy electrons orbiting in an outer shell. The mass of the atom is mostly concentrated within the nucleus where the protons and the neutrons reside. Although the electrons have a higher kinetic

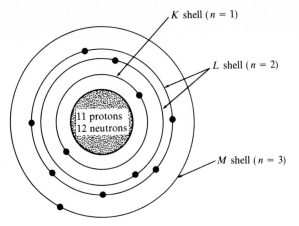

FIGURE 20-2
The atomic structure of sodium, atomic number 11, showing the electrons in the K, L, and M quantum shells. (*From Askeland [8]. Courtesy PWS-Kent Publishing Company.*)

energy than the protons and neutrons, they possess little mass. These characteristics are illustrated for the element sodium in Fig. 20-2 and summarized in Table 20-1. Additional definitions useful in discussions of atomic relationships are given in Table 20-2.

The orbital electrons are confined to discrete energy levels that are defined as electron shells. In an early model of the atom, each orbital shell is identified by the letters K, L, M, N, etc., with the K shell being the closest to the nucleus. The K shell holds the electrons having the least energy and these electrons are the ones most tightly bound to the nucleus. An electrically neutral atom will possess the same number of electrons as protons.

Later discoveries showed small energy differences and spin differences within each shell and additional quantum numbers were assigned as required for a complete description. These additional numbers will not be further discussed here. The interested reader, however, should consult Part III of this text for additional information.

TABLE 20-1
Atomic particles

Particle	Charge	Mass (amu*)	Comments
Proton	+1	1.007595	Nucleus of hydrogen atom
Neutron	0	1.008987	
Electron	−1	≈ 1/1840	
Positron	+1	≈ 1/1840	Emitted during disintegration of unstable neutron-deficient isotopes

*amu = atomic mass unit; 1 amu = 1.66×10^{-24} g.
Source: Reference 5.

TABLE 20-2
Atomic definitions and constants

Name	Symbol	Definition
Atomic number	Z	Number of protons in the nucleus
Mass number	A	Number of particles in the nucleus
Atomic mass unit	amu	One-sixteenth of the mass of the oxygen atom (1 amu = 1.66×10^{-24} g)
Angstrom	Å	1×10^{-10} m
Planck's constant	h	6.62620×10^{-34} J s
Avogadro's constant	N	6.02252×10^{23} mol^{-1} (atoms/at.wt.)
Speed of light in vacuum	c	2.997925×10^{8} m s^{-1}
Elementary charge	e	1.60210×10^{-19} C
Electron rest mass	m_e	9.1091×10^{-31} kg
Curie	Ci	3.700000×10^{10} dis/s
Joule	J	1.60210×10^{-19} eV
Roentgen	R	$2.57976(10)^{-4}$ C/kg

Source: Numerical values from *ASTM Metric Practice Guide*, 2nd ed. American Society for Testing and Materials, December 1966.

The nuclear and atomic structure, and the decay activity of the radioactive elements, are of significant interest in NDE since the electromagnetic radiation is emitted from these sources at characteristic energy levels established by the element itself. Penetrating radiation emitted by the radioactive sources originates from the disintegration process occurring in the nucleus of the atom and is known as gamma radiation.

Elements that emit penetrating radiation are called radioisotopes. The position of an element on the periodic chart is determined by its atomic number, i.e., the number of protons in the nucleus. It has been found that atoms of another element not having the same atomic number may have the same mass number as given by the total number of protons and neutrons in the nucleus. Further, atoms with the same proton number but differing neutron numbers are quite common. These latter atoms of an element with different mass number are isotopes. Within the isotope group of a particular element, some of the atoms have an unstable nucleus with the result that an ongoing disintegration process causes the release of electromagnetic radiation, i.e., gamma (γ) rays, as well as alpha (α) particles and beta (β) rays. These unstable isotopes that release the radiation are called radioisotopes.

The decay process is initiated when the ratio of protons to neutrons in the nucleus is either too large or too small for stability. Electromagnetic radiation from the nucleus generally occurs in conjunction with the emission of α or β particles.

Electronic transitions also lead to electromagnetic radiation. As an electron falls from one orbit (shell) to another lower-energy shell, the energy difference of the two shells is emitted as a quantum of electromagnetic radiation, and the wavelength of the emitted radiation is given by $E_1 - E_2 = hc/\lambda$, where E_1 and E_2 are the respective shell energies. In some cases, an electron jumping from one shell to another will cause successive jumps by electrons in adjacent shells resulting in radiation at several energy levels and at different wavelengths from that source. Depending on the value of E_1 and E_2, the radiation is x-ray, ultraviolet, visible, or infrared.

20-4 ELECTROMAGNETIC RADIATION SOURCES

Electromagnetic radiation may be excited using as a source either certain decaying isotopes for γ rays or vacuum tubes where high-energy electrons strike a target causing x-rays to be emitted. Both γ rays and x-rays are forms of electromagnetic radiation, indistinguishable except by the nature of the source.

20-4.1 Gamma-Ray Sources

Gamma rays are electromagnetic radiation emitted from an unstable source, i.e., an isotope. Each isotope will have characteristic nuclear energy levels and intensities for the emitted radiation. The γ-ray energy levels will remain constant for a particular isotope but the intensity will decay with time as indicated by the half-life. Table 20-3 lists pertinent characteristics of four isotopes most commonly used in NDE.

The symbol R/h m in Table 20-3 indicates radiation intensity with units of roentgen per hour per meter. A roentgen is a unit of measure defined to be the amount of energy absorbed from a beam of radiation passing through 1 cm³ of standard air. Intensity, therefore, is equated to the rate of energy absorption and is expressed in units of roentgen per hour 1 m from the source.

TABLE 20-3
Characteristics of common radiographic isotopic sources

Characteristic	Element			
	^{60}Co	^{192}Irs	^{137}Cs	^{170}Tm
Chemical form	Co	Ir	CsCl	Tm_2O_3
Half-life	5.27 yr	74.3 days	30.1 yr	129 days
γ rays (MeV)	1.33–1.17	0.31–0.47–0.60	0.66	0.084–0.052
R/h m Ci	1.35	0.55	0.34	0.0030
Typical source				
Curies	20	100	75	50
Diameter (mm)	2.5	2.5	10	2.5
(in)	0.1	0.1	0.4	0.1

Source: Reference 9.

Each of the isotopes listed in Table 20-3 possesses particular characteristics suitable for certain applications in NDE. The energy level of the emitted radiation is given in the table by the value in the column alongside γ rays (MeV). The intensity of the radiation is given by the value alongside R/h m Ci, where the curie (Ci) is the unit defining the specific activity of the isotope (see Table 20-2). For example, a ^{60}Co source with a strength of 20 Ci would have an intensity of 27 R/h m. In comparison, ^{192}Ir would be a very high-intensity source while ^{170}Tm would be a very low-energy source.

An active radiographic source is constantly disintegrating and therefore continually losing strength. This decay rate is indicated by the half-life as shown in Table 20-3, which expresses the time required for the source to lose half of its radioactivity in reference to the activity level at some initial time 0.

As it decays, the number of atoms in an isotope is decreasing. If the number of atoms at a certain time $t = 0$ is equal to N_0, then the number of atoms present at a later time N is given by

$$N = N_0 e^{-\lambda t} \tag{20-2}$$

where λ is the decay constant. At the end of the half-life period, $t = T_{1/2}$, one-half of the material would have disintegrated. Thus, with $N = N_0/2$ in Eq. (20-2),

$$\lambda T_{1/2} = \ln 2 \tag{20-2a}$$

and the decay constant is given by

$$\lambda = \frac{0.693}{T_{1/2}} \tag{20-2b}$$

Since the intensity of the radiation is directly related to the number of atoms present, Eq. (20-2) may be written

$$I = I_0 e^{-\lambda t} \tag{20-2c}$$

or, for an expression of the intensity at some time t in terms of the half-life $T_{1/2}$,

$$I = I_0 e^{-[0.693\, t/T_{1/2}]} \tag{20-2d}$$

Gamma-ray sources are typically constructed from a series of wafers or pellets of some source material, measuring approximately 2.5 mm (0.1 in) in diameter and 0.25 mm (0.01 in) in thickness. A typical 50 Ci source, then, may be made of a stack of 5 wafers each emitting radiation at a level of 10 Ci each. A source may be purchased with a specified activity level, e.g., 50 Ci, and will be used for some length of time. Since exposure is a function of intensity and time, it is clear that a source of diminishing activity level will require an increasing amount of time to achieve a proper exposure. At some point, the length of time becomes excessive and a new source is required.

20-4.2 Characteristics of x-ray Sources

X-rays are excited when high-speed electrons strike a suitable target causing an unstable condition at the target area that results in the release of quantum energy, or photons. The energy of the photon results from a transformation of the kinetic energy of the striking electron. Radiation excited in this manner is often called *bremsstrahlung*.

A general arrangement for a high-vacuum, x-ray tube and the associated circuitry is shown in Fig. 20-3. Two electrical circuits are involved in the operation of the x-ray tube. First, on the low-voltage (LV) side, the filament (cathode) is heated to raise the energy level of the conduction electrons to zero so that they physically leave the filament surface. The high-voltage (HV) circuit is responsible for accelerating the free electrons from the filament to the target.

The energy associated with this phenomenon is given by

$$\tfrac{1}{2}m_e v^2 = eV = hc/\lambda_{min} \tag{20-3}$$

where m_e = the mass of the electron
e = the charge on the electron
v = the velocity with which the electron strikes the target

The rapid deceleration of electrons at the target causes a continuous spectrum of emission of x-ray wavelengths down to λ_{min}. The target material is unimportant.

The tube voltage may be varied by the controlling/rectifying circuit as shown in Fig. 20-3. Since the controlling voltage may often be indicated by a voltmeter on the x-ray equipment, it is important to distinguish this voltage from

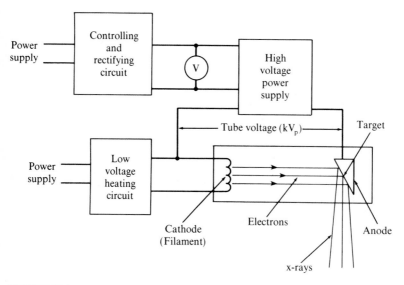

FIGURE 20-3
Layout of typical x-ray tube and circuit.

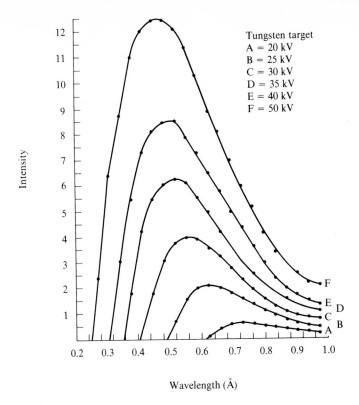

FIGURE 20-4
The continuous x-ray spectrum from a tungsten target at various tube voltages. (*From Ulrey [10]. Courtesy Physical Review.*)

the tube voltage that is being excited by the high-voltage (HV) circuit. Further details on x-ray tubes and circuits is given by Refs. 3 and 4.

The spectrum of the emitted radiation is affected by the tube voltage. As seen in Fig. 20-4, the maximum intensity shifts to the left as the tube voltage is increased [5, 10]. Further, the minimum wavelength of the quantum of radiation, where the intensity diminishes to zero, is also shifted to the left with the increased voltage. This minimum wavelength λ_{min} may be developed from Eq. (20-1) where

$$\lambda_{min} = \frac{hc}{eV} = \frac{12,399}{V} \text{ Å} \tag{20-4}$$

In the low-voltage heating circuit, an increase in tube current will increase the intensity of emitted radiation but will not shift the wavelength value for the maximum intensity [10].

The choice of target material only influences the line spectrum of the emitted radiation, which appears in conjunction with the continuous spectrum. A usual choice is tungsten, but copper, iron, and cobalt are also used. For typical

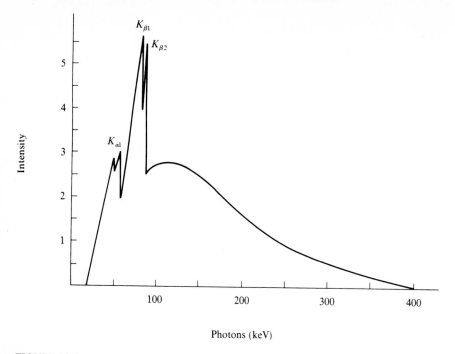

Photons (keV)

FIGURE 20-5

Energy spectrum of 400 kV$_p$ x-rays (constant potential, 4 mm aluminum filtration) showing characteristic lines from target material. (*From Halmshaw* [2]. *Courtesy Applied Science Publishers.*)

x-ray tube operation, an electron is accelerated from zero velocity at the heated
element by the high-voltage V across the plates.

The penetrating characteristics of the excited x-rays are a function of the
energy level, which is determined by the tube voltage and the target material.
Exposure times, as will be discussed later, are also affected by the intensity of the
x-rays. Halmshaw [2] has observed from Ulrey's data [10] that at lower tube
voltages, the maximum intensity occurs at a wavelength $\simeq 1.5$ times the minimum wavelength as given by Eq. (20-4).

A typical continuous energy spectrum for 400 kV$_p$ x-rays taken from
Halmshaw [2] is shown in Fig. 20-5. Shown are the characteristic peaks as well as
the minimum and maximum voltages of the beam. The maximum intensity of the
spectrum, neglecting the characteristic lines, occurs at about 133 kV or at
approximately one-third of the tube voltage of 400 kV. As noted by Halmshaw,
the common practice of using the tube voltage to describe the continuous
spectrum is not scientifically accurate. For example, where a 100,000 V accelerating voltage is used, this would be described at 100 kV x-rays or, in some
instances, 100 kV$_p$ for kilovoltage peak. The actual spectrum, however, will be a
function of the materials and other characteristics of the system. Commercially
available x-ray tubes are furnished with a rating chart showing the tube char-

TABLE 20-4
Frequently used units for photon energy and x-ray tube voltages

Unit	Application
eV, keV, MeV	X-ray photon energy in electron volts
kV_p	Corresponds to the upper voltage limit applied to the x-ray tube. Often called the operating kV* as indicated by voltage meters on x-ray machines
kV,* kV_{avg}	Commonly used for the x-ray energy. Note that $kV_{avg} \sim kV_p/3$ is often used as an estimate of the x-ray energy since, as seen in Fig. 20-5, the maximum x-ray intensity for 400 kV_p occurs near to 133 kV

*Industry practice has led to duplicate and contradictory usage of the term kV.

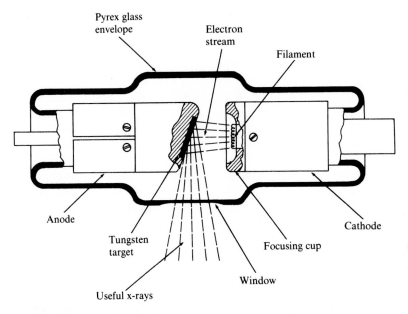

FIGURE 20-6
Collidge-type x-ray tube (*Courtesy General Electric Medical Systems.*)

acteristics. The rating chart has an appearance similar to that shown in Fig. 20-5. Table 20-4 further defines the relationships of tube voltages. x-ray energies, and x-ray intensities.

Details of the target area are shown in Figs. 20-6 and 20-7. Figure 20-6 shows the diverging pattern of the electrons emitted from the filament, the reflection at the target area, and the continued diverging pattern of the excited x-rays. The target angle θ represents the inclination of the target relative to the electron source stream, as shown in Fig. 20-7. Where the actual width of the struck target is H, the focal spot size is seen to be F. Later discussion will demonstrate the relationship of a small focal spot and increased sharpness in the

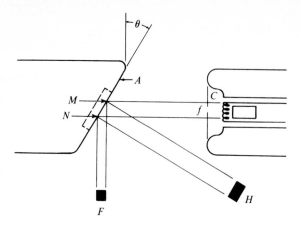

FIGURE 20-7
Principle of line focus x-ray tubes
[4].

radiographic image. Focal spots typically are square with sizes (i.e. the lateral dimension) ranging from 0.4–5 mm available in commercially produced equipment.

Later discussion will show that increased tube voltage and the resulting decrease in wavelength will increase the penetration ability for x-rays. Further, the increased intensity for greater tube current has already been noted. These advantages from increased electrical input, however, are not without associated costs, since the larger units generally require additional considerations such as more space and forced cooling. Available x-ray units may have maximum tube voltage ratings from 50 kV_p to over 1 MV_p and maximum tube currents from 2–25 mA (milliamperes) and above. Units using voltages less than 50 kV_p are normally air-cooled and will be smaller in physical size.

20-5 PARTICLE SOURCES

Streams of subatomic particles may be used for specialized inspections that may be difficult or impossible to accomplish with photons. Neutrons are of particular interest since their use has increased rapidly in recent years. Protons also have unique capabilities as described in the material to follow. A typical application of the particle sources is in thickness gauging as described in Refs. 4 and 11.

20-5.1 Thermal Neutron Sources

In the past, sources of thermal neutrons have been limited to nuclear reactors, which, of course, placed severe restrictions on use in general NDE. Recent years, however, have seen the development of moveable sources, which has enhanced their potential for industrial inspection applications.

The characteristics of several isotope sources of thermal neutrons have been listed by Berger [12] and Iddings [13]. Typical sources listed include ^{124}Sb-Be, ^{241}Am-Be, and ^{241}Am- ^{242}Cm-Be. Portable neutron source/moderator/collimator devices are now available for commercial inspection applications. Some have self-contained accelerators serving as the neutron source.

For each of the preceding sources, significant amounts of radiation other than neutrons are also emitted. Stray radiation from other sources has been a difficult problem to overcome. While this stray radiation increases the effort required to achieve clear, sharp images, a partial solution to the problem is available with the use of absorbers for the unwanted radiation. Collimator devices are also required at the neutron source since they tend to radiate spherically.

Thermal neutrons have energy levels in the range of 0.01–0.3 eV. Other neutron types available for use in NDE are cold, with an energy range below 0.01 eV, epithermal, with an energy range from 0.3–10,000 eV, and fast, having energies from 10 keV–20 MeV. Reference 14 contains a number of papers on neutron radiography.

20-5.2 Proton Sources

As reviewed by Koehler and Berger [15], sources for proton radiography generally are limited to accelerators, although machine and isotope sources are available. While there are over 100 such accelerators in the world suitable for this work, the use of protons in NDE is limited by the small number of sources. Beam control for these charged particles is accomplished using magnetic devices.

20-6 GEOMETRIC UNSHARPNESS

The layout shown in Fig. 20-1 illustrates a point source. With this arrangement, the image projected onto the recording plane (film) should be a sharp, albeit slightly enlarged, reproduction of the true shape of the object. True focal spots for sources used in conventional radiography are not point sources but rather may reach several millimeters in size. For these cases of a finite source size, the quality of the image detail becomes an important consideration. The term identifying the geometric relationship of the image quality is the geometric unsharpness U_g as given by

$$U_g = \frac{Ft}{L_o} \qquad (20\text{-}5)$$

where F = focal spot size
  t = distance from the object to the recording plane
  L_o = distance from the source to the object

The factors involved in establishing the geometric unsharpness are shown in Fig. 20-8.

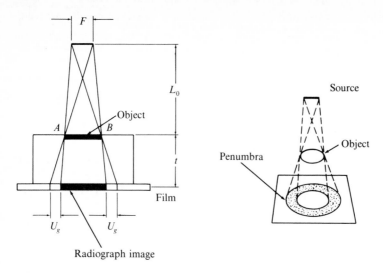

FIGURE 20-8
Geometric unsharpness in radiographic inspection.

20-7 INTENSITY CHANGE WITH DISTANCE

Since the distance from the source to the object is an important consideration in achieving a desired level of geometric unsharpness, the variation of radiation intensity with distance must be considered. For most electromagnetic sources, the source size is considerably smaller than the image plane, and for that condition, it is reasonable to assume point source radiation for the present discussion. In this case, the intensity will vary with changing distance according to the familiar inverse square law

$$IL^2 = C \qquad (20\text{-}6)$$

where I = intensity
L = distance to the image plane
C = constant

Comparing the intensity at distances L_2 and L_1, where $L_2 > L_1$, it is easy to see that the decreased intensity at distance L_2 will be given by

$$I_2 = I_1 \left[\frac{L_1}{L_2} \right]^2 \qquad (20\text{-}7)$$

20-8 ABSORPTION AND SCATTERING OF RADIATION ENERGY

The image of defects and other abnormalities projected onto the reference plane is a result of differences in the attenuation rates, or absorption, for various types of matter. Moreover, adjustment for proper exposure times and intensities is a function of the material attenuation. With this, then, it is necessary to briefly discuss some of the basic aspects of radiographic attenuation.

20-8.1 Structure of Crystalline Materials

The structure of crystalline, polymeric, and ceramic materials is covered in a number of basic references, e.g., Refs. 8 and 16. For brevity, the present descriptions will be limited to crystalline materials.

Metals are formed into crystalline structure with "metallic" bonding of the outer shells of electrons. The nature of metals is that only one, two, or three electrons are orbiting in the outer shell and the bonding process is one where these electrons are shared in an attempt to fill the outer shells. Other materials are formed by ionic, covalent, and van der Waals bonds, which define the relative placement of the atoms in the structure. Atomic characteristics for several elements used in engineering materials are given in Table 20-5.

A small, subgrain region of a metallic structure is illustrated in Fig. 20-9. The nuclei are shown separated by a space related by the fundamental atomic radius. Also shown is a single ray of x-ray photons approaching the metallic structure. Since the photons possess energy but have no mass or charge, it is

TABLE 20-5
Selected physical properties of elements

Element	Symbol	Atomic number	Atomic weight(g/at.wt)	Atomic radius(Å)	Density (g/cm³)
Aluminum	Al	13	26.98	1.43	2.699
Cadmium	Cd	48	112.41	1.5	8.65
Calcium	Ca	20	40.08	1.97	1.55
Carbon	C	6	12.01	0.71	2.225
Cesium	Cs	55	132.91	2.65	1.90
Chlorine	Cl	17	35.46	1.07	3.21×10^{-3}
Chromium	Cr	24	52.01	1.25	7.19
Cobalt	Co	27	58.94	1.25	8.85
Copper	Cu	29	63.54	1.28	8.96
Gold	Au	79	197.00	1.44	19.32
Hydrogen	H	1	1.01	0.46	0.09×10^{-3}
Iron	Fe	26	55.85	1.24	7.87
Lead	Pb	82	207.21	1.75	11.36
Magnesium	Mg	12	24.32	1.60	1.74
Nickel	Ni	28	58.71	1.25	8.9
Oxygen	O	8	16.00	0.6	1.43×10^{-3}
Tin	Sn	50	118.70	...	7.3
Titanium	Ti	22	47.9	1.47	4.51
Tungsten	W	74	183.86	1.37	19.3
Uranium	U	92	238.07	1.38	19.07
Zinc	Zn	30	65.38	1.33	7.13

Source: References 8 and 16.

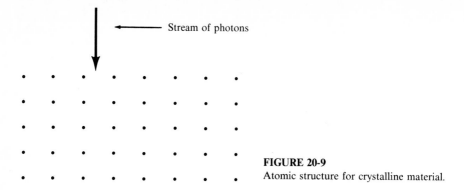

FIGURE 20-9
Atomic structure for crystalline material.

reasonable to perceive that they may be deflected and absorbed by the concentrated mass of the nucleus as well as the orbiting electrons in the shells. In reality, the x-ray photons exist in exceedingly large quantities and while some are absorbed and deflected, some pass through the material and exit the opposite side. It is the exiting photons that provide the imaging energy on the reference plane. Photons that either do not emerge from the material or that emerge at too small of an energy level to activate the film are considered to be scattered and therefore are not available for activating the recording medium. Thus, the intensity of the radiographic energy is said to be attenuated.

20-8.2 Scattering Mechanisms for Electromagnetic Radiation

Several mechanisms contribute to the total attenuation of radiographic energy. Following Refs. 1, 5, and 17, it is observed that electromagnetic radiation entering a sample of infinitesimal thickness dx with an intensity I will emerge from the sample with intensity $I - dI$, where dI represents the change in intensity due to absorption. The decrease in intensity dI through the thickness dx may be written

$$-dI = \mu I(x)\, dx \qquad (20\text{-}8)$$

where μ is the linear (total) absorption coefficient of the material and represents the fraction of energy removed from the beam per unit path length. The presence of the $I(x)$ term on the right hand side indicates x dependency of the intensity variable $I(x)$. Equation (20-8) may be rearranged to give

$$\frac{dI}{I(x)} = -\mu\, dx \qquad (20\text{-}8a)$$

which upon integration and mathematical manipulation yields

$$I = I_0 e^{-\mu x} \qquad (20\text{-}8b)$$

In the previous equation, I_0 is the intensity of the entering beam and I is the intensity of the beam exiting a material of thickness x. Considering only the scattering portion of the attenuation, the linear attenuation coefficient μ may be related to the atomic attenuation coefficient σ through the relationship

$$\mu = \frac{N\sigma\rho}{A} \qquad (20\text{-}9)$$

where the constants are defined in Table 20-2. Similarly, the mass attenuation coefficient μ/ρ is expressed by

$$\frac{\mu}{\rho} = \sigma \frac{N}{A} \qquad (20\text{-}9a)$$

The nature of the various attenuation coefficients is demonstrated in Example 20-1 and a partial list of attenuation coefficients from Hubbell [17] is given in Table 20-6 for various elements and in Table 20-7 for various materials.

Example 20-1. For x-rays at 0.10 MeV energy level, calculate the linear attenuation coefficient μ and the atomic attenuation coefficient σ if the linear attenuation coefficient μ/ρ is 0.4563 cm²/g at that energy level. Compare the atomic attenuation coefficient with the actual cross-sectional area of the copper atom.
The linear attenuation coefficient is

$$\mu = (\mu/\rho)\rho = (0.4563 \text{ cm}^2/\text{g})(8.96 \text{ g/cm}^3) = 4.088 \text{ cm}^{-1}$$

The atomic attenuation coefficient is given by

$$\sigma = \frac{\mu}{\rho}\frac{A}{N} = \frac{(0.4563 \text{ cm}^2/\text{g})(63.54 \text{ g/at.wt.})}{6.02252 \times 10^{23} \text{ atoms/at.wt.}}$$

$$= 4.81 \times 10^{-23} \text{ cm}^2/\text{atom}$$

The atomic cross-sectional area at the circumference is

$$a = \pi r^2 = \pi(1.28 \times 10^{-8})^2 \text{ cm}^2/\text{atom} = 0.51 \times 10^{-16} \text{ cm}^2/\text{atom}$$

The ratio of atomic attenuation coefficient to the atomic cross-sectional area is

$$\frac{\sigma}{a} = \frac{4.81 \times 10^{-23}}{0.51 \times 10^{-16}} = 9.43 \times 10^{-7}$$

which indicates that the scattering cross section is small compared to the total atomic cross-sectional area.

As additional information, Tables 20-6 and 20-7 show also the mass-energy absorption coefficient (μ_{en}/ρ), which is a measure of the average fractional amount of incident photon energy that is converted to charged particle kinetic energy during the atomic interaction. As described by Hubbell [17], this charged particle kinetic energy is therefore an approximation to the amount of energy made available for the production of effects associated with exposure to ionizing radiation. The reaction at the film or image plane is related to this ionizing radiation.

There are several mechanisms that contribute to the total atomic attenuation coefficient σ, and the relative effect of each is a function of the strength of the x-rays impinging on the material. The total attenuation coefficient may now be written as

$$\sigma = \sigma_{pe} + \sigma_s + \sigma_{pp} + \sigma_{pd} \qquad (20\text{-}10)$$

TABLE 20-6
Attenuation coefficients for selected elements*

Element	Atomic no.	Energy (MeV)	Attenuation coefficient	
			Mass μ/ρ (cm^2/g)	Energy μ_{en}/ρ (cm^2/g)
Magnesium	12			
		0.02	2.722	2.393
		0.05	0.3270	0.1432
		0.10	0.1684	0.03392
		0.15	0.1393	0.02762
		0.50	0.08646	0.2938
Aluminum	13			
		0.02	3.392	3.056
		0.05	0.3655	0.1816
		0.10	0.1701	0.03773
		0.15	0.1378	0.02823
		0.50	0.08446	0.02870
Titanium	22			
		0.05	1.204	10.967
		0.10	0.2711	0.1308
		0.15	0.1646	0.0540
		0.50	0.08191	0.02809
		1.00	0.05891	0.02561
		2.00	0.04180	0.02162
		4.00	0.03173	0.01907
Iron	26			
		0.05	1.944	1.630
		0.10	0.3701	0.2181
		0.15	0.1960	0.0797
		0.50	0.08413	0.02922
		1.00	0.05994	0.02604
		2.00	0.04265	0.02195
		4.00	0.03311	0.01984
Copper	29			
		0.10	0.4563	0.2952
		0.15	0.2210	0.1030
		0.50	0.0860	0.02943
		1.00	0.05990	0.02563
		2.00	0.04204	0.02156
		4.00	0.03318	0.01981

where the individual attenuation coefficients are identified as follows:

σ_{pe} = photoelectric effect
σ_s = scattering
σ_{pp} = pair production
σ_{pd} = photodisintegration

The characteristics of the first three contributors to the energy loss are illustrated in Figs. 20-10 through 20-13. Photodisintegration is a minor effect in NDE and is not discussed here.

TABLE 20-6

Element	Atomic no.	Energy (MeV)	Attenuation coefficient	
			Mass μ/ρ (cm^2/g)	Energy μ_{en}/ρ (cm^2/g)
Zinc	30			
		0.10	0.4950	0.3292
		0.15	0.2335	0.1134
		0.50	0.08450	0.02992
		1.00	0.05942	0.02582
		2.00	0.04236	0.02170
		4.00	0.03360	0.02005
Tin	50			
		0.10	1.6680	0.1.200
		0.15	0.601	0.417
		0.50	0.0940	0.040
		1.00	0.0580	0.0260
		2.00	0.041	0.028
		4.00	0.036	0.021
Tungsten	74			
		0.10	4.438	2.254
		0.15	1.581	0.9833
		0.50	0.1378	0.07722
		1.00	0.06616	0.0360
		2.00	0.04432	0.02286
		4.00	0.04037	0.02368
Lead	82			
		0.10	5.550	0.229
		0.15	2.014	1.135
		0.50	0.1613	0.09564
		1.00	0.07103	0.03787
		2.00	0.04607	0.02407
		4.00	0.04197	0.02463

*Attenuation coefficients based on earlier tabulations as given by Ref. 18 are furnished in Appendix B. The primary differences occur in the lower-energy values.
Source: Reprinted with permission from J. H. Hubbell, "Photon Mass Attenuation and Energy-Absorption Coefficients from 1 keV to 20 MeV," *Int. J. Appl. Radiat. Isot.*, vol. 33, Copyright 1982, Pergamon Press.

   The photoelectric effect as illustrated in Fig. 20-10 is mostly felt at relatively low-energy levels and the effect is negligible for materials having a low atomic number. In this process, the energy of the photon E_0 is totally transferred to an electron in some shell of the atom with the result that the electron may be moved to a different shell or ejected completely. At low photon energies, it is absorbed easily within the atom and the kinetic energy of the ejected electron is merely the difference between the energy of the photon and the binding energy of that shell, i.e., $E_0 - E_b$. As the photon energy is increased and reaches that of the binding energy of a particular shell of electrons, however, the absorption in-

TABLE 20-7
Attenuation coefficients for selected materials

Material	Energy (keV)	Attenuation Coefficient Mass μ/ρ (cm²/g)	Attenuation Coefficient Energy μ_{er}/ρ (cm²/g)
Concrete	5.0	171.8	163.1
	10.0	26.19	24.67
	50.0	0.3918	0.2048
	100.0	0.1781	0.03014
	500.0	0.08767	0.02984
Glass, lead			
	5.0	582.3	574.5
	10.0	102.8	98.82
	50.0	6.134	5.152
	100.0	4.216	1.686
	500.0	0.1429	0.07927
Glass, Pyrex			
	5.0	124.0	121.3
	10.0	16.78	16.17
	50.0	0.3004	0.1220
	100.0	0.1656	0.03193
	500.0	0.08696	0.02958
Nylon $(C_{12}H_{22}O_3N_2)N$			
	5.0	23.94	23.42
	10.0	2.996	2.67
	50.0	0.2066	0.02876
	100.0	0.1653	0.0237
	500.0	0.0950	0.0339
Polyethylene $(C_2H_4)N$			
	5.0	15.85	15.40
	10.0	2.023	1.717
	50.0	0.2081	0.02410
	100.0	0.1719	0.02420
	500.0	0.0945	0.03388

creases rather abruptly. This occurs for each of the electron shells and, for the K shell, is designated as the K absorption edge. A further increase in energy will result in a diminishing of the contribution of the photoelectric effect approximately with the cube of the energy. At some point, its effect becomes negligible.

The scattering coefficient σ_s is composed of two contributors, namely Compton scattering and coherent scattering, as illustrated in Figs. 20-11 and 20-12. As explained by Compton, the incident photon with an energy $E_0 = h\nu$ upon striking a relatively free electron causes a momentum transfer within the system of the photon and the electron. The struck electron will exit its shell at an angle with an energy E_e, while the photon will be deflected at an angle opposite to the electron and will exit the atom with an energy of $E_0 - E_e$. In effect, conservation of momentum and energy govern the process.

Coherent scattering, also known as Rayleigh scattering, occurs when the photon does not experience an energy shift upon striking the electron. This effect

TABLE 20-7

Material	Energy (keV)	Attenuation Coefficient	
		Mass μ/ρ (cm²/g)	Energy μ_{er}/ρ (cm²/g)
Polymethyl methacrylate $(C_5H_8O_2)N$	5.0	26.18	25.65
	10.0	3.273	2.944
	50.0	0.2069	0.0302
	100.0	0.1640	0.0263
	500.0	0.09408	0.03204
Polystyrene $(C_8H_8)N$	5.0	17.04	16.59
	10.0	2.150	1.849
	50.0	0.1982	0.02387
	100.0	0.1624	0.02293
	500.0	0.09376	0.03194
Polyvinyl chloride $(C_2H_3Cl)N$	5.0	225.4	214.9
	10.0	32.96	31.59
	50.0	0.4528	0.2596
	100.0	0.1883	0.04822
	500.0	0.0898	0.03057
Teflon $(C_2F_4)N$	5.0	52.82	52.10
	10.0	6.654	6.236
	50.0	0.2123	0.04861
	100.0	0.1499	0.02327
	500.0	0.0838	0.02852

Source: Reprinted with permission from J. H. Hubbell, "Photon Mass Attenuation and Energy-Absorption Coefficients from 1 keV to 20 MeV," *Int. J. Appl. Radiat. Isot.*, vol. 33, Copyright 1982, Pergamon Press.

is more pronounced for higher atomic number materials. As illustrated in Fig. 20-12, this type of scattering occurs when the incident photon excites the atomic electrons at a resonance oscillation and the photon leaves the atom merely deflected with negligible loss of energy. Coherent scattering is directly related to the incident angle of the incident beam and never contributes more than 20% of the total scattering losses. It is not considered to be an important process in metallic materials.

Above 1.02 MeV, the photon may disintegrate into two particles, an electron and a positron. This event is called pair production, as seen in Fig. 20-13 where the ejected energy of the electron is designated to be E_- and that of the positron E_+ and the energy balance is also shown. The process again conserves energy and momentum. Note that 0.511 MeV $= m_e c^2$ from the Einstein equation $E = mc^2$.

The combined effects of these absorption coefficients for iron are shown in Fig. 20-14, which is taken from Halmshaw [2]. The very large contribution at

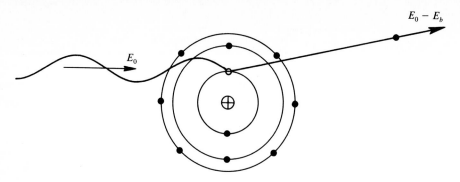

FIGURE 20-10
Photoelectric interaction of an incident photon with an orbital electron. (*From Emigh [1], The Nondestructive Testing Handbook, vol. 3, Radiography and Radiation Testing, 2nd ed., 1985, American Society for Nondestructive Testing.*)

lower-energy levels for the photoelectric effect pe is seen at the left end of the figure. Also notable are the lesser yet still significant effects from Compton (incoherent) and Rayleigh (coherent) scattering. The total linear absorption coefficient for iron at these energy levels is shown by the solid line labeled T. The effect of pair production pp begins at 1.02 MeV and becomes significant above 10 MeV, causing an increase in the total absorption.

The effect of increased tube voltage on total linear absorption is shown in Fig. 20-15, after Halmshaw [2], for a variety of elements used in engineering materials. In all cases, the initial decrease in absorption for increased tube voltages is seen, along with the characteristic absorption edges for lead and tungsten. Copper, lead, and tungsten are seen to follow iron in having an increase

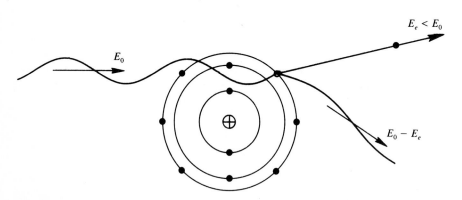

FIGURE 20-11
Compton scattering in which an incident photon ejects an electron and a lower-energy scattered photon. (*From Emigh [1], The Nondestructive Testing Handbook, vol. 3, Radiography and Radiation Testing, 2nd ed., 1985, American Society for Nondestructive Testing.*)

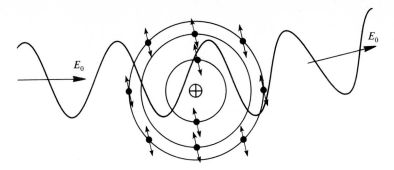

FIGURE 20-12
Coherent scattering of a photon without loss of energy. (*From Emigh [1], The Nondestructive Testing Handbook, vol. 3, Radiography and Radiation Testing, 2nd ed., 1985, American Society for Nondestructive Testing.*)

in absorption as the voltage is increased beyond a level yielding a minimum absorption.

A further relationship of the mass attenuation coefficient μ/ρ, x-ray energy, and atomic number is seen by the curves in Fig. 20-16 that were plotted from Ref. 18. For the lower-energy levels, the increased attenuation with increasing atomic number is clearly seen. At increasingly higher energies, however, the effect becomes minimal to where, at 1.5 MeV, the absorption coefficient is virtually the same for materials of all atomic numbers. Also shown on the figure are mass absorption coefficients for thermal neutrons that will be discussed in the following material.

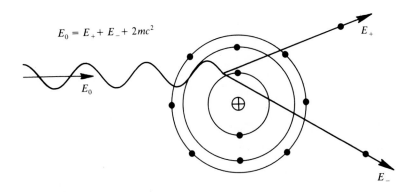

FIGURE 20-13
Pair production of an electron and a positron from an incident photon. (*From Emigh [1], The Nondestructive Testing Handbook, vol. 3, Radiography and Radiation Testing, 2nd ed., 1985, American Society for Nondestructive Testing.*)

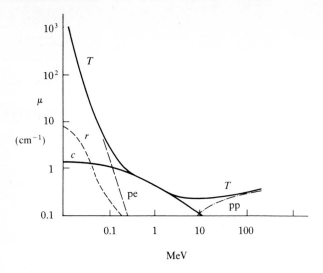

FIGURE 20-14
Curves of linear absorption coefficients for iron and energy (MeV). c = Compton scatter, pe = photoelectric, pp = pair production, r = Rayleigh scattering, and T = total absorption. (*From Halmshaw [2], Courtesy Applied Science Publishers.*)

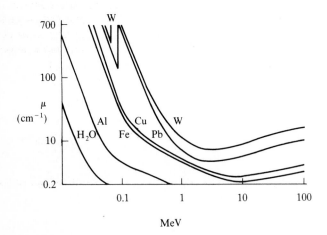

FIGURE 20-15
Total linear absorption coefficients for different materials. (*From Halmshaw [2]. Courtesy Applied Science Publishers.*)

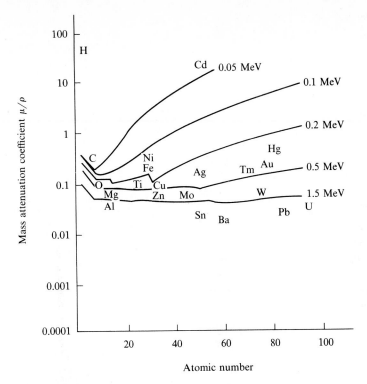

FIGURE 20-16

Mass attenuation coefficients for x-rays at various energy levels and for thermal neutrons. Atomic symbol positions represent μ_m for thermal neutrons and lines represent μ for x-rays.

Rather than determining the proper exposure using the various attenuation coefficients, standard industrial practice is to use a table of radiographic equivalence factors (REF) as given in Table 20-8, from Ref. 3. For each energy level there is one standard REF of 1.0 to which all others are compared. For example, for 100 kV x-rays, the table shows that copper is 18 times more attenuating than the standard, aluminum. Similar comparisons are made for the higher-energy levels although the standard metal shifts to steel. More discussion on these REFs will be presented in the inspection chapter to follow. These REF values are intended only as a source of initial estimate for determining the exposure for a specific instance. Radiographic equivalence factors calculated from the attenuation coefficients may not always agree with the REFs given in Table 20-8, particularly at the lower-energy levels.

20-8.3 Scattering Mechanisms for Particle Radiation

Several investigators have observed that the scattering characteristics of particle radiation are distinctly different from those of electromagnetic sources. This

TABLE 20-8
Approximate radiographic equivalence factors*

Material	x-rays								γ rays			
	50 kV	100 kV	150 kV	220 kV	400 kV	1000 kV	2000 kV	4–25 MeV	192 Ir	137 Cs	60 Co	Ra
Magnesium	0.6	0.6	0.5	0.08	...	...	...	...				
Aluminum	1.0	1.0	0.12	0.18	...	...	...	...				
2024 (aluminum) alloy	2.2	1.6	0.16	0.22	...	...	...	...	0.35	0.35	0.35	0.40
Titanium	...	...	0.45	0.35	...	...	...	...	0.35	0.35	0.35	
Steel	...	12.	1.0	1.0	1.0	1.0	1.0	1.0	1.0	1.0	1.0	1.0
18-8 (steel) alloy	...	12	1.0	1.0	1.0	1.0	1.0	1.0	1.0	1.0	1.0	1.0
Copper	...	18.	1.6	1.4	1.4	...	...	1.3	1.1	1.1	1.1	1.1
Zinc	...	...	1.4	1.3	1.3	...	...	1.2	1.1	1.0	1.0	1.0
Brass†	...	...	1.4	1.3	1.3	1.2	1.2	1.2	1.1	1.1	1.1	1.0
Inconel X alloy-coated	...	16.	1.4	1.4	1.3	1.3	1.3	1.3	1.1	1.1	1.1	1.1
Zirconium	...	...	2.3	2.0	...	1.0	...	...	1.3	1.3	1.3	1.3
Lead	...	...	14.	12.	...	5.0	2.5	3.0	4.0	3.2	2.3	2.0
Uranium	...	...	...	25.	...	...	...	3.9	12.6	5.6	3.4	

*Aluminum is the standard metal at 50 and 100 kV and steel at the higher voltages and gamma rays. The thickness of another metal is multiplied by the corresponding factor to obtain the approximate equivalent thickness of the standard metal. The exposure applying t̯ ̯us thickness of the standard metal is used. *Example:* To radiograph 1.27 cm (0.5 inch) of copper at 220 kV, multiply 1.27 cm (0.5 inch) by the factor 1.4, obtaining an equivalent thickness of 1.78 cm (0.7 inch) of steel.

†Tin or lead alloyed in the brass will increase these factors.

Source: References 3 (Reprinted courtesy of Eastman Kodak Company) 19 (*Nondestructive Testing Handbook,* vol. 3, *Radiography and Radiation Testing,* 2nd ed., American Society for Nondestructive Testing, 1985).

creates opportunities for inspection that simply do not exist for traditional electromagnetic radiation.

McGonnagle [4], Berger [12], and Iddings [13] have reported on the scattering characteristics of various neutrons. As shown in Fig. 20-16, plotted from data reported in Ref. 4, there appears to be no relationship between atomic number and the mass attenuation coefficient. In fact, hydrogen, which has the lowest atomic number, shows a very high neutron absorption. Moreover, the attenuation coefficient for carbon is only slightly lower than for mercury, which has a considerably larger atomic number. These attenuation patterns allow the inspection of rather unique combinations of materials, as will be discussed in a later section.

Figure 20-17 shows the transmission efficiency for x-rays and protons, from Koehler and Berger [15], as a function of density.

Example 20-2. Use the mass attenuation coefficients as given in Table 20-6 to estimate the radiographic equivalence factors (REF) for steel and aluminum for x-rays excited at a tube voltage of 450 kV. Compare these results with the REF obtained for these materials using the mass-energy attenuation coefficient and with the values given in Tables 20-6 and 20-7.

From Eq. (20-8b)

$$I = I_0 e^{-\mu x}$$

For aluminum and steel,

$$I_{Al} = I_0 e^{-\mu t (Al)}$$

$$I_{st} = I_0 e^{-\mu t (st)}$$

where t is the material thickness. Since the REF values relate equivalent intensities

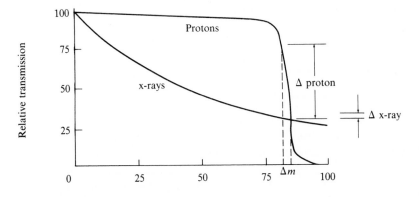

FIGURE 20-17
Typical relative transmission curves for monoenergetic protons (upper curve) and x-rays (lower curve). (*From Koehler and Berger [15]. Courtesy Academic Press.*)

on the recording plane,

$$\mu_{Al}t_{Al} = \mu_{st}t_{st}$$

and

$$t_{Al} = \frac{\mu_{st}}{\mu_{Al}}t_{st}$$

The energy of the emitted x-rays is approximately equal to one-third of the tube voltage. Therefore, for 150 kV x-rays

$$t_{Al} = \frac{(0.1960 \text{ cm}^2/\text{g})(7.87 \text{ g/cm}^3)}{(0.1378 \text{ cm}^2/\text{g})(2.699 \text{ g/cm}^3)}t_{st} = 4.15t_{st}$$

and for the mass-energy attenuation coefficient

$$t_{Al} = \frac{(0.0797 \text{ cm}^2/\text{g})(7.87 \text{ g/cm}^3)}{(0.02823 \text{ cm}^2/\text{g})(2.699 \text{ g/cm}^3)}t_{st} = 8.23t_{st}$$

Therefore, for the mass attenuation coefficient

$$t_{st} = 0.24t_{Al}$$

and for the mass-energy attenuation coefficient

$$t_{st} = 0.12 \ t_{Al}$$

Table 20-8 shows a REF for aluminum at 150 kV to be 0.12, which compares very well with the value obtained with the mass-energy coefficient and reasonably well with the mass attenuation coefficient. The lack of better agreement for the mass attenuation coefficient is most likely due to an inaccuracy in the one-third approximation for the tube voltage and x-ray voltage. For further comparisons, the reader is directed to the problems at the end of this part.

Where the x-ray transmission shows a gradual decline with increased density, the protons behave in a decidedly different manner. Here, the transmission change is very slight until a critical point is reached where the transmission decreases rapidly. In fact, where the lower scale represents the total range of the protons for a given material, it is observed that most of the attenuation occurs when the beam passes approximately 90% of the range thickness. The significance of this feature will be discussed in more detail in a later section.

20-9   SECONDARY RADIATION

In the previous discussion on scatter, the statement was made that any photon that was deflected in a collision with an atom was considered lost forever. With this assumption, the full contribution of a photon to the radiographic inspection process is felt only if the photon traverses the part being inspected without any loss of energy. While this assumption simplified the previous discussion, it must nonetheless be modified at this time.

The energy level is quite high for the photons entering the part being inspected and it is reasonable to expect that a photon could strike part of the atomic structure and be deflected while still retaining a large amount of its original energy. In that case, then, the deflected photon could strike succeeding atoms with the possibility that it could eventually emerge from the part with sufficient energy to activate the recording media. Since this deflected photon has not been considered in the calculations of the attenuation coefficients previously discussed, any interaction of it with the recording media would not be expected. The obvious result, then, would be a true exposure density greater than predicted by the previous discussion. Radiation from these deflected photons is called secondary radiation.

20-9.1 Broad-Beam and Narrow-Beam Geometry

The alignment of the stream of photons entering the inspected item strongly affects the amount of secondary radiation. Photon streams striking the part at normal incidence, e.g., Fig. 20-9, obviously would be less prone to excite secondary radiation than streams entering at an oblique angle, as would occur on the outer fringes of the arrangement shown in Fig. 20-1.

Narrow-beam geometry is used to define the normally incident photons as shown in Fig. 20-9. Clearly, all sources of very small size would radiate some photons at oblique angles and thus excite some secondary radiation on the outer fringes of the beam. Where a small source size has been shown earlier to decrease the amount of geometric unsharpness, the smaller size would have a deleterious effect from the standpoint of exciting amounts of secondary radiation. Photon streams originating from point sources, as shown in Fig. 20-1, are said to represent broad-beam geometry.

Calculated absorption coefficients and radiographic equivalence factors are typically assumed to obey the narrow-beam concept while actual empirical data would necessarily be based on broad-beam data. These factors offer an additional explanation of the occasional lack of good agreement between the observed results and calculated expectations when the calculations are based on the REF tables.

Earlier discussion on the geometric unsharpness indicated that greater distances from the source to the object would increase the sharpness of the image. This is because at larger distances, the width of the source becomes smaller compared to the source-to-object distance and the size of the shadow off of the edges is decreased. Increasing the distance from the source to the test sample also has the positive effect of decreasing the amount of secondary scatter due to the fact that the circumferential energy front of the radiation becomes increasingly flat at larger distances from the source, thus approaching a narrow-beam configuration.

20-9.2 Build-Up Factor

The amount of secondary scatter radiation is indicated by the build-up factor, as described by Halmshaw [2]. The build-up factor is defined

$$\text{build-up factor} = \left(1 + \frac{I_S}{I_D}\right) \tag{20-11}$$

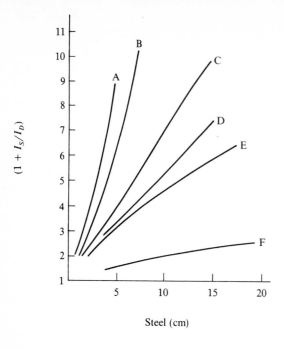

FIGURE 20-18
Experimental build-up factor curves for steel plates. A, 200 kV x-rays; B, 400 kV x-rays; C, 1 MV x-rays; D, 2 MV x-rays; E, 5 MV x-rays; F, 15 MV x-rays. (*From Halmshaw* [2], *Courtesy Applied Science Publishers.*)

where I_S = scattered radiation component
$\quad\;\; I_D$ = direct intensity

The scattered radiation component I_S represents energy that reaches a point on the film by any path other than a direct path.

The variation of the build-up factor with x-ray energy is shown by Fig. 20-18, from Halmshaw [2]. In this illustration, a value of 1 for the build-up factor indicates no scattered radiation, while increased scatter results in a larger value for the build-up factor. The lower-energy x-rays, i.e., those with longer wavelengths, are seen to suffer much greater scattering than those having higher-energy contents.

20-10 SUMMARY

This chapter has discussed in a general context the characteristics of the various types of energy used in radiographic NDE. Differences were observed not only in the excitation methods but also in the attenuation characteristics. This was noted to occur both for different materials as well as for different sources. The material to follow will describe applications of these types of radiographic energy to NDE.

CHAPTER 21

RADIOGRAPHIC
IMAGING

21-1 INTRODUCTION

The accurate reproduction of a recognizable image on some medium is a primary requirement for radiographic NDE. Typically, this image is obtained at the recording plane using film, visual display systems, or digitized systems. Since the basic principles of obtaining a satisfactory image are somewhat similar for all processes, the primary emphasis at this point will continue to be on film images. Some discussion at the end of the chapter will describe some fundamental advantages of the other techniques.

21-2 RADIOGRAPHIC IMAGES

The appearance of a distinguishable image on a radiograph is dependent on several factors, the most important of which is a difference in radiation intensity on various parts of the image plane. The difference in intensity, then, must be a function of the abnormality within the part being inspected. Intensity differences may be created by a number of circumstances, including a hidden object having an attenuation coefficient μ' that is different from the attenuation coefficient of the remaining material μ, as shown in Fig. 21-1. Consider in Fig. 21-1 that the specimen having an attenuation coefficient μ and a thickness t contains an object with an attenuation μ' and thickness mt, where m is a decimal fraction less than 1. In the region containing the defect, the thickness of the remaining material with attenuation coefficient μ is $m_1 + m_2$. It can be shown that the ratio of the

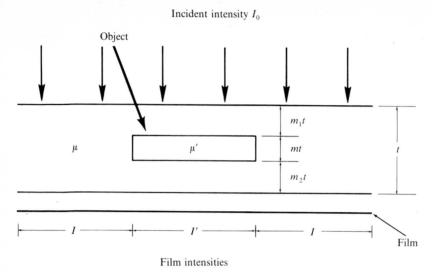

FIGURE 21-1
Image formation in radiography.

intensity for energy traversing the object to that of energy aside from the object is

$$\frac{I'}{I} = e^{-tm(\mu'-\mu)} \tag{21-1}$$

where I' = intensity on film for energy passing through the object
 I = intensity on film for energy passing through the full material
 μ = attenuation coefficient of test material
 μ' = attenuation coefficient of object material
 t = thickness of test material
 m = decimal fraction for ratio of thickness of object and sample thickness t

Equation (21-1) clearly indicates that where the attenuation of the object is less than that of the test material, the intensity I' will be greater than I. Also, the converse argument may be shown. Points 1 and 2 on the image plane represent adjacent locations on the film, which are for later discussion.

21-3 IMAGE RECORDING

The recording and processing of the image received on the recording plane has been receiving considerable interest in recent years. The traditional film recording systems provide an advantage in the fact that a hard, direct photograph type copy of the exposure is available for viewing and record keeping purposes. However, the time advantages for more real-time image display and analysis systems has placed considerable emphasis on fluorescent screens with television-like monitors as well as digitized and computer aided tomographic systems.

21-3.1 Film Radiography

Radiographic film is similar to photographic film in that there is a center carrier called the film base that is made of a thin sheet of polyester type material. This normally is transparent and serves only as the carrier for the chemically reactive materials that form the emulsion. The emulsion is the material that reacts to the radiation and, thereby, contributes to the production of the image. Emulsion consisting of a silver halide recording media with a binder of gelatin is applied to both sides of the base. In addition, a protective layer may be applied over the emulsion.

The silver halide is a granular material as indicated in Fig. 21-2. The grain size, designated as the graininess, has a significant effect on the exposure rate as well as the detail resolution ability of the film. As a result of the incident radiographic energy, each grain, acting independently, will undergo a chemical reaction. Upon processing, the grains that have been exposed will be darkened. The silver halide is removed from the unexposed grains in the processing leaving a transparent area, as shown in Fig. 21-2. Thus, the darkness, or density, of the film is directly a function of the exposure of the film grains.

The response of the film during exposure is indicated by the density obtained after processing. Film exposure E is a matter of time and intensity, i.e.,

$$E = IT \tag{21-2}$$

where I is the intensity of the radiation on the film and T is the time of exposure. A measure of the amount of exposure seen by the developed film is the

COARSE GRAIN   FINE GRAIN

EXPOSED   UNEXPOSED   EXPOSED   UNEXPOSED

FIGURE 21-2
Representative film grain size. Large (coarse) grain on left and fine grain on right. Exposed grains shown black. (*From General Dynamics* [20]. *Courtesy American Society for Nondestructive Testing.*)

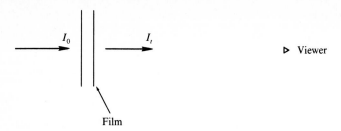

FIGURE 21-3
Light transmission through exposed radiograph.

transmission density D, which is given by

$$D = \log \frac{I_0}{I_t} \qquad (21\text{-}3)$$

where the light intensities are as indicated in Fig. 21-3. In reading the film, it is held to where a light source of intensity I_0 strikes the film as shown. The transmitted intensity I_t will be seen by the viewer looking from right to left at the opposite face of the film.

Film response is typically nonlinear as shown by the logarithmic exposure curves given in Fig. 21-4. Two films are shown, each having a different response

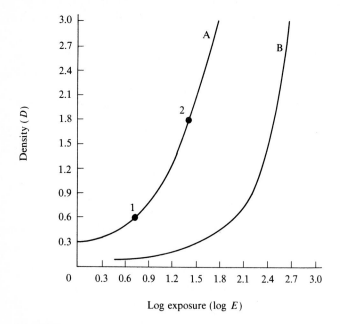

FIGURE 21-4
Film characteristics curves. High-speed film to left, low-speed film to right. (*From General Dynamics* [20]. *Courtesy American Society for Nondestructive Testing.*)

rate. The density change with exposure is the log relative exposure, as given by Eq. (21-4), where E_0 is a standard exposure at some arbitrary time T_0:

$$\log(RE) = \log(E) = \log\frac{E}{E_0} \tag{21-4}$$

It is convenient at this time to normalize E_0 to be unity and to refer to the log relative exposure as the log exposure, $\log(E)$, where, from Eq. (21-4), $\log(E) = \log(RE)$. For each of the two films depicted in Fig. 21-4, the response of the silver halide particles in the early part of the exposure is seen to be very slow. As the exposure increases, the response of the film increases logarithmically and the change in density becomes very rapid.

The exposure is an important factor in achieving good image density as well as contrast. Longer exposures simply darken the film, creating a higher density. Contrast between adjacent areas on the film, however, is a function of the film response curve and the ratios of the intensities for the adjacent areas. The amount of the exposure determines the density D, while the contrast is determined by the film characteristics as well as the exposure differential for the objects being radiographed. Following Halmshaw, the contrast at a density D is defined to be the slope of the film characteristic curve, i.e.,

$$G_D = \frac{dD}{d(\log E)} \quad \text{(measured at density } D\text{)} \tag{21-5}$$

G_D, then, varies with density and is a measure of the film contrast available at a chosen density. The relationship of exposure E, density D, and film contrast G_D is demonstrated in Example 2-1.

> **Example 21-1.** A radiographic test arrangement was such that an I'/I value of 1.25 was obtained for the adjacent particles 1 and 2 shown in Fig. 21-1. Using film A in Fig. 21-4, the first film exposure had a nominal density of 0.6. A second exposure achieved a density of 1.8. Compare the contrast for the adjacent dark and light areas represented by particles 1 and 2.
>
> The two densities are shown at points 1 and 2, respectively, in Fig. 21-4. The contrast at the two points would therefore be given by the slopes
>
> $$(G_D)_1 = 0.82$$
>
> and $$(G_D)_2 = 3$$
>
> Thus, the higher value for the contrast at exposure 2 indicates that a defect with the given characteristics would be more readily apparent with exposure 2 as compared to exposure 1.

The speed of initial response of the film to exposure is another important film parameter. For example, in Fig. 21-4 the two curves depict one film that could be called a high-speed film and another that is a lower-speed film. Film A would be the high-speed film since the grains of that film would begin reacting to the exposure considerably sooner than those of the other film. Realizing that exposure is the product of time and intensity, the effect of film speed is rather significant. For a constant intensity, for example, the grains of film A would produce the required density before the grains of film B. Where time consumed in the inspection is important, the film speed also becomes important. Similarly, a faster speed film might allow a suitable exposure to be achieved with a lower-power unit. Realizing that no advantage ever comes without a cost, it must be pointed out that generally the faster speed films have larger grains and, therefore, may not be able to produce the required detail.

Film development is also important in producing a quality image. Certainly, factors such as contrast and definition may be affected by the development process. It is also important to recognize that film graininess and speed may also be manipulated in the development. Further information on film development is furnished in the references.

Film handling during the exposure process contributes significantly to the quality of the radiograph. During the exposure, there is a considerable amount of stray energy that has been scattered within the test sample and may find its way to the film. Although low in intensity, this scattered radiation can fog the film and destroy the detail that otherwise would be achieved. A countermeasure to this problem involves the use of absorbing screens that are placed next to the film during the exposure. The film and the screens are placed in a film cassette that is then placed adjacent to the part being radiographed. Efforts to improve the performance of screens and films so as to create images with greater detail and sharpness are described by Moores [21].

A lack of definition due to film response characteristics is known as film unsharpness, as reviewed by Halmshaw [2]. Referring to Fig. 21-1, it is realized that maximum definition occurs when the film accurately reproduces the edge of the discontinuity as an edge on the film. In actuality, the change in density on the film caused by the edge of the object occurs over some finite distance. The width of this gradient is known as the film unsharpness U_f, as shown in Fig. 21-5.

Overall image sharpness is due both to film unsharpness U_f and geometric unsharpness U_g. Citing work of earlier researchers, Halmshaw [2] reports that total image unsharpness U_t may be adequately described by

$$U_t = \left(U_f^2 + U_g^2\right)^{1/2} \tag{21-6}$$

Additional discussion on film and geometric unsharpness will be given in the section following on source-to-film distance.

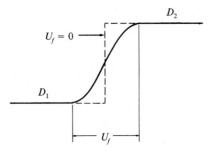

FIGURE 21-5
Film unsharpness.

21-3.2 Real-Time Radiography and Image Data Analysis

Real-time radiography represents a significant improvement in the NDE process since there can be an overall decrease in the elapsed time from the inspection to a decision on the appropriate action. Not only does this time period have safety implications, it also may affect the cost of continued operation and the cost of repair. It is for these reasons that the efforts in real-time radiography, as reviewed by Halmshaw [22] and by Bossi, Oien, and Mengers [23], are finding expanded application.

The analysis of accumulated data has long been a source of difficulty in radiographic NDE. Inefficiency results because decisions on the presence or absence of a defect and the correct classification of other indications has long been based on experience rather than quantified judgement. Image data analysis schemes have been described by a number of investigators, including Jacoby [24], De Meeser and Aerts [25], and Packer [26].

21-3.3 Photon Tomography

Tomography is a unique technique for imaging in that full three-dimensional data on scan fields can be obtained for a large number of objects. The technique uses narrow beam projection of a source that is moved in relation to the object. The responses from the object are stored for later analysis when the full image may be reconstructed. An application of this technique that is well known in the medical industry is CAT scan, for computer-aided tomography. Computed tomography (CT) is more commonly used in NDE applications. A number of applications have been reported by Snow and Morris [27], Gilboy and Foster [28], and Frank and Kohutek [29], among others.

21-4 SOURCE-TO-FILM DISTANCE

The distance from the source to the recording plane affects the image quality through the geometric unsharpness. Recalling from Eq. (20-5) that the geometric unsharpness is directly proportional to the focal spot size and the thickness of the specimen and inversely proportional to the source-to-film distance (SFD), it is clear that image sharpness will require that the source-to-object distance be maximized while the object-to-recording-plane distance be minimized.

Achieving a sharp image involves compromises. Where the geometric unsharpness is decreased with a greater SFD, inverse square law indicates that the exposure time must be lengthened. If the exposure time is decreased through the use of higher-energy radiation, Halmshaw reports that this results in an increase in the film unsharpness [2]. Halmshaw further observes, however, that a good estimate of a proper SFD for high sensitivity inspection might be obtained by increasing the SFD until $U_g = U_f$, which, when Eq. (20-5) is applied, yields a ratio of source-to-film distance and specimen thickness of

$$\frac{\text{SFD}}{t} = \frac{F}{U_f} + 1 \tag{21-7}$$

Where a plate 6 mm (0.236 in) in thickness is to be radiographed with a source having a focal spot size of 3 mm (0.118 in) and a 0.5 mm (0.02 in) level of unsharpness is desired, the resulting SFD/t is 7. More complete discussions on the parameters affecting image sharpness and source-to-film distance are given in Ref. 19 as well as Halmshaw [2].

21-5 DETERMINING RADIOGRAPHIC EXPOSURE

Obtaining a satisfactory radiograph is now recognized to involve both material and geometric considerations as well as to require knowledge of the source and film characteristics. These factors are summarized in exposure charts as shown in Figs. 21-6 through 21-9.

Figures 21-6 and 21-7 depict x-ray inspection of aluminum and steel, respectively. The thickness of the material to be inspected is given on the horizontal axis and the exposure is given on the vertical axis. Exposure (E) defines the x-ray intensity that is a function of the filament current in milliamperes, and the time of exposure in seconds. Thus,

$$E = AT \tag{21-8}$$

where A = filament current (mA)
T = time (s)

The previously discussed relationship of x-ray energy level and penetrating ability is demonstrated in these figures, where, for an assumed value of tube current, an increase in the x-ray energy level, as shown by the slanted lines, increases the thickness of the material that can be penetrated. Moreover, at a constant energy level, an increase in the tube current also increases the depth of material that can be penetrated. The differences in Figs. 21-6 and 21-7 are related to the fact that the steel is more difficult to penetrate than the aluminum, i.e., greater energy levels are required to penetrate equal amounts of steel as compared to aluminum. The information shown in these two curves is related through the previously discussed attenuation constants as well as the radiographic equivalence factors.

The parameters appearing in the box in the lower right corner of the figures give special parameters that describe the x-ray machine as well as the film and exposure characteristics. Typically, exposure curves must be established for each machine and film combination [3]. More discussion on exposure charts is given in a number of sources including Refs. 1–4, 7, 19, and 20.

Example 21-2. A steel plate 25.4 mm (1 in) thick needs to be inspected for internal defects and radiography has been proposed. Two instruments are available. One instrument has a tube voltage of 160 kV. The maximum filament current is 4 mA

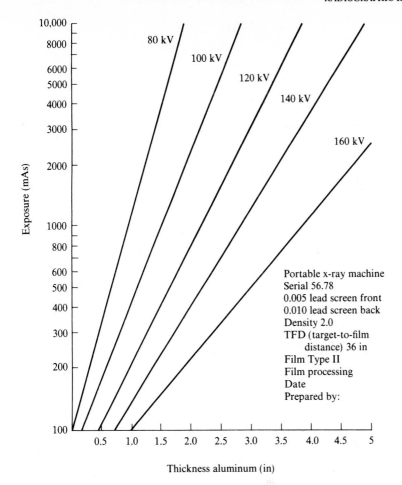

FIGURE 21-6
X-ray exposure chart for aluminum with portable equipment. (*From General Dynamics [20]. Courtesy American Society for Nondestructive Testing.*)

and the focal spot size is 0.4 mm. The other instrument has a maximum tube voltage of 220 kV. The maximum filament current for this machine is 6 mA and the focal spot size is 2.2 mm. Compare the inspection time as well as the image quality to be expected from these two machines. Assume that the parameters given in the corners of Figs. 21-6 and 21-7 apply here.

Machine Number 1: Figure 21-7 shows that at 160 kV, the exposure required for the sample would be 30,000 mA s. With the maximum 4 mA tube current available, the required inspection time would be

$$T = \frac{30{,}000 \text{ mA s}}{4 \text{ mA}} = 125 \text{ min}$$

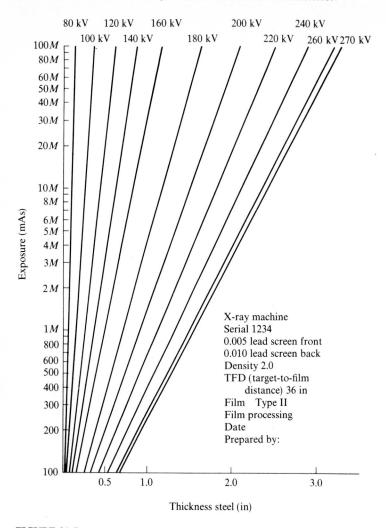

FIGURE 21-7

X-ray exposure chart for steel with permanent equipment. (*From General Dynamics [20]. Courtesy American Society for Nondestructive Testing.*)

The factors involved in the geometric unsharpness would be the side length of the focal spot as well as the source and object distances. The maximum distance that the object could be from the film is 1 in and the 36 in source-to-film distance yields a minimum source-to-object distance of 35 in. With this,

$$U_g = \frac{(0.015 \text{ in}) (1 \text{ in})}{35 \text{ in}} = 0.428 \times 10^{-3} \text{ in}$$

Machine Number 2: The corresponding values for machine number 2 yield

$$E = 700 \text{ mA s} \quad \text{and} \quad T = \frac{700 \text{ mA s}}{6 \text{ mA}} = 1.9 \text{ min}$$

Iridium 192 exposure factors for

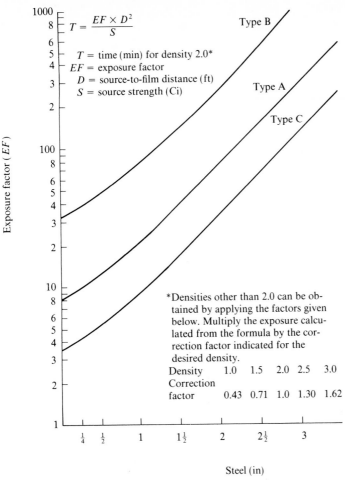

FIGURE 21-8

[192]Ir exposure chart for steel. (*From General Dynamics* [20]. *Courtesy American Society for Nondestructive Testing.*)

$$ \text{and} \qquad U_g = \frac{(0.087 \text{ in})(1 \text{ in})}{35 \text{ in}} = 2.5 \times 10^{-3} \text{ in} $$

Comparing the two machines, it is easy to see that the time required for the inspection with the smaller machine is unreasonable. For the larger machine, the time is satisfactory as is the geometric unsharpness.

Exposure information for ^{192}Ir and steel is given in Fig. 21-8. In this case, the strength of the source must be determined for each exposure. X-ray exposures for graphite composites are given in Fig. 21-9. It is noteworthy to observe the lower tube voltages that are used in comparison to those used for steel and

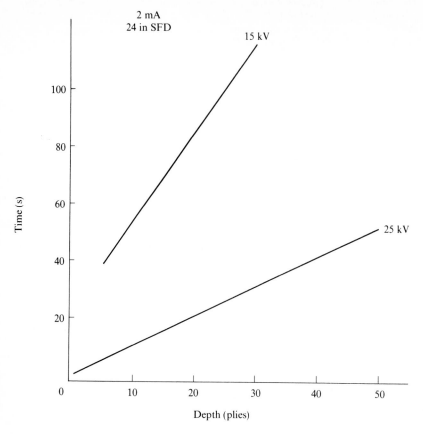

FIGURE 21-9
X-ray exposure for graphite-epoxy composites. (25 wt.% graphite, 75 wt.% epoxy). (*Courtesy General Dynamics.*)

aluminum. This is due to the lower absorption of the graphite composite material.

21-6   RADIOGRAPHIC SENSITIVITY

Radiographic sensitivity is a function of the ability of the system to produce a detailed image of the object as well as the ability of the recognition system to detect the presence of that object. For the present, the emphasis will continue to be on the characteristics of the system to produce a sharp, recognizable image on the recognition plane, without limiting the argument to any specific type of recognition system.

Following the presentation of Halmshaw [2], the thickness sensitivity, or contrast sensitivity, S for a material with a linear absorption coefficient of μ is expressed as the percentage of the total thickness x:

$$S = \frac{\Delta x}{x} 100 = \frac{2.3 \Delta D \left[1 + (I_S/I_D)\right]}{\mu G_D x} 100\% \tag{21-9}$$

where Δx = smallest observable thickness
  x = specimen thickness
  ΔD = density change on recognition media (e.g., film)
  I_S = intensity of scattered radiation
  I_D = intensity of direct radiation
  μ = linear absorption coefficient
  G_D = film contrast

For the Δx to be the minimum thickness change that can be detected, ΔD would be the corresponding minimum density difference that is recognizable by the system. For film radiography, this would obviously depend on the human eye. Halmshaw states that there is ample evidence that this equation agrees very well with experimental data with values for ΔD ranging from 0.006–0.01.

An extension of Eq. (21-9) is given by Halmshaw [2] to express the detail sensitivity and the crack sensitivity. For a linear crack of width W and length d, oriented at an angle θ, as shown in Fig. 21-10, the sensitivity is

$$Wd = \frac{2.3 \Delta D F}{G_D \mu} (d \sin \theta + W \cos \theta + U_T) \left(1 + \frac{I_S}{I_D}\right) \tag{21-10}$$

where F = form factor
  U_T = total unsharpness = $(U_g^2 + U_F^2)^{1/2}$

The form factor F can be normally assumed to be equal to 0.6. However, where $W(\tan \theta) > U_T$, then $F = 1$ may be used.

The most favorable orientation of a defect is so that the maximum distance of the anomalous material will be penetrated by the radiographic beam. Figure 21-11 shows how a possibly large defect could remain undetected if the beam were projected in the direction of the thinnest dimension of the defect. For

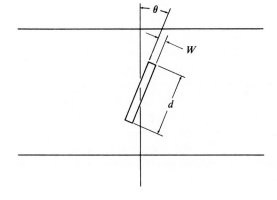

FIGURE 21-10
Crack dimensions for crack size sensitivity determination.

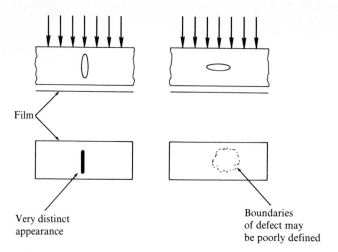

Film

Very distinct
appearance

Boundaries
of defect may
be poorly defined

FIGURE 21-11
Effect of object orientation on radiographic inspection.

volumetric defects such as voids, porosity, etc., this may not be a source of difficulty. For tightly closed cracks, however, the problem could be serious. Obviously, an inspection from more than one orientation minimizes the likelihood of defects being missed because of orientation.

21-7 EVALUATING IMAGE QUALITY AND SENSITIVITY

An indication of image quality often is an estimation of the smallest indicator that would be detectable. This is called an image quality index (IQI) and is usually established through the use of a device called a pentrameter. Other names for the pentrameter are penny and plaque. There are a myriad of image evaluation devices available, as specified in great detail in the various international inspection codes. These are described in several of the listed references. No attempt will be made here to describe them in a general context. A discussion using a typical pentrameter as an example will be given.

The pentrameter shown in Fig 21-12 is the type commonly used in industrial applications for evaluating weld or plate inspection quality. It must be made of either the same material as is to be inspected or a material that is radiographically equivalent. The thickness of the pentrameter is given by the dimension t while the ID number, fixed in lead on the surface of the pentrameter, indicates the maximum thickness of the material being inspected for which the pentrameter may be used. Three holes are in the pentrameter, labeled $1T$, $2T$, and $4T$, where the $1T$ hole diameter is equal to the pentrameter thickness. The diameters of the $2T$ and $4T$ holes are, respectively, twice and four times that of the $1T$ hole.

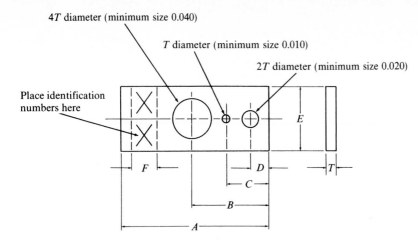

4T diameter (minimum size 0.040)

T diameter (minimum size 0.010)

2T diameter (minimum size 0.020)

Place identification numbers here

(*a*) Design for IQIs to but not including 180

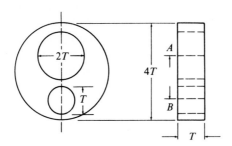

(*b*) Design for IQIs 180 and over

Table of Dimensions of IQI (Note 1)

NOTE 1—All dimensions in inches (Note 6).
NOTE 2—Tolerances for IQI thickness and hole diameter.
NOTE 3—XX identification number.
NOTE 4—IQIs No. 5 through 9 are not 1T, 2T, and 4T.
NOTE 5—Holes shall be true and normal to the IQI. Do not chamfer.
NOTE 6—To convert inch dimensions to metric, multiply by 25.4.

Number	A	B	C	D	E	F	Tolerances (Note 2)
5–20	1.500	0.750	0.438	0.250	0.500	0.250	±0.0005
	±0.015	±0.015	±0.015	±0.015	±0.015	±0.030	
21–59							±0.0025
60–179	2.250	1.375	0.750	0.375	1.000	0.375	±0.005
	±0.030	±0.030	±0.030	±0.030	±0.030	±0.030	
180–up	1.330T	0.830T	...	...	...	...	±0.010
	±0.005	±0.005					

FIGURE 21-12
Image quality indicator (IQI) or pentrameter, ASTM E 1025-84, made from the same material as the piece being inspected. (*Courtesy American Society for Testing and Materials.*)

TABLE 21-1
Image quality levels for hole-type pentrameters

Quality level	t (% of T_m *)	Perceptible hole diameter
1-2T	1	2T
2-1T	2	1T
2-2T	2	2T
2-4T	2	4T

*T_m is the thickness of the material being inspected.

In performing the radiographic exposure, the pentrameter must be situated such that it will be in the path of the beam, somewhat near to the area being inspected. If there is a risk that the pentrameter may be situated over a defective area, two shots must be taken.

The quality level and the related sensitivity are specified as given in Table 21-1. The quality level is specified with an alphanumeric symbol such as N-MT where N represents the maximum percent thickness of the material being tested for which the pentrameter is valid. M represents the minimum required perceptible hole diameter. For the shot to meet the specified requirements, the MT hole must be distinguishable in the radiograph. The sensitivity for the hole type pentrameter, as given by Halmshaw [2], is

$$\text{sensitivity} = \frac{\text{visible hole}}{\substack{\text{thickness of the specimen} \\ \text{being inspected}}} \times 100\% \qquad (21\text{-}11)$$

A typical pentrameter specification gives both the percent thickness of the material, e.g., 2, and the minimum hole that must be visible, e.g., 2T. A pentrameter giving these specifications would be a 2-2T unit.

While it is recognized by Halmshaw [2] that there is no simple relationship between the sensitivity as expressed by the IQI and that expressed by Eqs. (21-9) and (21-10), it is argued that a direct relationship would be a reasonable expectation. Dölle and Lemmer furnish additional results on experimental and theoretical studies of unsharpness and sensitivity [30].

21-8 SUMMARY

The intent of this section has been to introduce the reader to the basic requirements for obtaining a radiographic image. With this approach, much detail has been omitted. The various factors, such as exposure time and intensity, as well as contrast and image quality, have been discussed in sufficient detail to acquaint the reader with these principles.

CHAPTER
22

RADIOGRAPHIC
INSPECTION
SYSTEMS

22-1 INTRODUCTION

The preceding chapters have emphasized the physical principles of radiation sources, absorption, and geometry. This approach allowed a condensed and somewhat simplified discussion before the introduction of many of the factors that are quite important to the execution of a radiographic inspection. In the present chapter, some additional elaboration will be given on some of these additional factors and typical radiographic systems will be briefly described.

22-2 FILTERS AND SCREENS

Since a narrow-beam configuration is in actuality impossible to achieve and in fact often difficult to approach, several devices are available that minimize the effect of the broad-beam excited secondary radiation, i.e., scatter. Some of the more common of these items are the various filters and screens [2–4, 19]. Filters are typically placed near to the source, while screens may be placed in the film cassettes, on either side of the film sheet.

The purpose of the filter near to the x-ray source is to remove from the beam spectrum those energy levels that are more likely to cause scatter in the test specimen. For example, the longer wavelength, i.e., lower-energy, x-rays are less penetrating and, as shown in Fig. 20-18, produce greater scatter. A filter that removes the lower voltage x-rays from the spectrum is said to create a "hard" beam.

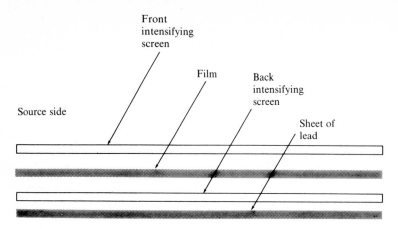

FIGURE 22-1
Typical film packet showing placement of screens.

Lead foil screens placed adjacent to the film as shown in Fig. 22-1 tend to reduce scatter from all sources. Some secondary radiation effects occur in the film itself where deflected photons may decrease the sharpness of the image. Lead screens also serve as intensifiers, thereby reducing the required exposure. In most cases, exposure characteristics for filter and screen combinations are established empirically for a particular film and x-ray machine. Moores [21] describes recent applications of screen and film combinations for image quality improvement.

22-3   TYPICAL SOURCE HANDLING DEVICE FOR ISOTOPES

An isotope source is constantly emitting radiation and cannot be turned off. Control of these sources is accomplished by moving it in and out of a storage container that is capable of confining the radiation energy within its walls. While there are a variety of these devices that serve specific purposes and have several levels of automated control, only one simplified system will be described here, in the interest of brevity.

The device shown in Fig. 22-2, from Iddings [9], consists of a movement controlling assembly, i.e., the crank, which connects to the source and moves it in and out of its safe storage container. When exiting the container, the source is moved along a tube until it reaches a collimator that controls the direction of the beam of photons. In its simplest form, the collimator is positioned over the item to be inspected, at the proper distance from the test surface, and the source is moved out of the container to the collimator. The time in the collimator controls the exposure time and the characteristics of the particular source controls the

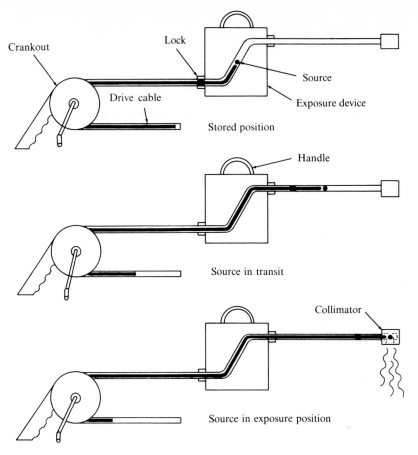

FIGURE 22-2
Source handling for radioactive isotopes. (*From Iddings* [9]. *The Nondestructive Testing Handbook,
vol. 3, Radiography and Radiation Testing, 2nd ed., 1985 American Society for Nondestructive Testing.*)

energy of the emitted radiation. Once the exposure is complete, the source is
retrieved to safety within the container, which is typically called a storage shield,
or camera.

Since the amount of radiation emitted by the source is typically of sufficient
strength to cause serious injury to operators, very strict controls are exercised
over the operation of these devices.

The size of a typical source container may be so that it can be easily carried
by one individual. Others, however, designed for automated inspection, may
require special handling. Pipeline inspection is most frequently accomplished
using a "pig" or "crawler," which travels inside the pipe through the length of
the pipeline and is activated by remote control. These devices may be too large
and heavy to be easily carried and will require special effort.

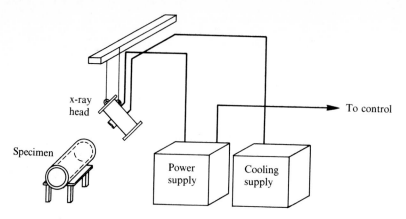

FIGURE 22-3
Typical x-ray inspection room.

22-4   X-RAY MACHINES

Because of the larger size of a typical x-ray head, the need for external power and a continuous cooling operation, x-ray machines are more frequently used in a room where suitable cranes and hoists are available for manipulating the inspecting system. A schematic diagram of an x-ray room equipment installation is shown in Fig. 22-3. For a 300 kVp unit, the x-ray head may weigh around 40 kg (88 lb). This is suspended from the hoist as shown. Either mechanical or electrical manipulation controls may be available. Connected to the head through cables and tubes are the external power supply and control unit as well as the cooling system. An x-ray room will be fully shielded with lead, concrete, or some combination of materials in order to provide shielding from stray radiation. Smaller power units may rest on the floor and be encased in a box lined with lead sheathing.

Portable x-ray machines having a wide range of energy levels are available for field use. Lower-energy units that do not require external cooling can generally be hand carried to the job site. Crawler-type units of 300 kV energy level are common for field use. Larger units with energy levels of 1.5 MeV which can be taken to the job site are also available [31]. In this case, however, the need for forced cooling the greater weight associated with the higher-energy levels increases the number of components involved. While this portability may not match that of other NDE techniques, it does offer the advantages of x-ray inspection in remote locations.

22-5   NEUTRON PARTICLE MACHINES

The application of neutrons in NDE has been significantly broadened by the development of neutron radiography systems that could be taken to the job site. This overcame a significant disadvantage of neutron radiography where previ-

FIGURE 22-4
Moveable accelerator/neutron generator. (*From Dance* [*32*], *Courtesy LTV Aerospace and Defense Company.*)

ously the part to be inspected needed to be taken to a nuclear reactor in order to be inspected. Figure 22-4 shows a moveable accelerator/neutron generator tube situated underneath an aircraft wing panel. The manipulation unit and the accelerator are mounted on a trailer and the power and control units are located nearby. Again, this represents limited portability but does offer the advantages of neutron radiography in previously impossible situations.

22-6   PROTON SOURCES

At the present, proton sources are limited to nuclear reactors and accelerators.

22-7   RADIOGRAPHIC INSPECTION FACILITIES

In common with most NDE systems, radiographic inspection is required in the laboratory, the shop, and the field. Generally, these requirements demand some differences in the equipment and facilities needed to support the inspection.

Unique to radiography is the strong emphasis on protection from stray radiation. Protection techniques, however, are well established in the industry and pose no particular problem.

Field inspection with x-rays most frequently will be accomplished with isotopes that normally are easily transported. In the laboratory or shop, the larger power tube units are usually supported with auxiliary power supply and cooling capacity. Isotopes may be used in the shop much as they are in field inspections. In addition, where film is used for the recording media, chemical processing equipment and space is needed. Video monitoring and digital signal analysis systems may require power to support the equipment.

Since the items to be radiographed may range in size from specimens smaller than a coin to a full size aircraft, the facilities must provide for the placement of the source and must accommodate the item being inspected. For large items that must be inspected with maximum sharpness, additional space will be required simply to achieve a suitable object distance and source L/D ratio. Additionally, inspection in populated surroundings may require shielding of the radiation using concrete, lead, or some combination of materials.

22-8 SUMMARY

Because of the wide range of capabilities of radiographic inspection, the variations in sizes of objects to be inspected, the somewhat restricted movement of higher-power sources, and the need for personnel protection, the facilities demands for this technique may be greater than for the others. The material in this chapter is not meant to give a complete discussion of machines, accessories, and facilities, but, rather to introduce the reader to the fundamental areas of concern. More detailed information is available from a number of the references.

CHAPTER
23

RADIOGRAPHIC INSPECTIONS

23-1 INTRODUCTION

The present chapter is intended to illustrate and discuss only a few applications of radiographic inspection in order that typical examples may be understood. There are, of course, a very large number of inspection techniques and applications that are beyond the scope of the present effort. Adequate discussion of specific applications is available from a number of reference sources.

At this stage, the reader is probably aware of the myriad of test variations that can affect a radiographic inspection. Not only is secondary scattering a potential problem source, but irregularity in the absorption of the materials because of chemical and metallurgical structure affects as well as thickness differences also may yield unexpected results. The effects of the secondary scattering may be minimized by using suitable films and screens and by selecting the optimum energy level.

For film radiography, the optimum exposure is often determined by trial and error, based on both the exposure charts and the experience of the operator. With real-time, video display units, however, the optimum tube voltage and exposure time, for example, may be determined with minimum loss of time.

The present examples have been selected to show optimum results that were achieved by experienced radiographers. Typically, there is no fixed formula for any inspection and each must be approached with deliberation and care.

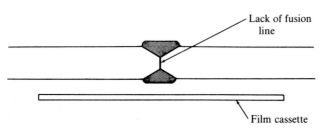

FIGURE 23-1
X-ray inspection of a double-V weld with a lack of fusion defect.

23-2   RADIOGRAPHIC APPEARANCE OF COMMONLY OCCURRING ANOMALIES

The density of defect indications on film or other recording media is inversely related to the total linear density of the material in the beam path. Decreased densities in the appearances of the indications on the radiographs will therefore indicate more dense material along the total beam path.

As previously described, the detectability of discontinuities by radiography is very much dependent on the shape and orientation of the anomaly as well as the density. In the material to follow, the effect of shape first will be discussed followed by a discussion of the orientation effects. Since defects found in welds and castings represent a large portion of the typical defects detected with radiography, these items will be used as typical examples. The present discussion will assume electromagnetic waves, i.e., either x-rays or γ rays, but comments related to orientation might well apply to other types of sources.

A frequently occurring weld defect is the lack of fusion example shown in Fig. 23-1. In this case, the failure to completely join the original butt ends of the plate has created a thin discontinuity perpendicular to the plate surface. In a radiograph obtained from the arrangement shown in Fig. 23-1, the lack of fusion defect would appear as a thin dark line following a straight path along the plate contact region. It would, of course, be either continuous or intermittent, depending on the pattern of the defect. The weld bead would appear less dense in the radiograph due to its greater thickness. Where the weld bead was ground flat, there would, of course, be no change in section and therefore no difference in appearance of the weld bead and the plate metal.

Other common weld defects, such as cracks, porosity, inclusions, etc., will also appear along the weld line but in a pattern less regular than that occurring for the lack of fusion defect. Porosity, i.e., a gas hole in the metal, would appear on the radiograph as a spot or a field of spots darker than the surrounding material. Inclusions of foreign matter, on the other hand, may appear on the radiograph as either darker or lighter spots or fields, depending on the density of the foreign material relative to that of the weld material. Both tungsten and copper, for example, would appear lighter than the surrounding material in a radiograph of a steel weld because of the greater density of the two metals compared to iron. Since these two metals are used in many electric welding rods, it is not uncommon to find traces of them showing in the weld radiographs.

Corrosion either in a flat or circular section may be readily detected using radiography, since the loss of material affects the exposure of the film. The

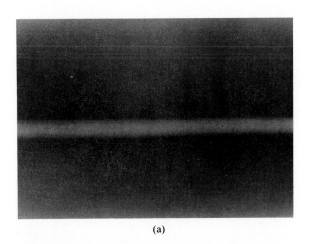

(a)

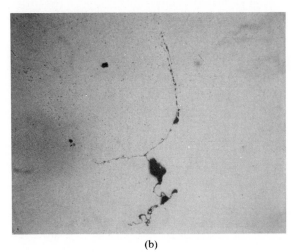

(b)

FIGURE 23-2
(*a*) Radiograph showing cracks in welded plate. aluminum 5451, 6.4 mm (0.250 in) thick, 80 kV, 5 mA, 32 s, 1 m (40 in) SFD, 3.0 mm × 3.0 mm focal spot size, class 1 film, no screens. (*Radiograph Furnished by NDE, Inc., Fort Worth, Texas.*) (*b*) Photomicrograph of defect region from the weld in (*a*) showing cracks and inclusion. 50 × , unetched. (*Photomicrograph furnished by NDE, Inc., Fort Worth, Texas.*)

appearance on a radiograph of material loss from corrosion would be as a dark field with the thinnest areas appearing the darkest.

23-3 INSPECTION OF UNIFORM FLAT SECTIONS

A flat plate or bar or a weld joining two flat plates or bars, are the most common examples for uniform flat section inspection. The typical inspection arrangement given in Fig. 23-1 shows the source located a suitable distance above the center line of the weld. The film would be placed in a carrier that would be located adjacent to the underside of the weld. Normally, the film carrier is fitted with screens to minimize the effects of secondary scatter.

Weld cracks with occasional inclusions are shown in Fig. 23-2(a) for a weld in a 6.4 mm (0.25 in) thick aluminum plate. Shown near to the center of the weld bead is an irregular field running longitudinally along the weld. This dark line is the crack field where the darker appearance of the cracks in the radiograph results from the lesser absorption of the air gap between the crack faces. Figure 23.2(b) shows the defects detected in the weld.

23-4 INSPECTION OF CIRCULAR SECTIONS

Often the geometry of circular sections as found in pipes, pipe fittings, piping systems, and pressure vessels makes them geometrically more suitable for radiographic inspection than flat sections. Figure 23-3 shows a source located at the

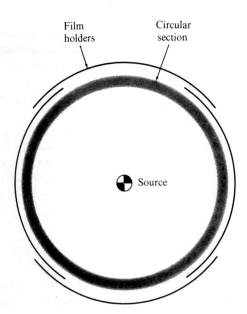

Film holders

Circular section

Source

FIGURE 23-3
Radiographic inspection of a circular section with internal source and circular film placement.

axial center line of a circular section with the film wrapped circumferentially around the outside of the pipe. Where large sections as shown would require multiple film sheets, smaller sections may be shot with one film sheet wrapped around the section.

The arrangement shown could be used for locating weld flaws as well as pipe wall defects such as longitudinal manufacturing defects, porosity, and loss of section. A plane perpendicular to the central axis of the part and intersecting the source would have the maximum intensity and least amount of secondary

FIGURE 23-4

(*a*) Radiograph showing a shrinkage defect in sand casting. 8630 cast iron, 31.8 mm (1.25 in) wall thickness, ^{192}Ir, 42 Ci, 40 s, central axis source, 203.2 mm (8 in) SFD, focal spot size 0.1 in × 0.1 in, class 1 film, lead screens, 0.13 mm (0.005 in) front, 0.26 mm (0.010 in) back. (*Radiograph furnished by NDE, Inc., Fort Worth, Texas.*) (*b*) Sand casting containing shrinkage defects shown in (*a*). (*Photograph furnished by NDE, Inc., Fort Worth, Texas.*) (*c*) Shrinkage defects found in sand casting shown in (*b*). (*Photomicrograph furnished by NDE, Inc., Fort Worth, Texas.*)

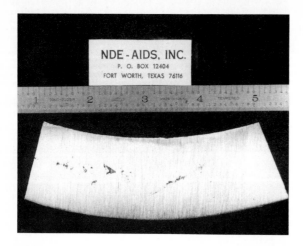

FIGURE 23-4
Continued.

radiation and geometric distortion. For large sections where multiple film sheets were used, the shot will need to be repeated because film edges will overlap causing areas to be missed.

The radiograph shown in Fig. 23-4(*a*) shows porosity in a sand cast pipe fitting shot with an internal source with the arrangement shown in Fig. 23-3. The casting is showing in Fig. 23-4(*b*) and the defects that were found are shown in Fig. 23-4(*c*).

Circular objects may also be radiographed with flat film located external to the pipe as shown in Figs. 23-5 and 23-6. In Fig. 23-5, the source-to-film distance varies at different beam angles but the source-to-object distance remains constant. The radiation beam essentially traverses a constant pipe wall thickness but the source-to-film distance changes away from the perpendicular from the source to the film. A large source-to-film distance is used in this case in order to minimize the geometric distortion. Multiple shots would be required in order to achieve a full 360° inspection of the weld or section wall.

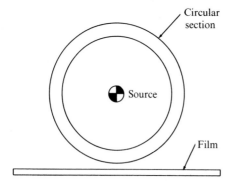

FIGURE 23-5
Radiographic inspection of circular section with internal source and flat external film.

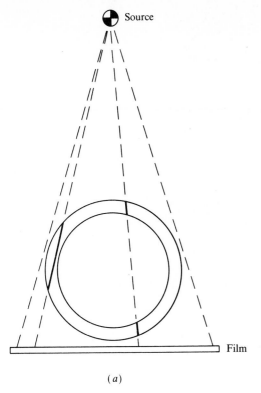

Source

Film

(a)

FIGURE 23-6
(a) End view of the setup for radiographic inspection of a circular section with external source and flat external film [7]. (b) Side view of the setup for radiographic inspection of a circular section with external source and flat external film [7].

The situation is only slightly altered with the arrangement shown in Fig. 23-6. In the center, slightly to either side of the perpendicular from the source to the film, the radiographic beam encounters a cross section with only slightly varying dimensions. At some distance from the normal, however, the wall thickness being traversed becomes considerably larger and the source-to-film distance is increasing with the result that the exposure is reduced. When this arrangement is used, several shots are required in order to achieve a full coverage of the section of interest.

It should be noted that whether a defect is located in the top or the bottom section cannot be determined from a single radiograph with the arrangement shown in Fig. 23-6 and multiple shots used to establish defect locations are time consuming and expensive. Orienting the circular section in Fig. 23-6(b) at an angle to the beam axis in the vertical plane will permit the beam to traverse both the top and the bottom of the weld, thereby revealing the exact location of a detected defect. A radiograph obtained with this arrangement for a high-pressure brass tubing is shown in Fig. 23-7. A small pin-hole defect is seen in the weld.

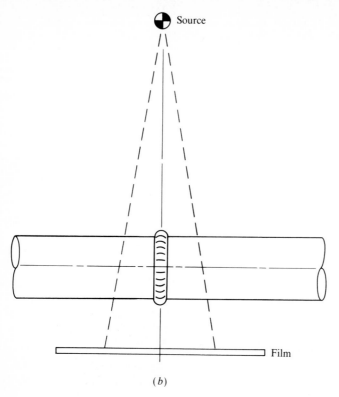

(b)

FIGURE 23-6
Continued.

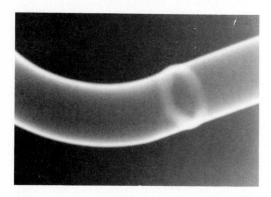

FIGURE 23-7
Radiograph of a weld in brass tubing. AMS 5557 brass tubing and AMS 5680 weld wire. Tubing wall thickness 1.5–3 mm (0.060–0.120 in), 180 kV, 10 mA, 16 s, 1 m (40 in) S F D , f o c a l s p o t s i z e 0.4 mm × 0.4 mm, class 1 film, lead screens, 0.13 mm (0.005 in) front, 0.26 mm (0.010 in) back. (*Radiograph furnished by NDE, Inc., Fort Worth, Texas.*)

Source

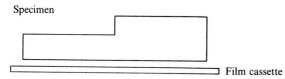

Specimen

Film cassette

FIGURE 23-8
Examination of an irregular cross section.

23-5 INSPECTION OF IRREGULAR SHAPES

Radiographic inspection of objects with varying cross sections represents an additional complication since the total absorption will be greater in the thicker parts of the section. Varying thicknesses may be present in a variety of parts such as castings and T-welds, as well as parts machined with sloping sides.

An example of a section with varying cross section is shown in Fig. 23-8. Exposed as shown, the thinner section would appear darker on the film than the thicker section. In a simple case such as this, the specimen could be inspected either with two exposures, each determined by the appropriate thickness, or with one shot using the double-loading technique to be described.

The first alternative would be to simply fill the area above the thinner section with radiographically equivalent material and establish the exposure based on the full thickness, as shown in Fig. 23-9(a). Another method is to double-load the film packet. The intent of this approach is to place in the film packet two films with different speeds. The faster film would be placed on top and the slower film underneath. Exposed as shown in Fig. 23-9(b), the faster film would be overexposed in the thin section and yet be properly exposed in the thicker section. In the thin section, the x-ray energy striking the slow film must first pass through the top, fast film layer before striking the underlying, slower film. With film speeds properly balanced, a suitable exposure may be achieved in each section with a single exposure.

Source

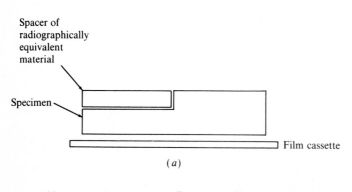

Spacer of
radiographically
equivalent
material

Specimen

Film cassette

(a)

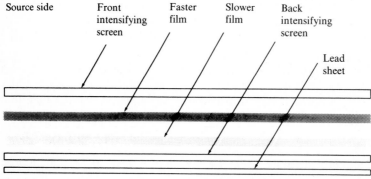

Source side

Front
intensifying
screen

Faster
film

Slower
film

Back
intensifying
screen

Lead
sheet

Back side

(b)

FIGURE 23-9
(a) Use of masking material for radiographic inspection of irregular cross sections. (b) Double film method for radiographic inspection of irregular cross sections.

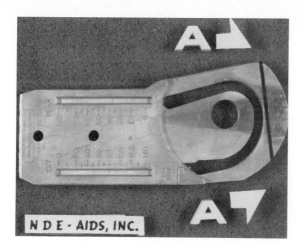

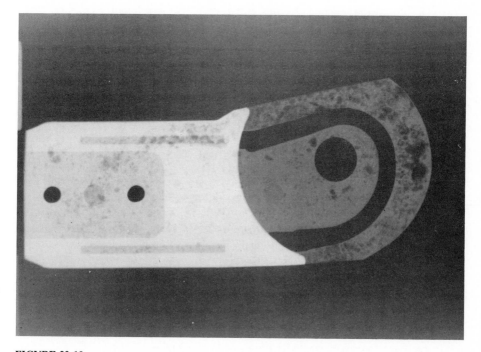

FIGURE 23-10

(a) Aluminum permanent mold casting, AA 383.0 material. (*Photograph furnished by NDE, Inc., Fort Worth, Texas.*) (b) Radiograph of an aluminum permanent mold casting showing gas pockets. Material thickness 6.4 mm (0.250 in), 80 kV, 5 mA, 30 s, 1 m (40 in) SFD, focal spot size 3.0 mm × 3.0 mm (0.118 in × 0.118 in), class 1 film, no screens. *Radiograph furnished by NDE, Inc., Fort Worth, Texas.*) (c) Photograph showing gas pockets found in the aluminum casting shown in (a). (*Photomicrograph furnished by NDE, Inc., Fort Worth, Texas.*)

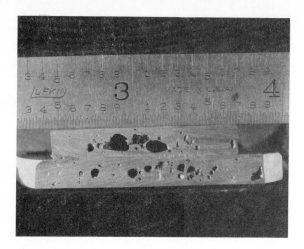

FIGURE 23-10
Continued.

An aluminum casting having different thicknesses and inspected with radiography is shown in Fig. 23-10(*a*). In the thin section used to establish the radiographic exposure, the material thickness was 6.4 mm (0.250 in). The radiograph in Fig. 23-10(*b*) clearly shows casting defects in the form of gas pockets in the thin section as well as the effect on the radiograph of the additional material in the thicker section. The defects found are shown in Fig. 23-10(*c*).

A nylon casting that has been radiographed for internal defects is shown in Fig. 23-11(*a*). The thickness of the thick and thin sections is approximately 12 (0.5 in) and 2 mm (0.08 in), respectively. The radiograph in Fig. 23-11(*b*) shows porosity due to turbulance during the material flow from the thin to the thick sections. A slice of the part, seen in Fig. 23-11(*c*), shows the defects.

Example 23-1. The weld specimen shown in Fig. 23-2 is approximately 6.4 mm (0.250 in) thick. It was inspected using 80 kV x-rays. Using Fig. 21-6, estimate a proper exposure for this specimen and compare your results with the data furnished with the radiograph. Was the source-to-film distance satisfactory to achieve a low level of unsharpness?

For 80 kV x-rays, Fig. 21-6 indicates that 185 mA s would be required for a proper exposure. Yet the data from the radiograph show that the exposure took 32 s at 5 mA yielding 160 mA s. While the 185 mA s exposure might be a good first attempt, the radiograph exposed at that level would be too dark. Dropping back on the time of exposure would lead to a good radiograph.

There are several possible explanations for the disagreement in the estimated and actual exposures. First, there is no assurance that the film used in the exposure is the same as that used for the exposure chart. Further, the exposure chart states that lead screens were used and there is no indication that there were screens used in obtaining the radiograph.

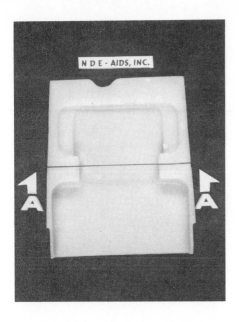

FIGURE 23-11
(*a*) Nylon type 6 casting. (*Photograph furnished by NDE, Inc., Fort Worth, Texas.*) (*b*) Radiograph of a nylon casting showing turbulence resulting from material flowing from the thin to the thick sections. Material thickness 12 mm (0.500 in) in thick part, 50 kV, 10 mA, 120 s, 1m (40 in) SFD, focal spot size 3.0 mm × 3.0 mm (0.118 in × 0.118 in), class 1 film, no screens. (*Radiograph furnished by NDE, Inc., Fort Worth, Texas.*) (*c*) Photograph showing turbulence in the casting shown in (*a*). (*Photograph furnished by NDE, Inc., Fort Worth, Texas.*)

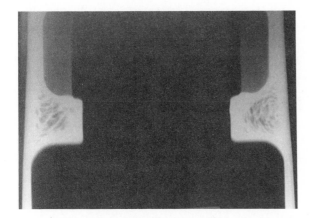

The suggested minimum SFD for an unsharpness of 0.5 mm, according to Eq. (21-6), would be seven times the plate thickness or $\text{SFD} = 6.4 \text{ mm} \dfrac{3 \text{ mm}}{0.5 \text{ mm}} + 1 =$ 44.8 mm (1.76 in). The actual SFD is seen to be 40 in (1.02 m), which is considerably beyond the value just calculated. However, the calculated SFD applies only directly under the beam and the larger SFD used in practice decreases the geometric unsharpness on the periphery of the beam.

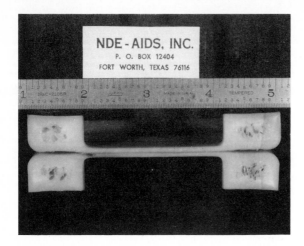

FIGURE 23-11
Continued.

23-6 EFFECTS OF MATERIAL PROPERTY VARIATIONS ON RADIOGRAPHIC INSPECTIONS

Dissimilarities in material properties may affect radiographic attenuation and thus create some difficulties in the interpretation of radiographs. In one such instance, Kruzic [33] has described the existence of the problem for nuclear reactor pressure vessel recirculating inlet and core spray nozzles. In this case, the linear images that may occur at the boundaries of Inconel and carbon or stainless steel welds could easily be interpreted as nonfusion defects when, in fact, they arise from the abrupt change in density at the interface between the two metals.

In order to investigate the problem, a series of radiographs were obtained using a setup similar to that illustrated in Fig. 23-12. The wall thickness of the steel parent metal at the weld area was 25.4 mm (1.0 in). Because of the differences in the absorption of the steel and the Inconel, a model with an equivalent steel weld thickness of 35.1 mm (1.38 in) was used for exposure calculations.

Where the ^{192}Ir source is shown in Fig. 23-12, similar inspections were made with a 300 kV x-ray head in the same location as the isotopic source. Double-loaded cassettes were used with 0.25 mm (0.010 in) front and back lead screens in place adjacent to the film. Three different class 1 films were used. The ^{192}Ir source had an intensity of 28 Ci or 1036 GBq (giga becquere.) The focal spot size was 2.24 mm (0.088 in). The x-ray head was a half-wave rectified type with 3 mA tube current and a focal spot of 2.54 mm (0.100 in).

Figure 23-13 shows a sample radiograph of the weld section with the change in density marks indicated. Where nonfusion images appear sharp, distinct, and generally linear, the change in density images normally appears as a series of parallel lines. They may be somewhat indistinct and fuzzy, yet, the

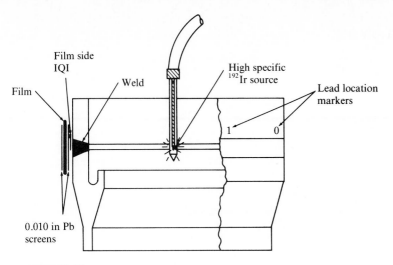

FIGURE 23-12
Setup for panoramic exposure weld in nozzle assembly. (*From Kruzic* [*33*]. *Reprinted with permission from Materials Evaluation, vol. 44, no. 11, 1986, The American Society for Nondestructive Testing, Columbus, OH.*)

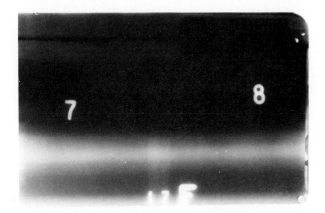

FIGURE 23-13
Radiograph of weld area showing material density change lines. (*From Kruzic* [*33*]. *Reprinted with permission from Materials Evaluation, vol. 44, no. 11, 1986, The American Society for Nondestructive Testing, Columbus, OH.*)

interpretation of radiographic images to be either a harmless change in density or serious nonfusion in the weld requires further investigation.

23-7 SUMMARY

Numerous examples of radiographic inspection have been given in this chapter. The examples have been chosen so as to indicate the breadth of nondestructive inspections available with the various sources of radiographic energy. Certainly the applications are not limited to the examples chosen, but rather are far too numerous to describe in a review presentation such as is given here. Further information is available to the reader from the numerous references.

CHAPTER
24

RADIOGRAPHIC TECHNIQUES FOR MATERIALS ANALYSIS AND STRESS MEASUREMENT

24-1 INTRODUCTION

Materials analysis and stress measurement are important applications of radiographic technology. In most cases, either beam attenuation or diffraction is used in order to investigate the parameter of interest. Attenuation has already been covered and will not be repeated here, but a brief overview of diffraction techniques seems fitting.

24-2 DIFFRACTION TECHNIQUES

X-ray diffraction techniques recognize the fact that perfectly monochromatic and perfectly parallel x-rays when striking crystallographic planes will be diffracted in an orderly fashion of reinforcement and cancellation so as to yield information about the spacing of the planes of the crystal. The spacing of the planes can be used as an indicator of stress level in the material, amongst other things.

The principles of x-ray diffraction techniques are described by Cullity [34]. Initially, it may be visualized that the incident energy striking the material surface, as shown in Fig. 24-1, consists of perfectly parallel and perfectly

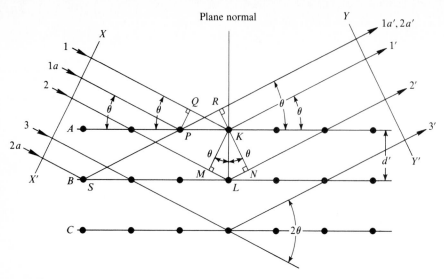

FIGURE 24-1
Diffraction of x-rays by a crystal. (*From Cullity* [*34*]. *Courtesy Addison-Wesley.*)

monochromatic x-rays. It is further assumed that they are perfectly in phase when passing plane XX'. A set of identical planes of atoms in a crystalline material is shown to be separated by a distance d'. The emitted x-rays are then shown to be leaving the material and passing through plane YY', where the diffraction angle θ is the same as for the incident wave.

Where the x-rays passing the reference plane YY' are perfectly in phase, they will be mutually reinforced and the additive magnitude of their intensities will increase. Conversely, x-rays not in phase will serve to cancel each other with the result that the emitted x-rays that are in phase will appear quite distinctly from the out of phase portion of the energy. Recognizing that where the planes in the region of interest will all be equally spaced at distance d', the diffraction characteristics of the reinforced beams are expressed by Bragg's Law

$$n\lambda = 2d'\sin\theta \qquad (24\text{-}1)$$

where n is the order of refraction that expresses the number of wavelengths in the path differences between the rays scattered by adjacent planes. The order of refraction n must be an integer number. Thus, where the diffracted angle θ and the wavelength of the x-rays λ may be established in a test apparatus, the planar spacing may be directly determined.

As noted by Cullity, diffraction is essentially a scattering phenomenon and the diffracted beam may be defined as one composed of a large number of scattered x-rays mutually reinforcing each other. Not only is the x-ray energy diffracted upon striking the material, but a considerable portion will be scattered as discussed in the previous chapters. Since the atoms will scatter the incident energy in all directions, each diffracted ray that is detectable must be a result of the combined reinforcement. The remaining portions of the scattered radiation are considerably weaker simply because of the lack of reinforcement. The result, then, is that the spacing for crystallographic planes may be measured by observing both the wavelength and the diffracted angle of the detected energy.

24-3 X-RAY MEASUREMENT OF STRESS

For a uniaxial stress field, Cullity gives the following relationship of stress and planar spacing:

$$\sigma_y = -\frac{E}{\nu} \frac{d_n - d_0}{d_0} \qquad (24\text{-}2)$$

where E and ν are as previously defined

d_n = spacing of crystallographic planes which that reflect at normal incidence under stress

d_0 = spacing of planes under no stress conditions

The expression for biaxial stress is similar. With the planar spacing known, then, the stress level in the material may be directly calculated.

Crystalline materials such as metals are composed of numerous grains each having the same crystallographic orientation within the grain. For isotropic material, the grains will be randomly oriented within the material. Thus, the planes in the crystal, while maintaining their constant spacing, will be oriented at a variety of directions relative to the dimensions of the sample being discussed.

Rather than using the incident-reflected (diffracted) beam method just described, a more common method for determining the planar spacing for stress measurement and material studies is the back-reflection method where an x-ray beam strikes the surface at normal incidence and the diffracted beams leave the material at an angle of 2θ from each other. Each diffracted beam forms a cone with the apex on the material and the normal, incident beam as the central axis of the cone. Thus, on the base of the cone, which is parallel to the material surface and might be represented by x-ray film, a series of concentric circles or rings would appear. The radius of these rings, called Debye rings, are representative of the planar spacing in the material. Where stressing the material results in a distortion of the planar spacing, this would be measurable by observing the change in radii of the Debye rings with the applied stress.

Numerous reviews exist for residual and applied stress measurement using the x-ray diffraction technique. Among these are the reports given by Weiss [35], Ruud [36], and Borgonovi [37].

A report on x-ray techniques specifically applied to the measurements of residual weld stresses has been given by Ruud, DiMascio, and Melcher [38]. Four test specimens of 12 in (0.305 m) diameter, schedule 80, austenitic stainless steel, girth-welded steel pipe were used in this study. The inside diameter was 11.376 in (0.289 m) and the wall thickness was 0.687 in (17.4 mm). Stress measurements were made on the inside surface of the weld with the x-ray source located in the pipe. Scans were made from $\frac{1}{2}$ in on either side of the weld fusion line and the scans were repeated at four locations around the circumference. Tensile stresses in the range of 10–40 ksi (68.9–276 MN/m^2) were found at the weld center line with high cold-worked induced compressive stresses found at $\frac{1}{2}$ in (12.7 mm) from the center line.

24-4 MATERIAL STUDIES USING RADIOGRAPHIC TECHNIQUES

Because of the unique planar spacing and atomic arrangements of the various elements, x-ray diffraction techniques are particularly useful in material studies. Besides the stress measurement capability previously described, x-ray techniques are useful in the analysis of the elements present in materials as well as in the determination of the percent phases present. In addition to relevant information in Refs. 35 and 36, Jenkins [39] presents a review of several applications of x-ray analyses to material studies. A description of photon attenuation methods for materials analysis has been given by Kouris, Spyrou, and Jackson [40].

X-ray techniques have been widely used for a number of years to study texture in crystalline material. In order to gather the data, however, it is necessary that the specimens be metallurgically mounted and polished. This fact prevents the inclusion of these methods in a nondestructive classification. Neutron scattering techniques, however, may be applied for texture determination in steel plates with no significant preparation. Allen and Sayers [41] report neutron scattering studies of 25.4 mm (1 in) thick mild steel plate. The neutrons enter the plate on one side and pass through the full thickness, thus giving an indication of the average texture in the material. With this technique, differences in texture induced by cold rolling were detectable.

24-5 SUMMARY

It has been the intent of this chapter to describe briefly some of the applications of radiographic techniques to determination of material properties and stress conditions. Since x-ray techniques have been used for this purpose longer than the other NDE methods, a vast amount of data exist showing x-ray analysis of materials. This represents a very useful data source when material properties are an important consideration.

CHAPTER
25

OTHER
APPLICATIONS OF
RADIOGRAPHY
TO NDE

25-1 INTRODUCTION

"Other applications of radiography to NDE" is meant to describe those inspections that are not accomplished with conventional x-radiography. These applications cover a wide range of areas: some in common usage, such as neutron radiography; others, like microfocus radiography, which are more specialized, and proton radiography, which represents unique, highly specialized techniques.

The through transmission and image production characteristics of radiography make the technique particularly amenable to application in some very specialized ways. In one case, the penetrating capability of radiography combined with a large computor memory and ordinary optical principles provide the basis for computor tomography (CT). In another case, the use of very small sources makes possible microfocus radiography which is useful for detecting smaller defects. Radiography can also be used in a stop-action mode with high power, short duration pulses in flash radiography. Other specialized applications of radiography include scanning electron microscopes where the secondary effects resulting from the radiation excited on a target material being studied are used as the investigative tool. Finally, the significant differences in the absorption characteristics of neutrons and protons, as compared to photons, creates a field of inspection capabilities far beyond that of x-radiography alone.

25-2 INSPECTIONS USING COMPUTED TOMOGRAPHY

Computed tomography utilizes scatter data from radiographic scans obtained at a large number of orientations of source, object, and receiver. The source and detector are moved in each scan plane as shown in Fig. 25-1. Additionally, the scan planes may be rotated until a full scan global field is obtained. The scatter data are digitized and stored and the results appear as a reconstructed image.

Engineering applications of computed tomography have been increasing in number in recent years, as described by Refs. 27–29, 42, and 43. A very typical application has been for low density, light element castings since the low absorption of these materials decreases the overall time requirement for obtaining the data set.

Gilboy and Foster [28] reported that a 30 h scan was required to obtain global data for a reconstruction of a light alloy pump casting, using 1.5 MeV x-rays. The thickest section of this part was 100 mm (3.94 in).

More recent work reported by Ellinger [42] and Johnson [43] give examples of CT inspections of industrial castings of automotive engine blocks and aircraft engine parts, respectively. Defects sought include thin walls as well as porosity, voids, inclusions, etc.

A typical tomogram of a gray iron engine block, as reported by Ellinger [42], is shown in Fig. 25-2. Using a 100 Ci source of ^{60}Co, 5 min was required to complete one scan and approximately 1 h was needed to complete the 13 section scans necessary to inspect one block. The straight lines on the left wall section

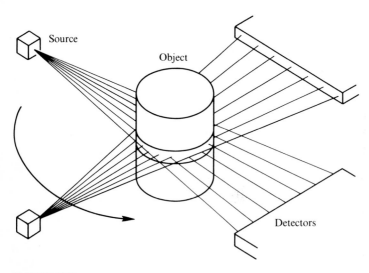

FIGURE 25-1
Scanning setup for computer-aided tomography (*From Kohutek and Frank [29]. Courtesy American Society of Mechanical Engineers.*)

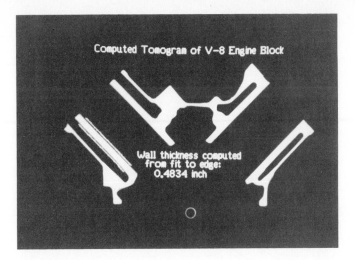

FIGURE 25-2
Tomogram of a gray iron engine block casting. (*Courtesy Scientific Measurements Systems, Inc.*)

show the overlay used for determining wall thickness. This technique is incorporated in an casting production line where earlier the blocks had been sectioned by a slow, saw cutting technique. With the CT method, there is considerable overall time saving compared to the destructive sawing techique.

Results for a study of tomographic weld inspection have been reported by Kohutek and Frank [29]. Using the scan arrangement shown in Fig. 25-3 results in a typical tomogram showing cracking and slag inclusions as shown in Fig. 25-4 for a steel weld sample measuring 2.4 mm thick by 304.8 mm long (0.095 in by 12 in). The slag inclusions appear as dark images toward the left of the tomogram and the cracking is indicated by the darker appearance of the field toward the right. These results were obtained using ^{192}Ir sources.

25-3 MICROFOCUS RADIOGRAPHY

Detecting flaws less than 10 μm in size in brittle materials such as ceramics is enhanced using microfocus radiography. Briefly, this technique uses a very small (microfocal) source, as shown in Fig. 25-5, which significantly decreases the size of the penumbra and therefore reduces the sharpness of the image. The image quality is further enhanced through the use of projection radiography where the enlarged image detail increases the resolution of the process. Radiation traveling along a path nearly perpendicular to the object and image plane will suffer very little scatter, while radiation on either side of the perpendicular will be more scattered and thereby not image the recording plane. Results on microfocus radiography are reported by a number of researchers including Berger and Kupperman [44], Parish [45], Bagnall and Kutzian [46], and Baaklini and Roth

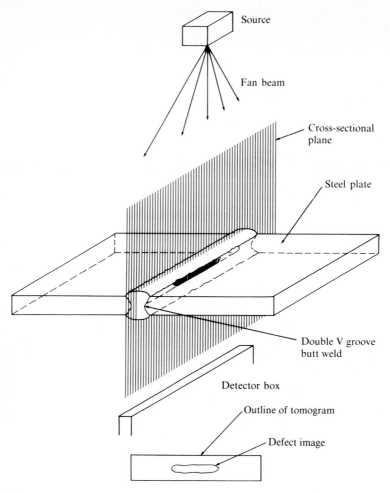

FIGURE 25-3

Computerized tomography techique for weld inspection. (*From Kohutek and Frank [29]. Courtesy American Society of Mechanical Engineers.*)

[47]. Typically, the resolution ability of a microfocus x-ray unit will be approximately equal to the size of the source. Parish reports tube characteristics ranging from 30–80 kV with currents of 0.5 mA at 80 kV and 1.0 mA at 50 kV or less and source sizes from 10–20 μm. Resolution of approximately 6 μm has been reported by Parish using fine wire mesh as the object.

Baaklini and Roth [47] have shown that the use of microfocus x-radiography will substantially increase the detection capability of surface and internal voids in structural ceramics. Their system had a 10 μm focal spot and tube characteristics of 30–60 kV and 0.25–0.32 mA. The magnification factor was 2.5 and SFD was 300 mm. Exposure times varied from 5–20 min. Using a variety of

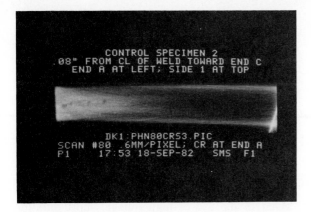

FIGURE 25-4
Tomogram of weldment showing cracking and slag inclusion. (*From Kohutek and Frank* [*29*]. *Courtesy American Society of Mechanical Engineers.*)

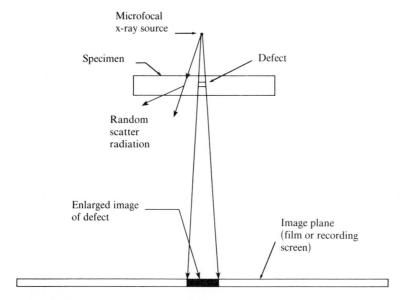

FIGURE 25-5
Principle of projection radiography in microfocus x-ray. The small, point source placed near to the specimen projects an enlarged, sharp image on a plane located a significant distance from the specimen. Contrast is improved due to random scatter in the specimen of radiation projected away from the perpendicular to the image plane.

ceramic materials as their test specimens, they reported a 90% probability of detection with a 95% confidence level for internal voids having dimensions on the order of 1–2% of the material thickness. Sample thicknesses ranged from 2–7 mm.

Where designing with brittle materials such as the high-strength ceramics is based on the minimum detectable flaw size for the material, the ability to detect very small flaws using microfocus x-radiography certainly increases the usefulness of these materials.

25-4 FLASH RADIOGRAPHY

Flash radiography is a label given to the use of high-voltage, short duration pulses to record images from moving objects. The summary given here is largely taken from Refs. 48 and 49, which should be consulted for more detailed information and specific technical references. Several unique capabilities are available through flash radiography. For example, where items are in motion, an image may be sequentially recorded at very short time intervals between the exposures. A very descriptive example of a use of flash radiography is in studies of the piercing of armor by a projectile. In this case, the behavior of the material being pierced and the effect on the projectile may be observed throughout the piercing process with flash radiography. Other applications more typically useful in NDE will be described in the material to follow.

The differences between flash radiography and conventional radiography rest more in specialized equipment rather than basic methods. In order to stop motion, very short duration pulses are necessary. Pulse durations in actual practice range from milliseconds for vibration investigations to nanoseconds for ballistic or shock exposure. Since the basic exposure of the recording medium is a direct function of exposure time, the short exposure time for flash radiography demands a current and voltage that is larger than it is for its conventional counterpart. In addition to the short duration pulse, flash radiography may also require rapid, sequential pulsing in order to create a multiple image. These requirements of short pulses and rapid firing create the need for the specialized equipment that sets apart flash radiography.

The x-ray tubes that are used in flash radiography are specially designed to decrease the rise time for the tube activity and also to increase the activity level. The electron current for flash radiography tubes is considerably higher than for conventional radiography. In the 100–2000 kV range, this high current capacity is typically achieved with a high vacuum, sealed-off configuration in which the cathode tip is pointed like a needle. The pointed shape at the end increases the electron current density on the surface of the cathode.

Pulses used to fire the x-ray tube ideallly should be narrow and rectangular in shape. These are typically achieved by using capacitor discharge equipment for the lower energies. Above 400 kV, the pulses are obtained from either a pulse transformer for the intermediate voltages or from a Marx–Surge generator for voltages in the MV range. Useful pulse durations range from 20–70 ns to 1 μs. At

350 kV, this type of equipment can give 1000 pulses per second with a pulse train of 60–100 pulses. Cineradiography, used for highly detailed recording of events, requires up to 2 MV at rates of up to 1×10^8 frames per second.

Due to the relatively low intensity and short duration of the exposure, special requirements are placed on the recording medium. For film, the need is for a fast film that because of its inherently large grain size, produces a coarser detail of the image. Screens often are used to enhance the image on the recording screen. Exposure speeds of over 100 times that of conventional films are reported for flash radiography exposures. Due to the rapidity with which the film process occurs, however, the use of these high speed films requires considerable care in order to obtain the correct density and contrast.

Since flash radiography is often used to record destructive events such as explosions and projectile penetration, there is a need to protect the image from destruction during the event that is being recorded. Where some protection may be obtained with protective armor for the radiographic equipment, optical techniques also may be used to place the imaging equipment at a safe distance from the event. Optical methods also offer the advantage of being able to store the image on either photographic film, video tape, or in computor memory.

Flash radiography has in recent years begun to show applications for industrial nondestructive evaluation. One excellent example of an industrial application is in the investigation of the metal flow in the casting of metal parts. Irregular flow patterns in the casting process that may result in defects may be located with this technique. Additionally, x-ray diffraction investigations of material stress, texture, etc., may be accomplished with flash radiography.

25-5   REAL-TIME RADIOGRAPHY

The improved resolution of microfocus radiography has been applied to real-time circumstances to greatly improve the performance of real-time systems in industrial applications. One such example is weld inspection as described by Munro et al. [50]. The improved resolution is only one advantage of microfocus x-ray for the real-time systems. Microfocus systems possess the ability to enlarge an image. This feature can be used to create magnification of up to 10 times for areas of interest in the part being inspected. Used in a production line, however, the increased magnification has the deleterious effect of decreasing the viewing area and slowing the inspection process.

An application of real-time radiography to the inspection of laser welds or electron beam welds in thin sheets and pipe having a metal thickness of approximately 1 mm is described by Munro et al. [50]. Using microfocus systems emitting x-rays of not over 350 keV, the authors were able to produce a sharpness of 0.5 mm without geometric magnification and 0.1 mm with sixfold magnification. With this system, porosities in the range of 0.025–0.1 mm were detected. Approximately 1 s was required to complete the image. For steel thicknesses of 10 and 30 mm, the authors report IQI sensitivities of 1.6 and 1.1%, respectively, without geometric magnification. An inspection system for industrial pipe also is described.

25-6 ELECTRON RADIOGRAPHY

On occasion, the NDE engineer may find the need to analyze specimens with a scanning electron microscope (SEM). There are similarities between the SEM process and conventional radiography that make it useful to briefly introduce the subject here. For more detailed discussions, the reader is directed to the number of reference sources on the subject, e.g., Ref. 51. Additionally, Buchanan et al. [49] discuss the subject of electron radiography that may occur in the SEM process.

The images generated by a SEM result from high-energy electrons striking a target in a fashion similar to that which occurs in a conventional x-ray tube. Where the x-ray tube is specifically designed to produce x-rays when struck by the electrons, the scattering characteristics of the target in a SEM vary with the target material. In most cases, where the target might be a biological sample, such as a fly, or an engineering sample, such as a fracture surface, the signals of greatest interest are the secondary and backscattered electrons. These vary with differences in surface topography as the electron beam sweeps across the surface. The scattering characteristics of the specimen are used to recreate the very high resolution images that are readily associated with SEM work. Also of interest, however, is the field of x-rays that are emitted as a result of electron bombardment. These x-rays can be used to obtain both qualitative and quantitative compositional information from areas of the specimen as small as a few micrometers across.

25-7 INSPECTION WITH THERMAL NEUTRONS

Because of the unique absorption characteristics of neutrons, neutron radiography may be utilized to accomplish radiographic inspections not possible with electromagnetic radiation. An example of the effect of the different absorption characteristics of neutrons in various materials may be seen using the coated section from a steel pipeline as shown in Fig. 25-6(a). The $\frac{1}{2}$ in (12.7 mm) thick steel pipe section is coated with a thin 0.015 in (0.38 mm) coating of powdered epoxy material for corrosion protection. The sample seen in Fig. 25-6(a) shows a break in the coating material. The break has been imaged with neutron radiography as seen in Fig. 25-6(b). A conventional x-ray inspection of the steel section would not image the coating break.

Neutrons are particularly useful in the detection of corrosion where there has been a minimal loss of material yet the corroded substance contains a high amount of hydrogen. Situations such as this are often found in aluminum honeycomb panels, as shown in Fig. 25-7. As reported by Berger [52], an inspection of an aluminum skin–aluminum core, adhesively bonded honeycomb aircraft assembly showed the effects of water-induced corrosion. In this case, before the materials were bonded, a drop of water was inserted into several of the honeycomb cells. After being stored for a year, a neutron radiograph of the section showed corrosion effects at the cells where the water was induced. The most severe corrosion is indicated by the arrow in Fig. 25-7.

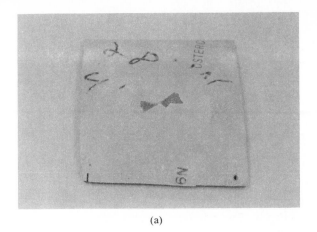

(a)

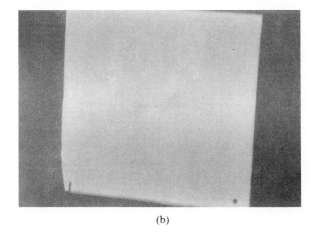

(b)

FIGURE 25-6
(*a*) Section from 12 mm (0.500 in) thick pipe wall with a break showing in 0.38 mm (0.015 in) thick powdered epoxy coating material. (*b*) Neutron radiograph showing the image of a break in epoxy coating material. (*Dr. Jon Reuscher, Texas A & M University, Nuclear Science Center.*)

Defects in adhesive bonding in structural joints may also be detected by neutron radiography. An inspection of a structural joint adhesively formed between metal inserts and a honeycomb material is shown in Fig. 25-8. The upper x-radiograph [Fig. 25-8(*a*)] clearly does not image the lack of adhesive in the bond areas, whereas the neutron radiograph in Fig. 25-8(*b*) shows the defective joint very clearly.

25-8 INSPECTION WITH PROTONS

Although protons have found limited application in nondestructive evaluation because of the small number of proton sources available, their unique characteristics deserve special mention at this point. Citing the work of West and Sherwood [53], Koehler and Berger [15] show an x-ray radiograph and a proton radiograph of a thin mechanical watch. In the proton radiograph, the teeth on the individual gears are clearly distinguishable as well as virtually all of the working parts. The

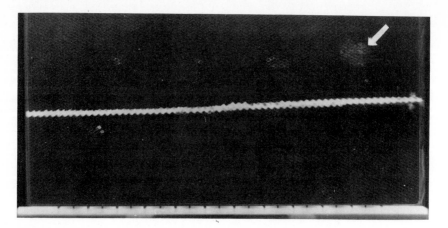

FIGURE 25-7
Thermal neutron radiograph showing corrosion in an aluminum honeycomb panel. The central white linear indication is the image of the adhesive. The broad white images, such as the one indicated by the arrow, show corrosion that resulted over a period of 1 yr from a drop of water inserted into the assembly. (*From D. Froom, US Air Force. Courtesy of Harold Berger* [52], *Industrial Quality Inc. Copyright ASTM. Reprinted with permission.*)

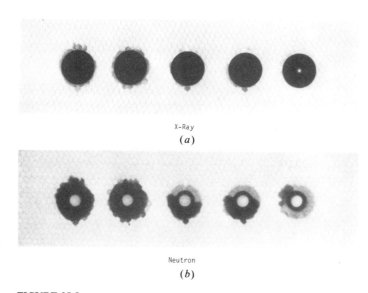

X-Ray
(*a*)

Neutron
(*b*)

FIGURE 25-8
(*a*) X-ray radiograph of flanged metal inserts bonded to honeycomb overlay. Adhesive deficient areas will not be indicated in this radiograph. (*Courtesy Dr. W. E. Dance, LTV Aerospace and Defense Co., Dallas, Texas.*) (*b*) Neutron radiograph of flanged metal inserts bonded to honeycomb overlay. Adhesive deficient areas are clearly indicated in this radiograph. (*Courtesy Dr. W. E. Dance, LTV Aerospace and Defense Co., Dallas, Texas.*)

x-ray radiograph, however, shows little detail because most of the area is either overexposed in the areas where there are few parts or underexposed where the photons were absorbed by the material used in the watch.

25-9 SUMMARY

Awareness of the capabilities of specialized techniques such as computed tomography, flash radiography, and microfocus x-ray is imperative for the NDE engineer. Compared to other NDE techiques, these specialized radiographic techniques are expensive. Typically, they are not available at most laboratories and inspection shops. Should the need to use them arise, however, many commercial NDE laboratories may have the equipment available to perform these services.

CHAPTER
26

SAFETY CONSIDERATIONS IN RADIOGRAPHIC INSPECTION

26-1 INTRODUCTION

The harmful effects of radiation on humans have been known for some time, with the result that the handling and inspection practices established by current standards are proven to be effective. As in any industrial environment, however, there are accidents with radiographic materials. These are infrequent, principally due to the well entrenched procedures for radiographic inspection. Because of the myriad of factors involved in radiation safety and the very complete reviews that are available in other sources, the present material will be very short, the intent being to give only a cursory view of the subject. The topics to be covered are the basics of exposure, protection, and monitoring. For more detail, the reader is directed to Ref. 54. Additionally, all licensed organizations involved in radiographic inspection are required to have a radiation safety officer who also can be consulted for advice.

26-2 RADIATION EXPOSURE AND DOSAGE

In the discussion of radiation exposure to humans, four terms need to be defined. These are summarized from Ref. 54:

Absorbed dose is the mean energy ionizing radiation absorbed by irradiated material per unit mass of the absorbing material. The unit expressing this quantity is the gray (Gy) where 1 Gy = 1 J/kg = 100 rad.

TABLE 26-1
Quality factors for various radiation types

Radiation type	QF
X-rays, γ rays, electrons, and β rays	1
Neutrons, energy < 10 keV	3
Neutrons, energy > 10 keV	10
Protons	10
α particles	20
Fission fragments, recoil nuclei	20

Source: Reference 54.

The *dose equivalent* provides a comparison scale common to all forms of radiation. The dose equivalent is the product of the absorbed dose and a constant (QF), which will be defined and is related to the type of radiation involved. A commonly used unit for this quantity is the rem. The SI unit for the dose equivalent is the sievert (Sv), which is equal to 100 rem = 1 J/kg.

Exposure is a function of the amount of measured radiation at some reference point relative to the x-ray device or the radioisotope. A commonly used unit for this is the roentgen, while the SI unit is the sievert.

The *quality factor* (*QF*) expresses the relative severity of exposure for the various forms of radiation. The QF has replaced the formerly used rbe for expressing this factor. For example, typical conservative values for QF, as given in Ref. 54, are furnished in Table 26-1.

Recognizing that radiation is often present in environments not being radiated with an active NDE source, the term permissible dose is an administrative recommendation for the purposes of minimizing the biological effects of radiation.

The Nuclear Regulatory Commission lists the following limits for human exposure to radiation. The maximum limit for whole body radiation per quarter calendar year is 12 mSv (1.25 rem). Specifically listed in the whole body group are the head, trunk, active bloodforming organs, the lens of the eyes, and the gonads. For the hands, forearms, feet, and ankles, the allowed value is 188 mSv (18.8 rem) per calendar quarter. For the skin of the whole body, the values are 75 mSv (7.5 rem) per calendar quarter.

Other considerations may be used to establish a more specific exposure limit for individuals. For example, exposure in previous years may affect the allowable exposure in the current year. Moreover, restricted and unrestricted operational areas may need to be defined based on level of exposure. Special considerations for minors and females are also listed.

As previously stated, this material is meant merely to introduce the subject of radiation effects on humans and expert sources such as those previously listed should be consulted before engaging in radiographic inspection.

26-3 RADIATION MONITORING

A variety of approved radiation monitoring devices are available from vendors. In some cases, an individual will be required to wear a film badge that constantly monitors the amount of radiation that is received. The amount of radiation emitted from a container or a room may be monitored with various instruments. The level of radiation in an area will determine the protection required for workers. Often, radiation levels may be reduced by enclosing the test area in a radiation absorbing container, which may be the size of a small box or large enough to accommodate an entire aircraft.

26-4 PROTECTION OF WORK SPACES

As may be observed from the discussion at the very beginning of the material on radiography, all matter absorbs radiation to some extent. One of the most frequently used techniques for controlling radiation in the work space is to enclose it within a highly absorbing container. It is not surprising that lead is commonly used as a shielding material. Under certain conditions, many typical building materials such as concrete blocks or even wooden doors may also be used as shielding.

Worker control may also be used to reduce the overall amount of exposure. For example, where an exposure rate at a work station is known to be 100 μSv/h and the permissible occupational exposure dose per week is 900 μSv, the maximum permissible work time at that station would be 9 h per week. Since the intensity of sources decreases with the inverse square law, as described earlier, the exposure may also be controlled by maintaining a suitable distance from the source.

26-5 LICENSING OF RADIOGRAPHIC WORK

Provisions as outlined in Ref. 55 specify that a firm must be licensed by the proper authority before it can obtain a radiographic source. In the United States, basic authority for control of isotopic sources rests with the Nuclear Regulatory Commission (NRC). The user must be licensed by either the NRC or other NRC approved regulatory body to use the source for inspection purposes. X-ray machines, on the other hand, may be purchased without licensing, but registration of a proper training procedure with the authoritative agency is required before operation can commence.

The Nuclear Regulatory Commission may delegate to the various states administrative responsibility for enforcing radiation safety in the workplace, provided that the state's regulations are more severe than those of the NRC. An *agreement state* is one that has assumed responsibility for radiographic matters, while one that has left the authority in the hands of the federal government is called an *NRC state*.

26-6 SUMMARY

The material in this chapter has had two purposes. One has been to impress upon the reader that there are important human safety considerations that must be met when involved in radiographic NDE. The second purpose has been to assure the person who may become involved in radiographic NDE that a safe environment may be established, provided that well proven procedures are followed. With these precautions, the advantages of radiography for materials inspection and evaluation may be made available without any sacrifice to human safety.

PROBLEMS
FOR
PART IV

Chapter 20

20-1. List the characteristics and describe one specialized application of each of the following radiographic sources: x-rays, γ rays, protons, and neutrons.

20-2. The total radiation absorbed by a material being inspected is a sum of the individual absorption processes as expressed by σ_{pe}, σ_s, σ_{pp}, and σ_{pd}. Obtain or estimate from the tables the mass attenuation coefficient μ/λ at energy levels of 0.02, 0.10, 1.0, and 10 MeV for x-rays in steel, aluminum, copper, lead, and tungsten. Compare in a qualitative sense the relative contribution of each of the absorption processes to the overall absorption and list the major absorption process or processes for each of the materials and at each energy level.

20-3. For x-ray tube voltages of 660, 450, and 150 kV$_p$, estimate the x-ray energy level for the x-rays having the maximum intensity.

20-4. An isotope source made of five ^{192}Ir wafers emits radiation energy with an intensity of 27.5 R/m h. Twenty-five days later, five more wafers are added. At the time that they are added, the strength of the new wafers is 50 Ci. At the time that the new 10 wafer capsule is assembled, what is its energy level and intensity?

20-5. A capsule containing ^{60}Co wafers is 31.75 mm (1.25 in) in diameter. It emits 1.35 R/h m/Ci. For a 20 Ci source, calculate the intensity of radiation along an axis from the surface of the capsule to a distance 15 m (49 ft).

20-6. Plot an activity curve (i.e., strength versus time) for ^{60}Co, ^{192}Ir, and ^{170}Tm. Assume an initial source strength as shown by the typical values in Table 20-3 and extend the plot until the source strength is 10% of the original value.

20-7. Use the radiographic equivalence factors to determine the approximate equivalent thickness of aluminum that would produce the same film exposure as a piece of 12 mm thick carbon steel. Assume 150 kV x-rays, tube voltage of 450 kV$_p$.

20-8. Calculate the radiographic equivalence factors (REF) at 150 kV$_p$ tube voltage for aluminum, steel, copper, lead, Pyrex glass, and polyethylene, using values from Tables 20-6 and 20-7. Use aluminum as the standard material. Compare your results for the metals with those given in Table 20-8. What circumstances may account for the discrepancies?

20-9. Calculate the radiographic equivalence factor (REF) for steel and titanium using absorption coefficients from Table 20-6 and those given in Appendix B. Compare the results obtained from the two tables. *Note:* Refs. 56 and 57 may be consulted for further discussion.

20-10. Red brass is composed of 85% copper and 15% zinc while a phosphor bronze is composed of 90% copper and 10% tin with a trace of phosphorus. Compare the linear attenuation coefficients for these two metals for 1.5 MeV x-rays.

20-11. Calculate the radiographic equivalence factors for titanium and steel based on the linear attenuation coefficients given in Appendix B for the x-ray energies listed in Table 20-8. Use interpolation to obtain your answer. How close do your calculated results agree with those given in Table 20-8? For further comparison, you may see Refs. 56 and 57.

20-12. When x-ray tube voltage is increased, what is the effect on wavelength, penetration, resolution, and radiographic unsharpness in a typical inspection?

Chapter 21

21-1. A 5 mm source is used to inspect a 25 mm thick steel plate. The distance from the source to the front of the plate is 610 mm. A 25 mm diameter flaw is located 6 mm below the plate surface, on the center line from the source. Sketch this inspection arrangement, indicating the image size on the film and the umbra and penumbra. Calculate the geometric unsharpness for these conditions. Repeat the calculations for a flaw located 23 mm below the plate surface and compare your results.

21-2. Work Problem 21-1 using a 19 mm source.

21-3. A properly exposed radiograph was obtained for an aluminum weld 3 in thick with a source 60 in from the film. The geometric unsharpness, however, was found to be unsatisfactory and the source-to-film distance was increased to 120 in. What would be the proper exposure time for this new placement compared to the original exposure time?

21-4. Compare the exposure time required for a nominal 12 in diameter steel pipe with an OD and wall thickness of 12.750 and 0.330 in (0.324 m and 8.382 mm), respectively, with that needed for a nominal 24 in pipe with corresponding dimensions of 24.0 and 0.375 in (0.6096 m and 9.52 mm). Assume that the internal source is located at the center line of the pipe and that the film is wrapped externally around the circumference.

21-5. A fabrication process in an aircraft manufacturing facility calls for two aluminum plates $1\frac{1}{2}$ in thick to be butt-welded with a double-V weld. The weld metal has been ground on each side to give a thickness of $1\frac{3}{4}$ in. Each weld must be radiographed for defects in the weld metal as in the heat-affected zone. If a satisfactory radiograph of the weld was obtained with a 15 s exposure using a portable x-ray device emitting x-rays at 140 kV, what would be the proper exposure time for an inspection of the heat-affected zone of the plate using the same equipment? Assume a source-to-film distance of 36 in for both tests.

21-6. Repeat Problem 21-5 for steel, assuming a 200 kV x-rays and using permanent equipment.

21-7. A radiographic inspection is required of a 76.2 mm (3 in) thick titanium bar. Using a 450 kV$_p$ tube voltage and the specifications given in Fig. 21-6 for aluminum, prepare a suitable test plan for this specimen.

21-8. ^{60}Co is reported to have a half-life of 5.3 yr. At the time of purchase, a proper radiograph for a particular sample could be obtained with an exposure of time T. After 2 yr what would be the relative exposure time required to obtain a comparable radiograph?

21-9. The half-life of ^{192}Ir is 75 days. If a proper radiograph were obtained with a ^{192}Ir source 10 days prior to the present, what would be the relative exposure for the identical item at the present time?

21-10. If a radiograph of a test sample shows a 2T hole and the outline of a 0.020 in pentrameter, what is the thickness of the specimen if it is stated that a 2-2T quality radiograph has been obtained? What is meant by the statement "2-2T quality radiograph?"

21-11. What is the qualitative term used to indicate the size of the smallest detail that can be seen in a radiograph?

21-12. Describe the relationship of radiographic film grain size and the speed of exposure and the definition quality.

Chapter 23

23-1. A high-pressure manifold connection is to be machined from leaded brass having a composition of 97% Cu and 3% Pb. The specifications require that before the block is machined it must be radiographed and certified that no defects greater than 0.24 mm in diameter exist internally. The dimensions of the block are 55 mm long, 80 mm wide, and 12 mm thick. Prepare a test plan to accomplish this inspection, giving the specific details on the most suitable equipment, the exposure time, the required film density, and the quality level of the exposure.

23-2. A pipeline crew is performing an initial radiographic inspection of an exposed line that is 180 km (112 mi) long and is made of nominal 12 in steel pipe having OD and wall thickness dimensions of 12.75 and 0.330 in (0.324 m and 8.382 mm), respectively. Each original unwelded pipe section was approximately 40 ft (12.2 m) in length. The ^{192}Ir source is moved internally in the pipe and the type A film is wrapped circumferentially, outside the pipe. A density 2 radiograph is required. Each day the crew can typically inspect 200 welds. Calculate the exposure time needed for each weld, first at the start of the inspection and then at the completion.

23-3. Welds in a polyethylene pipe are to be inspected using an internal source located at the pipe center line. If the pipe is 6.625 in (0.168 m) in diameter and has a wall thickness of 0.280 in (7.112 mm), would ^{192}Ir be a proper choice as a radiographic source? If not, what source would you recommend? Justify your answer.

23-4. A weld in an aluminum pipe is to be inspected using radiography. The pipe has an 8 in (0.203 m) OD and the wall thickness is 0.094 in (2.39 mm). If it is to be inspected with an external source and using the criteria given in Fig. 21-6, prepare a suitable test plan for this inspection.

23-5. Two thick-walled pipes are to be inspected using the same ^{60}Co source, which is to be located internally at the pipe center line. If the outside diameters of the two pipes are 8 and 12 in, respectively, compare the exposure times required for the two pipes.

23-6. A welded connection joins two 1 in (25.4 mm) thick steel plates which are oriented at an angle of 60° relative to each other. Propose a proper radiographic inspection of this weld zone considering the affect on the density of the varying section and a suitable placement of the pentrameter for image quality evaluation. Assume that you are to use a 240 kV x-ray machine having a tube current of 5 mA. Use the exposure specifications given in Fig. 21-7.

23-7. The information furnished with Fig. 23-11(b) showed that the radiograph was obtained with 50 kV x-rays. Use data available for aluminum and nylon from Tables 20-6 and 20-7, respectively, and estimate a radiographic equivalent factor for nylon relative to aluminum at 50 kV. Using this result, calculate an exposure for the nylon sample used in Fig. 23-11(b) and compare your results with the data furnished with the radiograph.

23-8. When using an ^{192}Ir source with a strength of 42 Ci, what working distance would be required to reduce the dose rate to 2 mR/h with no shielding? If the source and the immediate surrounding region were placed in a container having a lead sheet 3.175 mm (1/8 in) fully around the interior, what would be the new comparable working distance?

23-9. Using 45 Ci of ^{60}Co and 1 in (25.4 mm) thick lead sheet as a shield, what would be the dose rate at 25 ft (7.62 m) from the source.

23-10. Compare some typical characteristics and NDE applications of low-energy x-rays, e.g., 150 kV$_p$, and high-energy x-rays, e.g., 1.5 MeV.

REFERENCES
FOR
PART IV

1. Emigh, C. R., "Radiation and Particle Physics," *Nondestructive Testing Handbook, vol.* 3, *Radiography and Radiation Testing*, 2nd ed., American Society for Nondestructive Testing, Columbus, OH, 1985, sec. 1, pp. 62–90.
2. Halmshaw, R., *Industrial Radiology—Theory and Practice*, Applied Science Publishers, Englewood, NJ, 1981.
3. Quinn, R. A., and Sigl, C. C. (eds.), *Radiography in Modern Industry*, Eastman Kodak Company, Rochester, NY, 1980.
4. McGonnagle, W. J., *Nondestructive Testing*, Gordon and Breach, New York, 1961.
5. Semat, H., *Introduction to Atomic and Nuclear Physics*, 3rd ed., Rinehart, Company, New York, 1959.
6. Kaplan, I., *Nuclear Physics*, 2nd ed., Addison-Wesley, Reading, MA, 1962.
7. *NDT Radiography Manual*, FS 5.284:84036, Superintendent of Documents, Washington, DC, 1968.
8. Askeland, D. R., *The Science and Engineering of Materials*, Brooks-Cole, Monterey, CA, 1984.
9. Iddings, F. A., "Isotope Radiation Sources," *Nondestructive Testing Handbook*, vol. 3, *Radiography and Radiation Testing*, 2nd ed., American Society for Nondestructive Testing, Columbus, OH, 1985, sec. 3, 113–151.
10. Ulrey, C. T., "An Experimental Investigation of the Energy in the Continuous X-ray Spectrum of Certain Elements," *Physical Review*, vol. 11, no. 5, pp. 401–410, 1918.
11. Egerton, H. B. (ed.), *Nondestructive Testing*, Oxford University Press, London, 1969.
12. Berger, H., "Neutron Radiography," in R. S. Sharpe (ed.), *Research Techniques in Nondestructive Testing*, vol. I, Academic, London, 1970, chap. 9.
13. Iddings, F., "Utilization of a Low Voltage Accelerator for Neutron Radiography," *Proceedings of the Seventh Symposium on Nondestructive Evaluation of Components and Materials in Aerospace, Weapons Systems and Nuclear Applications*, April 23–25, Southwest Research Institute, San Antonio, TX, 1969, pp. 363–372.
14. Berger, H., *Practical Applications of Neutron Radiography and Gaging*, STP-586, American Society for Testing and Materials, Philadelphia, PA, 1976.
15. Koehler, A. M., and Berger, H., "Proton Radiography," in R. S. Sharpe (ed.), *Research Techniques in Nondestructive Testing*, vol. II, Academic, London, 1973, chap. 1.
16. Flinn, R. A., and Trojan, P. K., *Engineering Materials and Their Applications*, 3d ed., Houghton-Mifflin, Boston, MA, 1986.
17. Hubbell, J. H., "Photon Mass Attenuation and Energy-Absorption Coefficients from 1 keV to 20 MeV," *International Journal of Applied Radiation Isotopes*, vol. 33, pp 1269–1290, 1982.
18. "Attenuation Coefficient Tables," *Nondestructive Testing Handbook*, vol. 3, *Radiography and Radiation Testing*, 2nd ed., American Society for Nondestructive Testing, Columbus, OH, 1985, sec. 20, pp. 837–878.

19. Quinn, R., and Domanus, J. C., "Film and Paper Radiography," *Nondestructive Testing Handbook*, vol. 3, *Radiography and Radiation Testing*, 2nd ed., American Society for Nondestructive Testing, Columbus, OH, 1985, sec. 5, pp. 186–255.
20. *Nondestructive Testing, Radiographic Testing*, CT-6-6, *General Dynamics*, 2nd ed., American Society for Nondestructive Testing, Columbus, OH, 1983.
21. Moores, B. M., "Screen/Film Combinations for Radiography," in R. S. Sharpe (ed.), *Research Techniques in Nondestructive Testing*, vol. VI, Academic, London, 1982, chap. 7, 289–308.
22. Halmshaw, R., "An Analysis of the Performance of X-Ray Television-Fluoroscopic Equipment in Weld Inspection," *Materials Evaluation*, vol. 45, no. 11, pp. 1298–1302, November 1987.
23. Bossi, R., Oien, C., and Mengers, P., "Real-Time Radiography," *Nondestructive Testing Handbook*, vol. 3, *Radiography and Radiation Testing*, 2nd ed., American Society for Nondestructive Testing, Columbus, OH, 1985, sec. 14, pp, 593–640.
24. Jacoby, M. H., "Image Data Analysis," *Nondestructive Testing Handbook*, vol. 3, *Radiography and Radiation Testing*, 2nd ed., American Society for Nondestructive Testing, Columbus, OH, 1985, sec. 15, 641–673.
25. De Meester, P., and Aerts, W., "Analysis of Radiographic Image Recording Systems," in R. S. Sharpe, ed., *Research Techniques in Nondestructive Testing*, vol. V, Academic, London, 1982, chap. 1, pp. 1–52.
26. Packer, M. E., "The Use of Digital Image Processing in the Detection of Corrosion by Radiography," in R. S. Sharpe (ed.), *Research Techniques in Nondestructive Testing*, vol. V, Academic, London, 1982, chap. 2, pp. 53–74.
27. Snow, S. G., and Morris, R. A., "Radiation Gaging," *Nondestructive Testing Handbook*, vol. 3, *Radiography and Radiation Testing*, 2nd ed., American Society for Nondestructive Testing, Columbus, OH, 1985, sec. 16, pp. 674–704.
28. Gilboy, W. B., and Foster, J., "Industrial Applications of Computerized Tomography with X and Gamma Radiation," in R. S. Sharpe (ed.), *Research Techniques in Nondestructive Testing*, vol. VI, Academic, London, 1982, chap. 6, pp. 255–287.
29. Kohutek, T., and Frank, K., "Application of Photon Tomography to the Inspection of Bridge Weldments," Paper 85-WA/NDE-2, American Society for Mechanical Engineers, New York, 1985.
30. Dölle, H., and Lemmer, K., "Comparison of Codes and Standards for Radiographic Inspection and Experimental and Theoretical Studies on Unsharpness and Sensitivity Requirements," *Materials Evaluation*, vol. 43, no. 2, pp. 188–195, February 1985.
31. Lapidas, M. E., "Radiographic In-Service Inspection of Cast Austenitic Nuclear Plant Components," *Materials Evaluation*, vol. 44, no. 1, pp. 108–113, January 1986.
32. Dance, W. E., private communication, 1987.
33. Kruzic, R. W., "Difficulties of Radiographic Evaluation on Inconel Welds Joining Dissimilar Metals," *Materials Evaluation*, vol. 44, no. 11, pp. 1336–1339, October 1986.
34. Cullity, B. D., *Elements of X-ray Diffraction*, 2nd ed., Addison-Wesley, Reading, MA, 1978.
35. Weiss, V., "X-rays for Nondestructive Characterization of Materials Properties," in C. O. Ruud and R. E. Green, Jr. (eds.), *Nondestructive Methods for Material Property Determination*, Plenum, New York, 1984, pp. 3–19.
36. Rudd, C. O., "Application of a Positron Sensitive Scintillation Detector to Nondestructive X-ray Diffraction Characterization of Metallic Compounds," in C. O. Rudd and R. E. Green, Jr. (eds.), *Nondestructive Methods for Material Property Determination*, Plenum, New York, 1984, pp. 21–38.
37. Borgonovi, G. C., "Determination of Residual Stress from Two-dimensional Diffraction Patterns." in C. O. Ruud and R. E. Green, Jr. (eds.), *Nondestructive Methods for Material Property Determination*, Plenum, New York, 1984, pp. 47–57.
38. Ruud, C. O., DiMascio, P. S., and Melcher, D. M., "Nondestructive Residual-Stress Measurement on the Inside Surface of Stainless-Steel Pipe Weldments," *Experimental Mechanics*, vol. 24, no. 2, pp. 162–168, June 1984.
39. Jenkins, R., "X-ray Diffraction and Fluorescence," *Nondestructive Testing Handbook*, vol, 3, *Radiography and Radiation Testing*, 2nd ed., American Society for Nondestructive Testing, Columbus, OH, 1985, sec. 17, pp. 705–731.

40. Kouris, K., Spyrou, N. M., and Jackson, D. F., "Materials Analysis using Photon Attenuation Coefficients," in R. S. Sharpe (ed.), *Research Techniques in Nondestructive Testing*, vol. VI, Academic, London, 1982, chap. 5, pp. 211–254.

41. Allen, D. R., and Sayers, C. M., "Neutron Scattering Studies of Textures in Mild Steel," *NDT International*, vol. 14, no. 5, pp. 263–269, October 1981.

42. Ellinger, H., "Three-Dimensional Imaging Techniques using Industrial Computed Tomography," Scientific Measurements Systems, Austin, TX, Presentation at 1988 American Society for Nondestructive Testing, Spring Conference, Orlando, FL, 11–15 March 1988.

43. Johns, C., "Medical CAT-Scanning for Aircraft Engine Components," *The Leading Edge*, Fall 1987, pp. 27–31.

44. Berger, H., and Kupperman, D., "Microradiography to Characterize Structural Ceramics," *Materials Evaluation*, vol. 43, no. 2, pp. 201–205, February 1985.

45. Parish, R. W., "Microfocus X-ray Technology—A Review of Developments and Applications," *Review of Progress in Quantitative Nondestructive Evaluation*, vol. 5A, pp. 1–20, 1986.

46. Bagnall, M. J., and Kotzian, B., "Microfocus Radiography of Jet Engines," *Materials Evaluation*, vol. 44, no. 13, pp. 1466–1467, December 1986.

47. Baaklini, G. Y., and Roth, D. J., "Probability of Detection of Internal Voids in Structural Ceramics using Microfocus Radiography," *Journal of Materials Research*, vol. 1, no. 3, pp. 457–467, May/June 1986.

48. Charbonnier, F., "Flash Radiography," *Nondestructive Testing Handbook*, vol. 3, *Radiography and Radiation Testing*, 2nd ed., American Society for Nondestructive Testing, Columbus, OH, 1985 sec. 11, pp. 491–531.

49. Buchanan, R. et al., "Specialized Radiographic Methods," *Nondestructive Testing Handbook*, vol. 3, *Radiography and Radiation Testing*, 2nd ed., American Society for Nondestructive Testing, Columbus, OH, 1985, sec. 19, pp. 759–835.

50. Munro, J. J., McNulty, R. E., Nuding, W., Busse, H. P., Wiacker, H., Link, R., Sauerwein, K., and Grimm, R., "Weld Inspection by Real-Time Radiography," *Materials Evaluation*, vol. 45, no. 11, pp. 1303–1309.

51. Goldstein, J. I., Newbury, D. E., Echlin, P., Joy, D. C., Fiori, C., and Lifshin, E., *Scanning Electron Microscopy and X-Ray Microanalysis*, Plenum, New York, 1981.

52. Berger, H. "Neutron Radiographic Detection of Corrosion," in G. C. Moran and P. Labine (eds.), *Corrosion Monitoring in Industrial Plants Using Nondestructive Testing and Electromechanical Methods*, ASTM STP 908, American Society for Testing and Materials, Philadelphia, PA, 1986, pp. 5–16.

53. West, D., and Sherwood, A. C., "Radiographs Produced by Multiple Scattering of Charged Particles," Report AERE-R7190, UK Atomic Energy Establishment, Harwell, UK, 1972.

54. Burnett, W. D., "Radiation Protection," *Nondestructive Testing Handbook*, vol. 3, *Radiography and Radiation Testing*, 2nd ed., American Society for Nondestructive Testing, Columbus, OH, 1985, sec. 18, pp. 732–758.

55. "Standards for Protection against Radiation," Code of Federal Regulations, Title 10 (Energy), Part 20, US Government Printing Office, Washington, DC.

56. Nir-El, Y., "New Information on Radiographic Equivalents," *Materials Evaluation*, vol. 44, no. 13, pp. 1462–1463, December 1986.

57. Iddings, F. A., "Radiographic Equivalents," *Materials Evaluation*, vol. 43, no. 8, pp. 888–890, July 1986.

PART
V

PENETRANT TECHNIQUES IN NONDESTRUCTIVE EVALUATION

CHAPTER
27

PRINCIPLES
OF PENETRANTS

27-1 INTRODUCTION

Penetrant inspection utilizes the natural accumulation of a fluid around a discontinuity to create a recognizable indication of a crack or other surface opening. As will be described later, capillary action attracts the fluid to the discontinuity in a concentration heavier than in the surroundings. In order for the fluid concentration to be recognized, the background area must be of sufficient contrast to distinctly reveal the defect on the surface. The complete penetrant flaw detection system, therefore, consists of the fluid mechanics on the surface, as well as the recognition system that is used to detect the indication.

A typical example of the principles involved in penetrant inspection is the ready visibility of cracks in concrete slabs after a rain. When the slab is thoroughly wetted, the water seeps into the cracks. When the rain has been mostly dried by the sun and the thinner surface water has been evaporated, the heavier water concentration remaining around the cracks clearly reveals their location and shape.

The concrete slab example demonstrates several principles of penetrants. First, the slab initially is thoroughly wetted with a fluid that flows readily and evenly over its surface. Second, the fluid is drawn into the crack by a capillary action and, third, the excess fluid is removed from the surface creating a good contrast between the still moist area and the dry expanse of the slab. Finally, it is

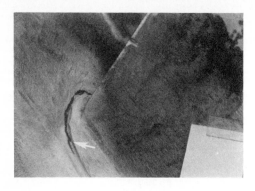

FIGURE 27-1
Visible penetrant indication at the weld for a pipe hanger lug. Surface prepared for inspection by grinding. (*From Reinhart* [*1*]. *Courtesy Critical Path.*)

important to note that penetrants are useful only for surface-breaking defects. A crack or an open pore that did not reach the surface would not be detected, regardless of its size.

The example of the cracked concrete slab describes a rather simplistic application of the principle of penetrants. To assume that all penetrant applications are similarly simple would be naive. Careful and thorough consideration of the surface and fluid properties and the flaw recognition systems to be used is important to the success of a penetrant inspection. Many test situations, however, are very straightforward, and the only instructions that may be required are those on the back of a penetrant spray can. The engineer must be aware, however, that many applications of penetrant inspection are far beyond the rudimentary sophistication of the spray can.

An example of a crack revealed with penetrant inspection is shown in Fig. 27-1. The black outline of the crack, as indicated by the arrow, is clearly evident. The ease in recognizing the defect is enhanced by the clean background surface that creates a good contrast for the penetrant material.

The material in the remainder of this section will introduce some of the physical principles of penetrants. Subsequent sections will then discuss in more detail the application of penetrant techniques for a variety of flaw detection tasks.

27-2 PRINCIPLES OF FLUID MECHANICS ON SURFACES

Penetrant inspection is directly concerned with the behavior of fluids on surfaces. Surface tension and capillarity are two areas of fluid mechanics that play significant roles in penetrants. Surface tension is a fluid property that affects the flow of the fluid on the surface to be inspected. Capillarity is the driving force in the movement of the penetrant materials along the solid surface, into the cracks, and, finally, out of the cracks and into the developer. These topics will now be reviewed briefly.

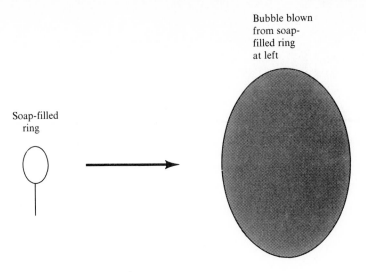

Bubble blown
from soap-
filled ring
at left

Soap-filled
ring

FIGURE 27-2
Bubble on right blown from the smaller ring on the left. The larger bubble has greater surface tension than the fluid in the smaller circle on the left.

27-2.1 Surface Tension

The physics of the surface forces exerted by a fluid film are described thoroughly in standard textbooks on physics and fluid mechanics, e.g., Refs. 2–4. For a fluid, a large soap bubble blown from a small ring may demonstrate some of the princicples of surface physics. As shown in Fig. 27-2(a), a small ring has been dipped in a soap solution so that a single two-sided film of soap is contained within the ring. Skillful blowing through the ring, however, can produce a bubble with a surface area many times larger than the surface in the ring. While the fluid would prefer to revert to the initial, smaller size of the ring, i.e., a smaller surface area, the pressure of the air in the bubble prevents this from occurring. Thus, the nonequilibrium fluid state of the bubble is being supported by the surface forces originating from the internal air pressure. These surface forces of the internal air are counteracted by the opposing surface forces of the fluid film of the soap solution. If the bubble were to be pierced, the fluid ideally would revert to its lowest-energy state, i.e., a spherical liquid drop. In this lowest-energy state, the surface forces constraining the fluid clearly are lower than those constraining the original bubble.

The parameters involved in the surface kinetics of the liquid drop may be described following the presentation of Prandtl and Tietjens [2]. Figure 27-3 shows a small element of the surface of the drop with the surface tension σ acting in the plane of the fluid. Surface tension is defined as the force acting on the surface and perpendicular to the length dl with units of force per unit length N/m. The radii of curvature for the surface are shown to be r_1 and r_2, where, for

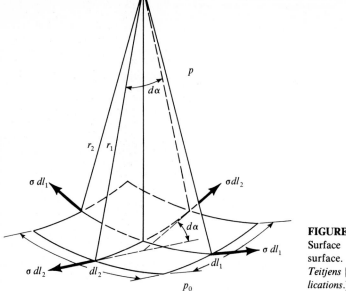

FIGURE 27-3
Surface tension on a curved surface. (*From Prandtl and Teitjens* [2]. *Courtesy Dover Publications.*)

generality, the two radii are assumed to be different. For the element having sides of length dl_1 and dl_2, the orthogonal forces along the two axes are $\sigma \, dl_1$ and $\sigma \, dl_2$, respectively. The resultant forces R_2 perpendicular to length dl_2 can be written as

$$R_2 = \sigma \, dl_2 \, da = \sigma \, dl_2 \frac{dl_1}{r_1} \tag{27-1a}$$

where σ is the surface tension of the liquid at a gas interface. Along the side dl_1, the resulting perpendicular force R_1 is

$$R_1 = \sigma \, dl_1 \frac{dl_2}{r_2} \tag{27-1b}$$

In an equilibrium state, the surface forces must balance, which for the present case requires that the sum of the surface tension forces be equal to the pressure differential forces across the surface boundary. Where the pressure inside the boundary, as shown in Fig. 27-3, is p and that outside the boundary is p_0, the resulting pressure force would be

$$(p - p_0) \, dl_1 \, dl_2 = \sigma \left(\frac{1}{r_1} + \frac{1}{r_2} \right) dl_1 \, dl_2 \tag{27-2a}$$

or

$$p - p_0 = \sigma \left(\frac{1}{r_1} + \frac{1}{r_2} \right) \tag{27-2b}$$

TABLE 27-1
Surface tensions of typical liquids at 20°C in contact with air.

Liquid	Surface tension σ (mN / m)
Benzene	28.9
Hexane	18.4
Kerosene	26.8
Lube oil	25 – 35
Methanol	22.6
Octane	21.8
Water	72.8

Source: Reference 5.

For the specific case of a spherical drop of a liquid where $r_1 = r_2 = r$, Eq. (27-2b) clearly reduces to

$$p - p_0 = \frac{2\sigma}{r} \qquad (27\text{-}2c)$$

It is interesting to observe at this point that the important parameters involved in the formation of a liquid drop are the internal and external pressures and the surface tension of the liquid. Thus, for identical pressure conditions, a liquid with a higher surface tension will form a drop with a larger radius. Surface tension, therefore, is a characteristic property of the liquid; Table 27-1 lists some typical values.

A comparison of the surface tension values for the liquids given in Table 27-1 shows that for drops of the same radius in air at atmospheric pressure, kerosene would have a lower internal pressure than would water.

27-2.2 Capillarity

Fluid capillarity is used in two phases of penetrant inspection. First, it is the mechanism that draws the penetrant into the crevice or opening that is being sought. Second, it is the same mechanism that then subsequently draws the penetrant out of the crevice and into the developer, thereby increasing the visibility of the anomaly.

Capillarity describes a balance of two forces acting on a fluid. One is the cohesive force that describes the mutual forces of attraction of the fluid particles. In contrast, the adhesive forces are those that act between the fluid particles and the solid interface. The dominance of either the cohesive or the adhesive forces determines whether or not the fluid flows easily and evenly over the solid and into the exposed surface anomalies.

The vector sums of the cohesive (C) forces and the adhesive (A) forces are shown for three different material conditions in Fig. 27-4 [3]. In each case, the

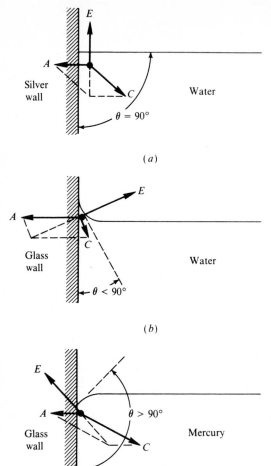

FIGURE 27-4
Effect of cohesive and adhesive forces at fluid, solid, and gas interfaces. (*From Sears and Zemansky [3]. Courtesy Addison-Wesley.*) (*a*) Cohesive forces balance. (*b*) Adhesive forces greater than cohesive forces. (*c*) Cohesive forces greater than adhesive forces.

contact boundaries are assumed to be gas/liquid, solid/liquid, and solid/gas. The adhesive forces A are shown in all cases to be directed perpendicular to the solid/liquid boundary with the magnitude varying due to the different adhesive forces for the specific materials. Similarly, the cohesive forces C are shown directed downward and into the liquid with the magnitude and direction again being a function of the particular fluid. In Fig. 27-4(*a*) for a silver wall in contact with a water and air interface, it is observed that the vector sum of forces A and C is directed downward. For particle equilibrium, the reaction force E acts upward, perpendicular to the surface. The contact angle of the liquid, θ, is the slope of the liquid relative to the solid wall. For the air/silver/water combination, $\theta = 90°$. A change in the solid material affects the contact angle as shown

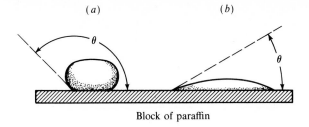

(a)   (b)

Block of paraffin

FIGURE 27-5
Fluid droplet on a surface showing
large and small contact angle θ.
(*From Sears and Zemansky [3].
Courtesy Addison-Wesley.*) (*a*)
Water free of detergent. (*b*) Water
with detergent added.

in Fig. 27-4(*b*). Here, the increased magnitude of the adhesive forces relative to
the cohesive forces gives the reaction force E a positive slope with $\theta < 90°$, as
shown. A different fluid, e.g., mercury, adjacent to a glass wall shows $\theta > 90°$,
mainly due to the greater cohesive forces for mercury as compared to water.

The angle θ defines the wetting ability of the fluid on a particular solid.
Figure 27-5 shows two extreme cases of the relationship of wetting ability of a
fluid on a solid and the contact angle θ. For reliable penetrant inspection, θ must
be less than 90° and should be as small as possible. Typically, θ is 10° or less for
penetrant materials [6]. It should be noted also that the wetting ability of a
particular fluid may be altered by the addition of wetting agents that increase the
ability of the fluid to flow over the solid surface.

The spreading coefficient (S) over a solid surface for a fluid having a
surface tension σ has been given by Alburger et al. [6]:

$$S = \sigma_{sg} - (\sigma + \sigma_{sl}) \tag{27-3}$$

where σ_{sg} = surface tension of the solid/gas interface
    σ_{sl} = surface tension of the solid/liquid interface

In order for a fluid to spread on the solid surface, the surface energy of the
solid/liquid interface must be at a lower level than that of the solid/gas
interface, or $\sigma_{sl} < \sigma_{sg}$.

The movement of the fluid into the surface-breaking defects is a function of
the contact angle θ as well as the surface tension of the fluid. This phenomenon
is demonstrated by the familiar capillary tube comparison, as shown in Fig. 27-6.
For a liquid that wets the interior of the tube at a contact angle $< 90°$, the
adhesive forces move the perimeter of the fluid up the tube wall until an
equilibrium height h is reached, as shown in Fig. 27-6(*a*). The lifting force P on
the column, a result of the surface tension at the perimeter of the fluid, is

$$P = 2\pi r\sigma \cos \theta \tag{27-4}$$

where r is the radius of the tube and σ is the surface tension of the fluid in
contact with the solid. The downward force W is a function of the weight of the
fluid column, i.e.,

$$W = \rho g \pi r^2 h \tag{27-5}$$

where ρ = density
    g = gravitational constant

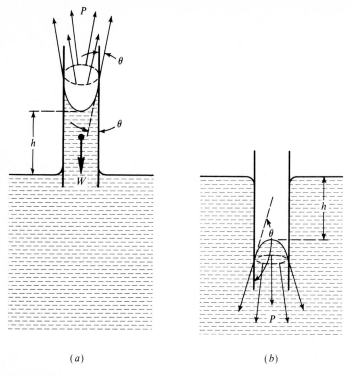

(a)   (b)

FIGURE 27-6
Capillarity. (*From Sears and Zemansky [3]. Courtesy Addison-Wesley.*) (*a*) *Liquid rises in tube for contact angle θ less than* 90°. (*b*) *Liquid is depressed in tube when contact angle θ is greater than* 90°.

Setting Eq. (27-4) equal to Eq. (27-5) and rearranging yields the height of the column as

$$h = \frac{2\sigma \cos \theta}{\rho g r} \tag{27-6}$$

Thus, the ability of the fluid to enter surface cavities is directly proportional to the surface tension σ and inversely proportional to the contact angle θ, the density ρ of the fluid, and the size of the opening r.

A satisfactory penetrant inspection is therefore dependent on the several factors as given in Eq. (27-6). The surface tension and the density of the fluid are, of course, dependent upon the type of penetrant material used. The contact angle is a function of the properties of both the fluid used in the inspection and the material being inspected. Further, the cleanliness of the inspected material surface affects the contact angle. The depth of the defect also is seen to affect capillarity as given in Eq. (27-6). While a smaller defect is seen to increase the amount of penetrant drawn into the cavity, it must be recognized that smaller flaws may be more difficult to detect in some circumstances. As discussed in a

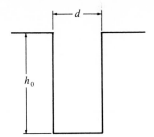

FIGURE 27-7
Blind cavity for capillary action.

number of sources, including Ref. 6, the actual excursion depth of penetrant into a closed crack will, in fact, be less than the value h given by Eq. (27-6) due to the internal pressure of the gas trapped at the crack tip. Fluid viscosity affects the time required to achieve a thorough wetting of the surface and a full penetration of the fluid in a cavity. Thus, the overall inspection time is very much determined by the penetrant viscosity.

The time required for the penetrant to fill the cavity is called the *dwell time.* Penetrant activity at a cavity has been described in Ref. 6 and an analysis of the relationship of dwell time and fluid properties has been given by Prokhorenko et al. [7, 8]. An expression by Prokhorenko et al. of the time needed to fill a cavity, as shown in Fig. 27-7, is

$$t = \frac{6\mu\varepsilon h_0^2 \Psi}{d\sigma \cos\theta}\left[\frac{\xi^2}{2} - (1 - \Psi)\left(\xi + \Psi \ln\frac{\Psi - \xi}{\Psi}\right)\right] \tag{27-7}$$

where
$$\Psi = \frac{2\sigma \cos\theta}{2\sigma \cos\theta + p_a d}; \qquad \xi = \frac{h}{h_0}$$

and

$$\varepsilon = \frac{2\tilde{k}^2(d/2)^2 \sinh(\tilde{k}d/2)}{\left[2(\tilde{k}d/2)^2 + 3\delta\right]\sinh(\tilde{k}d/2) - 3\delta(\tilde{k}d/2)\cosh(\tilde{k}d/2)}$$

For Eq. (27-7),

μ = coefficient of dynamic viscosity
σ = surface tension
h_0 = crack depth
h = wetted depth in crack
d = crack width
θ = fluid wetting angle
p_a = atmospheric pressure
$\tilde{k}, \delta$ = penetrant microstructural properties obtained experimentally [8]

Equation (27-7) may be used to compare the filling time for various crack sizes and fluids. It is important at this point to emphasize that the parameters $\tilde{k}$

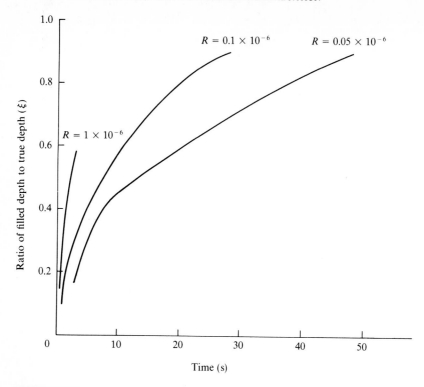

FIGURE 27-8
Fill rate for cavities of different width R (in meters). Depth D is constant and ξ represents the ratio of filled depth to true depth.

and δ are structural properties of the fluid that must be determined experimentally [8]. Further, the equation applies to a blind capillary, i.e., one with smooth walls and closed at the bottom end. While these conditions are not typically met in ordinary nondestructive testing circumstances, the equation is useful in comparing the actions of typical fluids and the reactions at typical crack sizes.

Calculated results from Ref. 7 are shown in Fig. 27-8 for water, which is a typical base for many penetrants. Where the parameter ξ represents the ratio of the filled cavity depth to the full depth, the three plotted curves show the increased time required for filling tighter cracks.

When the surface conditions of an actual crack are considered, the filling times are greatly increased. In fact, calculations by Prokhorenko et al. [7] show an increase of over 40% for the 5 μm crevice when some structural conditions that simulate actual crack conditions are considered.

The fluid properties also determine the smallest cavity in which the penetrant may flow, as described in Ref. 6, where the limiting defect width d_c is given as

$$d_c = \frac{K\sigma \cos \theta}{\Delta p} \tag{27-8}$$

where K = fluid infiltration rate, a function of the crack opening geometry
(spherical pore $K = 3.14156$, narrow crack $K = 2$)

σ = surface tension of fluid in air

θ = contact (wetting) angle of fluid on solid

Δp = capillary pressure

While Eq. (27-8) indicates that a lower surface tension decreases the size of the defect that may be infiltrated by the penetrant, Prokhorenko et al. caution that other properties as given in Eq. (27-7) may play a significant role in establishing the infiltration of the penetrant into the crevice [8].

27-3   OPTICAL PRINCIPLES IN DEFECT RECOGNITION

Optical recognition systems are used for the detection of surface anomalies arising from penetrant inspection. The full range of the light spectrum is used, from visible light to ultraviolet. Each type has characteristics that are uniquely utilized in penetrant inspection.

27-3.1   Visible Light Spectrum

Historically, there probably has been a greater evolution in the defect recognition systems than in any of the other principles involved in the penetrant technique. Initially, the only detection method used was the unaided observation by the eye of the human inspector. Thus, most early efforts in inspection improvements were concentrated on the visible light spectrum.

The wavelengths in the visible light spectrum are listed in Table 27-2. In general, a light source will contain not one pure color, but rather will have energy at a variety of colors, or wavelengths, where the perceived color of the source is determined by the wavelength that dominates the source spectrum. Where a typical normal distribution curve exists around the peak wavelength, it is easy to see that different hues of a particular color could exist. For example, where a

TABLE 27-2
Approximate wavelengths of visible light (nm).

Color	Wavelength
Violet	380–450
Blue	450–490
Green	490–560
Yellow	560–590
Orange	590–630
Red	630–760

Source: Reference 9.

source with a peak energy at 480 nm wavelength might be expected to be blue with a small amount of green, a source at 460 nm might be blue with a touch of violet. In most cases, the human eye could distinguish between the two light sources with wavelengths at 480 and 460 nm.

The characteristics of the human eye strongly affect the perception of brightness of an indication. As reviewed by Sears [10], a standard observer under good lighting conditions detects maximum luminosity at approximately 550 nm, which is at the upper region of the green color. The luminosity drops off significantly on either side of the peak, reaching one-half of the peak value at 510 and 610 nm. The 510 nm light would be in the lower region of green while the 610 nm light would be in the orange region. Under subdued lighting, the peak luminosity shifts to the left, with the peak being at 510 nm and the lower and upper one-half luminosity points at 470 and 560 nm, respectively. The 470 nm light would be in the blue region, while the 560 nm would be at the intersection of green and yellow.

As is clear from the foregoing, the nature of the light source strongly affects the perceived brightness of the colored region being observed. Natural light and artificial light might vary significantly in both the peak energy wavelength as well as the spread of the energy distribution curve. This phenomenon may be both a hindrance and an asset in penetrant inspection, as will be discussed in the material to follow.

27-3.2 Fluorescence and Ultraviolet Lighting

An improvement of significant magnitude occurred when particles that fluoresce under an ultraviolet (UV) light were introduced into the penetrant fluid. Fluores-

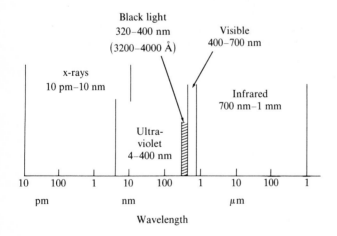

FIGURE 27-9
Partial electromagnetic spectrum showing ultraviolet, visible, and infrared ranges. (*Courtesy Magnaflux Corporation.*)

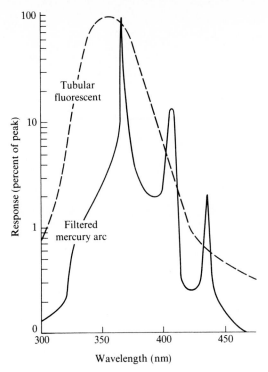

FIGURE 27-10

Emission of mercury arc and tubular fluorescent ultraviolet light. (*Courtesy Magnaflux Corporation.*)

cence describes the release of light energy by some substances when they are excited by external radiation such as ultraviolet light [11]. For the present case, when the particles in the fluid are struck by the incident ultraviolet light, they are raised to a higher-energy level. After being excited, each particle, then, returns to the original unexcited level with the emission of light having a wavelength longer than the original source. Thus, the emitted light is in the visible spectrum.

The partial electromagnetic spectrum given in Fig. 27-9 shows the location of the ultraviolet spectrum as well as that of the visible region previously discussed. Because the ultraviolet light spectrum is not normally seen by the human eye, ultraviolet lights have been labeled *black lights*. The advantage of this in penetrant inspection is that regions holding greater amounts of the fluorescent penetrant appear very bright. When the inspection is performed in very subdued light, the regions clear of penetrant material will appear black. With this scenario, the visibility of very small indications is greatly enhanced.

While the full ultraviolet spectrum ranges from wavelengths of 1–400 nm, the segment used in penetrant inspection is from 320–400 nm, which is adjacent to the visible spectrum. Typical spectral emission curves for commercial black lights are shown in Fig. 27-10. Peak emission is commonly at 365 nm. Both unfiltered, i.e., tubular fluorescent, and filtered curves are shown.

27-3.3 Laser Scanners

A significant improvement in both the reliability and sensitivity of penetrant inspection occurred with the introduction of laser scanning devices that are able to automatically scan large areas. Lasers, being monochromatic light sources, do not have the side energy wavelengths present. Thus, fluorescent indications excited by a laser are distinct in both form and color. A typical scanner might consist of a helium/cadmium (He-Cd) laser emitting a deep blue at a wavelength of 441.6 nm [12, 13]. When directed at typical fluorescent penetrant materials, a visible yellow light is emitted. Light filtering techniques may be used to eliminate the blue light from the background, leaving the dominant yellow emission from the anomaly containing the penetrant. With the monochromatic laser light being available in a number of possible wavelengths, more precise matching of excitation source and organic dyes in penetrant materials may lead to greater improvements in penetrant sensitivity and reliability.

27-4 BACKGROUND CONTROL—CONTRASTS

The major activity in the early development of the penetrant technique was on improvements in the conspicuity of the defect area as compared to the surroundings. Particular emphasis was placed on the correct lighting of the field being observed as well as studies on the sensitivity of the human eye to certain colors. Both natural light and artificial light in the visible spectrum were used in early investigations. Studies of contrasts lead to the development of inspection materials that yielded a striking contrast to the penetrant that had crept in the crevices, pores, etc., of the part being inspected.

The most common technique used to create a satisfactory background contrast for the penetrant inspection is the use of a developer to draw excess penetrant from the defect area. Developers are an absorbent material applied to the inspection surface in a thin layer after the penetrant has been cleaned from the surface but not from the crevices.

Capillarity is the driving force that the developer uses to remove the penetrant from the defect. Figure 27-11 shows that for a crevice completely filled with penetrant, the visible width is limited to the crack width d. After the

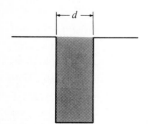

FIGURE 27-11
Cavity of width d filled with penetrant from capillary action.

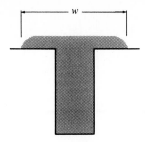

FIGURE 27-12
Width w of penetrant drawn from cavity by developer.

developer has been applied, the visible width of the defect is now w, as shown by the enlarged region in Fig. 27-12. Thus, the developer plays a significant role in the penetrant inspection process. The physical principles active during the development process are described by Refs. 6 and 8.

27-5 MATERIAL PROPERTIES AFFECTING PENETRANT INSPECTION

As a surface effect technique, it is abundantly clear that penetrant inspection is sensitive to the properties and condition of the surface being inspected. The wetting ability of the fluid is a function of the wetting angle that is unique to a particular material and fluid interface. Moreover, whether the surface of a particular material is wiped with a rag, cleaned with a solvent, or etched could impact the wetting ability of the fluid. The mechanical condition of the solid may also affect the inspectability. Where a smooth, flat surface of a defect-free material would yield a no-defects test result, a surface of the same material containing minor corrosion pitting or rolled laps, for example, could give false indications of defects. Another possibility in the scenario just cited is that assumed spurious indications could mask true defect indications.

There appears to be little documentation of the surface condition quality required for a satisfactory penetrant inspection. The rules and recommendations as listed in Table 27-3 should aid the inspection process.

TABLE 27-3
General rules governing material surface conditions for satisfactory penetrant inspection.

1. The surface must be such that the inspecting fluid flows evenly and easily.
2. Rough or uneven surfaces impede the inspection process. Fluid retained in surface cavities that are not cause for rejection can either give spurious indications or mask true defects, or both.
3. Rough or uneven surfaces are more difficult to inspect since extra care is required to properly clean the region to be inspected.
4. Materials with characteristically high porosity are difficult to inspect because of the large amount of fluid naturally retained in surface opening pores. Examples of these materials are concrete, some polymers, clay pottery, wood, and paper.

Some problems that might be encountered in inspecting a particular material may be overcome by choosing a different penetrant that may be less susceptible to the problems encountered.

27-6   SUMMARY

The physical principles and pertinent nomenclature for penetrant inspection have been introduced in this first section. With this basic information, the discussion will now proceed to sections where more definitive discussions will be given.

CHAPTER 28

TECHNIQUES FOR PENETRANT INSPECTION

28-1 INTRODUCTION

Penetrant inspection may be accomplished on a variety of materials, both ferrous and nonferrous. Glass as well as polymers and ceramics often may be inspected using penetrants. Irregularly shaped parts that may be very difficult to inspect with the other techniques may be in some cases conveniently inspected with penetrants. Disbonds in layered composite materials and leakage in pressure vessels and piping may also be detected using these methods. The need in penetrant inspection for surface exposure of the defects or anomalies, however, must always be recognized.

In the simplest case, penetrant inspection is both inexpensive and easy to perform. Visual dye penetrants may be used with no electrical power requirements. In some cases, the capital investment for penetrant inspection may be quite small and the required inspector training minimal. At the other extreme, the investment required to establish a fully automated laser scanning penetrant inspection may be considerable. Further, the engineering support necessary for such an system may also be significant.

The material in the present section will be derived from a number of sources, most notably Refs. 6 and 14–27, which give considerable detail on a variety of penetrant applications. The intent of this part of the textbook will be to provide an introduction to the basic techniques of penetrant inspection and to cite some examples showing application to particular tasks. The reader is directed to these references for additional detail.

28-2 BASIC TECHNIQUES FOR PENETRANT INSPECTION

Using the physical principles previously described, penetrant inspection is accomplished with a few fundamental steps. An overview of the fundamental steps as outlined in Fig. 28-1 and illustrated in Fig. 28-2 will be followed by a more detailed discussion of the various techniques and applications. The detailed discussion will describe several variations available to achieve specific test objectives and conditions.

One of the most important steps in the penetrant inspection process is the initial cleaning of the surface area to be inspected. Ideally, the crack that is being sought is in an area that is clean and free of extraneous cavities. It must be open to the surface in order for the penetrant to enter, as indicated by the sketch in Fig. 28-2(a). Scale, flakes, paint, simple dirt, grease, and other chemicals that are not cleaned from the surface will tend to accumulate penetrant that can either mask real indications or create indications of defects where none exists. The initial cleaning process may be at one of several levels of severity, doing what is

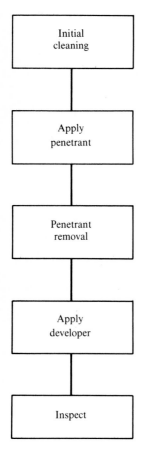

FIGURE 28-1
Basic steps in penetrant inspection.

Penetrant

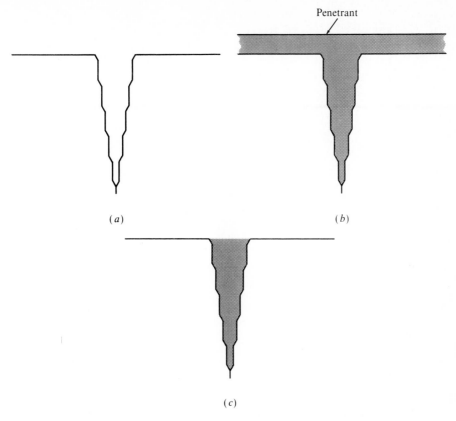

(a)

(b)

(c)

FIGURE 28-2
Typical fluid action in penetrant inspection. (a) Open defect. (b) Applied penetrant in the cavity and covering the surface. (c) Penetrant cleaned from the surface but remaining in the crack. (d) Enlarged width of penetrant due to action of the developer. (e) Detection of defect by visible inspection. Area may be illuminated with black light for fluorescent inspection or white light illumination for visible dye penetrants.

required to free the surface of spurious material. While too severe mechanical cleaning action could close the exposed crack or pore, cleaning that is too mild could leave undesirable residual material. A satisfactory combination of solvents, brushes, rags, etchants, etc., must be chosen for a particular inspection problem. Rags and paper towels that leave residue on the inspection surface must be avoided. It is imperative that the cleaned surface be adequately dried before the application of the penetrant since excess cleaning fluid would dilute the penetrant and diminish the brilliance of the indication.

The next step in the inspection process is the application of penetrant fluid to the cleaned surface. The fluid should spread freely and evenly over the surface and move into the crack, as shown in Fig. 28-2(b). As previously described, the dwell time, which is the amount of time required to move into the crack, will vary

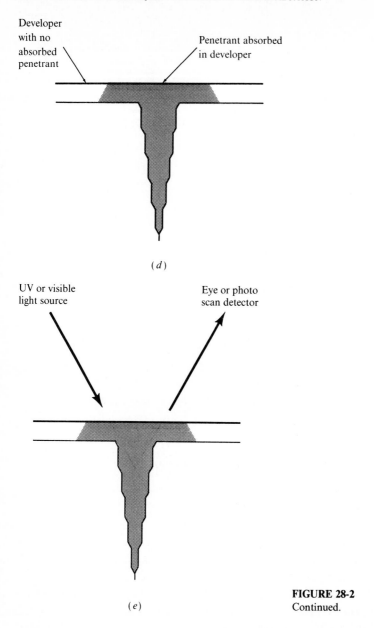

(d)

(e)

FIGURE 28-2
Continued.

depending on crack size and shape characteristics as well as the previously discussed physical characteristics of the fluid and also the environmental conditions such as temperature and surface inclination. The penetrant fluid consists of two components, the indicating particles and the vehicle that carries the particles into the crack. The indicating particles may respond in the visible light spectrum or they may fluoresce under ultraviolet or laser excitation. Penetrants may in general be either water-based or petroleum-based. Details regarding typical penetrant materials will be covered in the application sections to follow.

The penetrant removal step, i.e., the second cleaning, may be both the most difficult and the most important step in the entire process. The desired result is shown in Fig. 28-2(c) where the surface is completely clear of penetrant yet the crack retains all of the fluid that entered it and the meniscus is not depressed into the surface. Two specific problems may occur in this step. On the one hand, insufficient cleaning will leave a background of penetrant on the surface. With this, the defect will appear only slightly different from the background area and the contrast between the background and the flaw may not be sufficient for the defect to be recognized. On the other hand, cleaning too aggressively may remove the penetrant from the upper region of the defect with a result that the developer does not reach the penetrant and no defect is indicated. Often this second cleaning step will be repeated several times in order to remove sufficiently the background surface penetrant. Care must be exercised, however, so that over-cleaning will not occur.

Following this second cleaning step, the developer is spread over the surface to draw penetrant out of the crack and increase its visibility. As seen in Fig. 28-2(d), the excess penetrant that is drawn out of the defect increases the apparent width that makes the defect easier to identify. The developer is a powder-type material that uses capillary action to draw the penetrant out of the crevice. Another important function of the developer is that it covers the surface with a color that provides good visual contrast to the penetrant, thus increasing the conspicuity of the defect.

The last step in the process is the scanning of the surface in the search for indications, as shown in Fig. 28-2(e). The scan may be under visible light conditions or with ultraviolet or laser incident light and recognition may be with the human eye or with automated optical scanners.

28-3 PENETRANT MATERIALS AND PROCESSES

A typical penetrant inspection involves a variety of materials for cleaning and developing as well as the penetrant material itself. Additionally, the procedure may vary, depending on the object to be inspected and the test surroundings. In general, the fluids involved in the penetrant process are either petroleum- or water-based and the solvents or cleaners are dictated by the base of the penetrant fluid. In many instances, however, both petroleum- and water-based fluids are used. In this event, emulsifiers are needed to break down the petroleum base to where the fluid may be removed with water.

The penetrant inspection processes are typically classified by the indicating method, i.e., visible or fluorescent dye, and the method used to remove the excess penetrant from the test piece. Table 28-1 shows a generalized grouping of the various penetrant families for fluorescent materials. Visible dye penetrants would have a similar grouping of identifying terms.

TABLE 28-1
Penetrant inspection systems

System	Symbol	Emulsifier	Cleaner	Sensitivity
Water-washable or self-emulsifiable	WW	None	Water	Regular, high, and ultrahigh
Postemulsifiable	PE	None Hydrophilic Lipophilic	Water	Regular, high, and ultrahigh
Prewash	PWHE	Hydrophilic	Water	Regular, high, and ultrahigh
Solvent removable	SR	None	Petroleum Base	Regular and high

Source: Reference 15.

Fluorescent penetrant inspection methods are designated by the American Society for Testing and Materials as method A, while visible dye penetrants are designated as method B [17]. Within method A, the classifications are further grouped into process "families" that describe the method of removal, namely water washable (Procedure A-1), post-emulsifiable (Procedure A-2), and solvent-removable (Procedure A-3). Visible penetrants are similarly categorized as Procedure B-1, Procedure B-2, and Procedure B-3. Further, dual purpose penetrant materials are available that may be used with either method A or method B. The specific characteristics of the penetrant materials and the process families will be discussed in the material that follows.

28-3.1 Penetrants

The penetrant material consists of the indicating (tracer) dye plus the vehicle fluid. The indicating dye may give a color contrast to the surroundings, as is the case for visible dye penetrant methods, or a brightness contrast for the fluorescent dye penetrants. For visible light penetrants, the dye is usually red in color, while for fluorescent penetrants, the dye appears bright yellow-green under the ultraviolet light. The selection of petroleum- or water-based penetrants is mostly determined by the circumstances of the inspection being accomplished. On the one hand, water-based chemicals are not volatile, which lessens the chance of explosions or the inhalation of dangerous fumes by inspection personnel. Water-based chemicals, however, may have a deleterious effect on material surfaces, particularly for items made from steel.

Detection sensitivity is a significant factor in the choice of penetrant material. It is not usually a significant factor, however, in the selection of water washable, postemulsified, or petroleum solvent systems since most levels of sensitivity are typically available with each system.

The quantification of penetrant sensitivity has been of interest to a large number of investigators since the minimum flaw size that is detectable generally dictates the penetrant fluid to be used. Penetrant sensitivity, however, is frequently expressed in more general terms such as regular, high, and ultrahigh [15]. Penetrant sensitivity levels are designated in military standard MIL-I-25135, Rev. D, as 1/2, low, normal, high, and ultrahigh [18]. Fluorescent penetrants are available in the full range of sensitivity levels while visible dye penetrants are available at all levels except for the highest. The highest sensitivity, however, is achievable only with the water-washable and postemulsifiable systems, as will be discussed later.

It is important to recall at this point that the wetting ability of the fluid on the material being inspected is a significant factor in penetrant inspection. Because of this, sensitivity levels for a group of penetrants established empirically using a sample with a smooth mirror-like surface may not always correlate well with results obtained from a machined surface. Comparing data from the same test samples, however, has been shown to give a good ranking of penetrant sensitivity.

A review of several efforts to quantify penetrant inspection characteristics has been given by Packman, Hardy, and Malpani [19]. An early scheme presented by Lomerson compared the reproducibility of several penetrant materials when applied to a control set of cast turbine blades [20]. In these tests, a standard penetrant material was chosen and the performance of other penetrants was compared to the standard using a two-fold congruency analysis. Visual observations of the fluorescent indications were used.

For three trials, the reproducibility of flaw indications for the test penetrant was compared to the standard penetrant. A mean or average number of reproducible indications was determined for each penetrant. An expression of the sensitivity, then, was given to be the ratio of the mean of the number of reproducible indications obtained from the test penetrant over that of the standard penetrant. For the range of penetrants tested, the sensitivity ranged from < 50% to > 200%, the latter indicating the greater sensitivity. The measure of sensitivity in this series of tests would be the fact that some penetrants would indicate more defects in the control turbine blades than others, the implication being that the additional indications were obtained from defects that were smaller.

Additional experiments using the two-fold congruency test were performed by Hyam, who investigated the effects on the sensitivity of various other processing parameters such as the cleaning and emulsifying techniques [21]. While it is recognized that sensitivity results obtained for these series of tests might not necessarily be repeatable for another material and surface condition, the relative performance of various penetrant materials under the same conditions was clearly demonstrated.

Using standard test panels with known defect characteristics, Vaerman has reported results quantifying penetrant sensitivity [22]. Four standard test panels were used, each having a thin chromium layer over a nickel layer that was on a brass strip 2 mm (0.079 in) thick. Each panel was 35 mm (1.38 in) wide and 100 mm (3.94 in) long.

The total number of cracks in each panel was known and the cracks in each panel were the same size, with the width approximately one-twentieth of the crack depth. The crack depth was known to be equal to the thickness of the combined chromium and nickel layer. The cracks in the panel with the largest cracks were 2.5 μm (0.1 × 10^{-3} in) wide and approximately 50 μm (1.97 × 10^{-3} in) deep. For the panel with the smallest cracks, the crack dimensions were 0.5 um (0.02 × 10^{-3} in) wide and approximately 10 μm (0.4 × 10^{-3} in) deep. Thus, a particular penetrant might be expected to indicate uniformly all of the cracks for the panel with the largest dimensions but progressively fewer of the cracks in the panels with the smaller dimensions. For the smaller cracks, the results showed that the penetrant might indicate only part of the total length of the crack. In the most severe case, no cracks might be indicated.

Each of the panels was illuminated with a UV laser and detection was accomplished using a photodetector. A total of N scans was made for each panel perpendicular to the crack field and the number of total crack indications was recorded. Since the total number of cracks were known, the sensitivity (S) was determined to the percentage of detected cracks as expressed by

$$S = \frac{\text{number of detected cracks}}{N \times \text{number of actual cracks}} \times 100 \qquad (28\text{-}1)$$

A series of penetrant compounds from two different suppliers was used for comparative testing. Figure 28-3(a) shows the indications for the panel with the

(a)   (b)

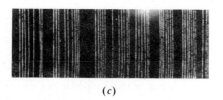

(c)

FIGURE 28-3
Test panels indicating penetrant sensitivity. (*From Vaerman [22]. Courtesy American Society for Nondestructive Testing and SNECMA, Evry, France.*) (a) Postemulsifiable process AP-13, 50 μm panel, 100% cracks detected. (b) Postemulsifiable process AP-13, 20 μm panel, 88.3% cracks detected. (c) Postemulsifiable process AP-4, 50 μm panel, 100% cracks detected.

largest size flaws and the compound designated AP-13. For these circumstances, all of the cracks were detected and the sensitivity was 100%. The results obtained with the same penetrant used on a panel with smaller cracks is shown in Fig. 28-3(b), where the sensitivity was found to be 88.3%. Another postemulsifiable penetrant (AP-4) also showed a 100% sensitivity for the largest cracks (Fig. 18-3c) but a more rapidly declining sensitivity for the smaller ones.

The results obtained in these experiments clearly indicate the differences in penetrant sensitivity that may exist for various materials. For the 50 μm panel, the one with the largest cracks, sensitivity for the various penetrant materials tested ranged from 100% to 67.6%. For the panel with the smallest cracks, i.e., the 10 μm panel, the maxium sensitivity was 67.6% with several penetrant materials indicating none of the cracks for a 0% sensitivity. The greatest sensitivity was generally obtained with postemulsifier penetrant systems while the least sensitivity came from the water-washable systems, as will be discussed in the section to follow. The reader is cautioned that these results, obtained under controlled, laboratory conditions, may not always be observed in actual practice due to process variations that occur in a shop or field environment.

The dynamic properties of penetrants migrating into cracks has been studied experimentally by Tanner, Ustruck and Packman [24]. Central to their work was the fact that they were able to measure the amount of penetrant dye that is contained in a crack. This was accomplished using a fluid bath of toluene as the control volume. The sample being studied was immersed in the toluene bath and the amount of dye released from cracks was determined by measuring the optical properties of the toluene that they previously had calibrated.

The first intent of their work was to measure the adsorption and desorption of the dye into the crack for two types of samples. For the first set of samples, the adsorption set, each plate was initially degreased and then immersed in penetrant material containing one of three different concentrations of dye, namely 50, 75, and 100% by volume in toluene. For the second set, the desorption set, the plates again were degreased but were then immersed in a 100% solution of dye. These plates then were immersed in the same 50 and 75% solutions of dye.

The results of these experiments are shown in Fig. 28-4. In both cases it can be seen that the second set of plates had a significantly higher level of dye in the cracks than did the first set of samples that were coated in the degreased condition with the 50 and 75% solutions. The authors suggest that the area within the curve could be used to discriminate between deep and shallow cracks in unknown samples. Further, they note that the final portion of the adsorption curve has a very steep slope, which infers that if solvent remains in a crack when the dye penetrant is applied, then the subsequently added dye would be diluted so that a much reduced indication of the crack would appear.

Penetrants may be applied in a number of ways including spray cans for localized application and immersion for complex shapes or for the simultaneous inspection of many items of the same part. Additionally, the penetrant fluid may be applied over the surface with a simple hose that gives an even, smooth flow. Spray techniques are adequate for application to small areas. Since the fluid

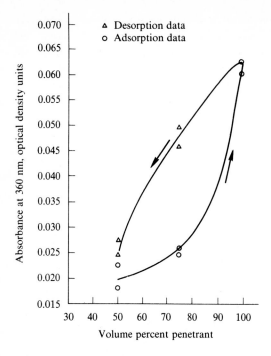

FIGURE 28-4
The adsorption–desorption hysteresis curve for penetrant FP-91 in contact with the chrome-plated cracked steel plate. The vertical scale indicates the amount of fluorescent dye in the cracks. (*From Tanner et al. [23]. Reprinted with permission from Materials Evaluation, vol. 38, no. 9, 1980, The American Society for Nondestructive Testing, Columbus, OH.*)

leaving the spray can cannot be recycled, it is an impractical technique for the inspection of large surface areas. Material usage is minimized with the immersion or recirculating hose techniques since only that penetrant that adheres to the part is consumed. Application techniques will be described in more detail in the material to follow.

28-3.2 Cleaners and Emulsifiers

Correct and adequate cleaning is crucial to a satisfactory penetrant inspection. Cleaning before the application of the penetrant assures that the penetrant may enter the crack or crevice; the penetrant removal step following the application of the penetrant must be correctly done to provide the correct viewing or scanning background for the defect detection process. Finally, a thorough removal of all of the penetrant inspection materials from the surface is usually required in order to return the item to service. In all cases, the parts should be thoroughly drained of extraneous material in order to decrease the amount of extra cleaning required.

Mechanical initial cleaning methods must be used with caution since shot peening, wire brushing, or sanding, for example, can smear or peen the surface layers so that the crack openings are closed to the penetrant. Whitehorn has investigated the effect on subsequent penetrant inspection of blasting of aluminum samples with a lignocellulose abrasive [24]. Lignocellulose abrasives are a wood-based blasting material often derived from various fruit stones such as plum, peach, and apricot. They have proved to be effective in the removal of paint from aircraft grade aluminums; the interest of the investigations was to determine the effect of the blasting on the detectability of surface-breaking defects with penetrants. The results clearly showed that very fine cracks could be closed off by the blasting and the effect was increased as the blasting air pressure was increased from 172 (25 psi) to 276 kPa (40 psi). The author felt, however, that the effect on cracks visible to the unaided eye was not serious. In any event, a light etch was found to restore the surface to the original, unblasted condition.

A cleaning fluid must act as a solvent for the material that is to be removed. For water-based penetrants, a simple water wash or rinse is suitable for the cleaning steps. For petroleum-based penetrants, there are two alternative methods for cleaning the test piece. The most direct approach is to use an oil- or chlorine-based solvent. Another method is to use an emulsifier that reacts with the oil-based penetrant to form a water-soluble substance, which then may be removed by water washing or rinsing.

The two types of emulsifiers are hydrophilic and lipophilic. Hydrophilic emulsifiers are composed of materials similar to common detergents, which react with the oil-based penetrant in a way that removes the penetrant from the surface in a scrubbing process that is enhanced by the final water rinse. Diffusion plays a minor role in the action of the hydrophilic emulsifiers. Lipophilic emulsifiers, on the other hand, are oil soluble and they diffuse into the penetrant, breaking down the structure so that the penetrant may be rinsed away with water.

Experimental results presented by Hosokawa and Hosoya [25] show that crack detectability may be affected by emulsification concentration and contact time. Booth et al. [26] have similarly described the effect of penetrant contamination in the emulsifier upon the effectiveness of the emulsification process. Penetrant levels of less than 2% by volume were shown to significantly reduce the effectiveness of the emulsifier. Methods for evaluating the contamination of emulsifiers are given in Ref. 25. In any case, it is important to recognize that care must be exercised in the mixing and application of emulsification chemicals.

A prewash technique may be used to minimize the amount of emulsifier required in the cleaning process [15]. For this step, the part that is coated with oil-based penetrant is water-rinsed first to remove excess penetrant from the surface. With the amount of penetrant on the surface significantly reduced, the amount of emulsifier required to complete the process is reduced. A further advantage of this technique is the ability to recover much of the penetrant from the rinse solution, thus diminishing the waste and pollution effects of the chemicals. In the application of the prewash hydrophilic emulsifier, care must be exercised in assuring the proper mixing of the chemicals so that removal by the emulsifier of penetrant from openings will not occur in the process [27].

Penetrants that use an emulsifier in the initial removal process are named *postemulsified* penetrants. These materials are distinguished from the prewash group in the fact that the postemulsified penetrants do not use the water rinse before the application of the emulsifier. Some oil-based penetrants may contain an emulsifying material that will react with rinse water to convert the penetrant from an oil- to a water-based material. These penetrants are labeled *self-emulsifying*.

There are a number of methods used in the cleaning process such as wiping with a cloth, dipping in a tank, rinsing with a hose, or some combination of these. The choice of cleaning process again depends on the many variables of the test, notably the size, shape, and material of the item as well as the test environment. Also, the cautions given for the penetrant removal stage would not generally apply for the final cleaning process that prepares the part for return to service. Similarly, the initial cleaning process does not require the same level of care and caution as does the penetrant removal step.

In the penetrant removal step, gentle wiping of the inspection area with a clean, dry cloth may be sufficient for very smooth surfaces. If some solvent is required, the cloth should be mildly wetted so that there is little likelihood of removal of the penetrant from the crevices. A cloth made from material containing a large amount of polymeric fibers may be unsatisfactory for this step since it will not absorb the solution in large enough quantities. Rags and paper towels that are prone to leave fibers on the test surface are similarly undesirable since that may obscure the area with spurious indications during the viewing process.

Larger items are often either sprayed with a gentle stream of the proper solvent or immersed into trays, tanks, or vats. In the penetrant removal phase, the need to adequately remove the penetrant from the viewing surface must not be allowed to result in too severe a cleaning process so that the penetrant also is removed from the crevices and pores.

Remnants of the cleaning fluid left upon the surface may seriously affect the brilliance of the penetrant indication [28]. Experimental investigations were conducted using 12 different cleaning solutions as contaminants in the penetrant. Contaminants varying from 2–10% by volume were introduced into the penetrant fluid. An absorbing filter paper was inserted into each contaminated penetrant and the filter paper was subsequently dried and studied for fluorescent brilliance. Six different penetrants were used from the investigations. In most cases, the smallest amount of contaminant, which was 2% by volume, showed a serious reduction in the fluorescent brilliance of the penetrant. Depending on the type of contaminant, the 2% amount caused a reduction in brilliance from 2% to 57% with the data generally spread between these two limits. The author's conclusion is that a 2% level of contamination is sufficient to seriously affect the results of a penetrant inspection.

28-3.3   Developers

Following the second cleaning step, i.e., the removal of excess penetrant, the developer material is used to enhance the conspicuity of the indication. As previously mentioned, there are two basic contributions of the developer. First, the developer must attract the penetrant out of the crevice by capillary action. This requires that the developer be highly absorptive of the penetrant fluid. Second, the developer must create a scanning or viewing background that greatly contrasts with the appearance of the penetrant. For fluorescent penetrants, the developer background should appear black when illuminated by the ultraviolet light. The penetrant material concentrated around the defect will brightly fluoresce and appear distinct from the black background of the developer. The developer for visible dye penetrants normally creates a white background that contrasts to the normal red appearance of the dye pulled out of the crack or pore.

The developer material may be one of several types, namely dry powder, aqueous (wet) powder-suspension, solvent-suspendible, plastic-film, and water soluble. Application may be accomplished by several techniques, depending on the size and shape of the part as well as the system to be used. Among these application techniques are spray, immersion, passing the part through a developer dust cloud chamber, fluidized beds, and electrostatic means. Where wet developers are used, the drying process may be speeded by moving the part to an oven after the developer has been applied. This must be done with some caution, however, as will be discussed later.

As described by Booth et al. [15], the developer application process may significantly affect the sensitivity of the inspection. The greatest sensitivity is obtained with solvent spray, plastic film spray, and water-soluble spray. The least sensitivity is obtained by the dry immersion and dust cloud methods.

Solvent suspendible developers may be applied to both the visible light and fluorescent penetrants. The enhanced performance of these developers is derived from the solvent action of the developer with the penetrant that aids the movement of the penetrant out of the defect. While the volatile solvent-based developers dry very quickly, thus speeding the inspection process, they must be used with some care since the fumes that they release may be both flammable and toxic. The plastic film developers are generally applied by spraying over the surface. After the material has dried, the film may be stripped from the surface to provide a permanent record of the inspection results.

Aqueous or wet powder-suspension developers are applied typically by immersion or spray immediately after the washing step and before the part is dried. These developers are typically used in high-volume inspections in industrial applications. After the developer is applied, the part is dried in a moderate temperature oven. With evaporation, the absorbent developer brings the penetrant out of the defect in the same manner as for the dry developers. Because the drying of the developer forms a very tight bond between the film and the surface, the aqueous developer is more sensitive to smaller cracks than the dry powder materials.

Dry powder developers may be applied directly to the part immediately after the penetrant removal process. These developer particles are generally white in appearance but when applied to the part being inspected they form a colorless film. Immersion of the part, passing it through an air agitated dust cloud, or a fluidized bed are typical means of applying dry developers. Electrostatic techniques are also used. The dry powder developers are the least sensitive in finding very small defects.

Water-soluble developers are clear solutions consisting of water-soluble crystalline substances dissolved in water. When dried on the surface, the materials recrystallize into the developer film. These developers are applied directly after the excess penetrant has been rinsed from the surface. Application may be by immersion, stream, or spraying. Water-soluble developers generally have good sensitivity, although not as high as the solvent sprays.

Care in the maintenance and application of all developer materials is essential in the execution of a satisfactory and reliable penetrant inspection. The proper mixture ratio of developer material and vehicle is important for the fluid-based developers. The purity of the developer must be consistently monitored since contamination may destroy its effectiveness. Moreover, layer thickness may affect the results since too thin a layer may not draw the penetrant out of the crack while too thick a layer may mask the indication. The dual role of the developer in providing the proper contrasting background and the capillary action for drawing the penetrant from the crevice places special emphasis on obtaining the correct developer thickness during the penetrant inspection. Glazkov and Bruevich [29] have investigated the reflection characteristics, i.e., the whiteness, of developers used in visible dye penetrant inspection. Parameters that were found to affect the whiteness of the developer were the thickness of the layer as well as the coefficients of diffuse reflection for both the developer material and the underlying material that would be the test sample. Using experimental data and curve fitting techniques, they established the mathematical model for the whiteness of the developer to be

$$\eta_c = \eta_{max}(1 - e^{-Xt}) \tag{28-2}$$

where η = reflection coefficient (whiteness) of the developer at thickness t
η_{max} = maximum attainable value of the reflection coefficient
X = light scattering index for the developer material

The results obtained for several values of the light scattering index showed for the highest value, ($X = 400$) the reflection coefficient η reached a maximum value at a layer thickness of approximately 15 μm (0.6×10^{-3} in) The material with a moderate light scattering index ($X = 160$) reached the maximum reflection coefficient value at a layer thickness of 20 μm (0.8×10^{-3} in), which represents a 33% increase in the required layer thickness for maximum reflection. Materials with lower values for the light scattering index required increasingly thicker layers to reach the maximum reflection, as given by Eq. (28-2). Although the results were obtained for visible dye penetrant materials, the necessity for adequate coverage of the test piece is nonetheless demonstrated for all penetrant developers.

The relationship of the capillary action properties of the developer and the penetrant and the time required to achieve complete wetting of the developer from penetrant fluid in a crevice has been studied by Prokhorenko [30]. For the case where the average size of the particles in the developer is considerably greater than the crevice opening, the parameters most affecting the migration of the penetrant from the crevice to the developer are found to be the dynamic spreading coefficient, the surface tension, and the adhesion tension. Where the particle sizes are less than or equal to the crevice opening, the surface tension most significantly affects the complete wetting characteristics, while the adhesion coefficient governs the incomplete wetting of the developer. Since the thickness of the developer layer determines the amount of migration of the penetrant that is required for complete migration, the results are therefore useful in analyzing the effect of the layer thickness on the adequacy of the penetrant indication.

28-4 DEFECT RECOGNITION

Penetrant indications are recognized by significant and localized increases in either the brilliance of the fluorescing dye or by the contrasts to the background for the visible dye. The recognition of an indication may be accomplished by either visual means or an optical scanner. For automated scanning techniques, the penetrant indication may be quantified by a signal-to-noise ratio. This ratio has no meaning, however, for visual techniques.

28-4.1 Visible Dye Penetrants

Visible dye penetrant techniques depend on visual recognition of the accumulation of penetrant around the defect. These applications may be accomplished in either artificial or natural light in the visible range. Since color contrast is the normal visual means for recognition of these indications, the proper choice of background material, i.e., developer, and indicating material, i.e., penetrant, will enhance the detectability of the indication. Typically, visible dye penetrants will use a red dye with a white background developer. Color contrasting penetrant materials are available in the water-washable, postemulsifiable, and solvent removable process but are not available in the ultrahigh sensitivity [15].

28-4.2 Penetrant Brilliance

Fluorescent dye penetrants utilize the brilliance of the fluorescing particles as the recognition effect. Under proper test conditions, the background is totally black, which creates the optimum conditions for recognition.

Previous discussions in this section have described several factors that affect the brilliance of the indication, namely contamination and dilution of the various chemicals. The present material will present in more detail the brilliance levels and sensitivity that might be expected.

Investigations of the relative brilliance obtained from several penetrant materials have been conducted by Schmidt and Robinson [31]. The authors note that since most penetrant materials are very absorbent of ultraviolet light, only the particles in the upper regions of the penetrant/developer layer that will receive UV light will fluoresce. For large cracks, much of the penetrant is drawn into or near the crack and away from the upper surface. This penetrant will not fluoresce because it is never reached by the UV light. Similarly, for small cracks, much of the penetrant dye remains at the surface and thus may produce a more

TABLE 28-2
**Brightness ratios for test penetrants MX-2 and MX-3
compared to standard penetrant MS-3**

Percentage concentration	MX-2 / MA-3	MX-1 / MA-3
10	1.94	1.66
8	1.85	1.70
4	2.33	1.91
2	3.11	2.32
1	4.15	2.92
0.4	2.47	1.47
D-20 panels	3.89	1.59

Source: Reference 31.

brilliant fluorescent indication. Thus, the brilliance of the indication may have little if any relationship to the depth of the crack.

The relative brightness of two different penetrant materials is compared to a standard penetrant in the tests of Schmidt and Robinson [31]. All of the penetrants were postemulsifiable type with designations MX-2 and MX-1 for the penetrants being compared and MA-3 for the standard. For their work, a set of penetrant and developer contentrations were prepared. Filter papers were immersed in the penetrant and developer mixture and brightness data were obtained using a fluorometer. Their results are shown in Table 28-2 along with data from a standard sensitivity panel designated D-20, which represents moderately fine sensitivity.

The authors first note that the higher concentrations give lower brightness ratios for both data sets. As the percentage of penetrant in the mixture is decreased, the brightness increases, reaching a peak at 1% penetrant/developer mixture. They further note that the peak brilliance compares favorably with that of the D-20 panel for the 1% mixture. In summary, they feel that the amount of penetrant in the penetrant developer mixture affects the overall sensitivity in that low concentrations, i.e., 1% or less, are better for very small cracks, while mixtures near 10% would optimize the detection capability for larger cracks.

28-4.3 Visual Scan—Lighting and Detection

Correct lighting conditions are of considerable importance for visual inspections, whether the visible dye or the fluorescent technique is used. In depth discussions of this topic are contained in a number of sources and will not be repeated here. A summary, largely based on Ref. 12, will be given for the present case.

The lighting requirements for visible dye penetrants are less strict than for the fluorescent dyes. In visible dye inspection, recognition of the presence of a defect is based on color contrast between the developer and the dye penetrant. Generally, incandescent, fluorescent, vapor arc, or natural sunlight lighting is satisfactory.

Although the spectral characteristics of the light source are not a significant factor, the appearance of the indication can be enhanced if the light source is deficient in the color of the penetrant but strong in the color of the developer. This will give a bright background with a dark appearance for the penetrant indication. Comfortable lighting levels of 300–550 lx (30–55fc) are satisfactory for most circumstances although levels of 1000 lx (100 fc) or greater might be required for inspecting critical areas for fine cracks.

The level of ambient visible light is of considerable importance when using fluorescent dye penetrants since the brightness contrast is the recognition mechanism that is used. Since the size of the crack that can be detected by the human observer is a function of this brightness contrast, very small cracks are best detected under very subdued lighting conditions of approximately 10 lx (1 fc) which is typically encountered in hooded inspection booths. At higher lighting levels, such as dim interior, bright interior, or bright outdoors, fluorescent penetrant may still be used but with a reduction on the size of the crack that may be detected. For an ambient lighting level of 10 lx (1 fc), the minimum intensity of the ultraviolet light is 0.3 W/m^2 for fine cracks and 0.1 W/m^2 for coarse cracks. For bright interior lighting of 1000 lx (100 fc) the minimum UV light intensity is 50 and 5 W/m^2 for fine and coarse cracks, respectively.

Ultraviolet inspection lamps are available from commercial vendors and have intensities generally ranging from 3–8 W/m^2 for tubular lights and from 70–90 W/m^2 for mercury vapor lamps. Other more specialized lights also may be available. Holden [32] has described the characteristics of the ultraviolet lights use in nondestructive testing.

It is well know that the human eye adapts, albeit somewhat slowly, to a wide range of lighting conditions. For example, an inspector moving from a brightly lit area to a dark inspection booth might require from 5–20 min for dark adaptation to occur. As observed in the previous chapter, the adjustment of the eye to the lower lighting conditions includes a shifting of its peak sensitivity to a shorter wavelength.

The size sensitivity of the human eye has been found to be satisfactory for most inspection applications. For example, Booth et al. [15] report that cracks with dimensions as small as 25 μm (0.001 in) have been regularly detected by visual techniques. Maximum reliability and sensitivity can only be achieved if the human inspector is operating at top performance. This, of course, is possible only if the vision of the inspector is monitored on a regular basis.

28-4.4 Automated Scan—Lighting and Detection

High speed inspection of large quantities of parts may be accomplished using automated lighting and scanning systems. While conventional ultraviolet lamps may be used as the excitation source, the employment of laser sources has greatly increases the speed and reliability of the inspection. Optical scanners similar to television cameras may be used as the detector. Photocells that show significant increases in the speed of operation also have been used.

Automated inspection using a television scanner is accomplished by using the TV scanner to illuminate the part with UV light. The field is then viewed with a television camera. The output of the camera is routed through a signal processing system that is calibrated to reject parts having recognizable defects. The rejection criteria are based on the magnitude of the optical signal received by the signal processing system from the television camera.

The operation of a typical automated laser scanning system is shown in Fig. 28-5. In this case, the part shown is scanned by the laser spot that is moved with the scanning mirror. Since the laser beam is very narrow, only a small area is illuminated at any one time. The fluorescence of any penetrant gathered around a defect is located with the photocell operating through the light collecting mirror.

Compared to the television scanning system, inspection with the laser system is much more effective for a variety of reasons, as reviewed by Borucki et al. [12]. First, the higher intensity of the laser beam results in fluorescent indications that are much brighter than those achieved with conventional UV light sources. Secondly, because of the greater capabilities in contrast ratios for the phototube, the detail discrimination is superior for the laser system. Also, the laser system does not suffer from optical "depth-of-field" problems that are associated with the television, giving a much sharper image. Lastly, the electronic circuitry for the laser system is simpler.

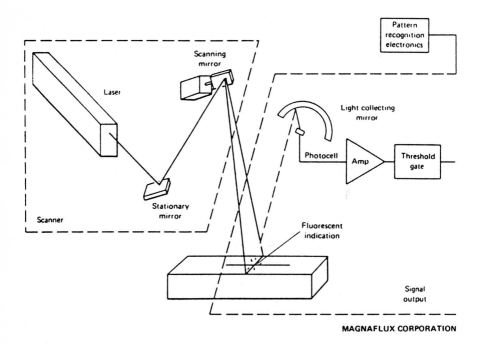

MAGNAFLUX CORPORATION

FIGURE 28-5
Flying spot laser scanning system for flourescent penetrant inspection. (*Courtesy Magnaflux Corporation.*)

FIGURE 28-6
Railroad coupler pin with crack enhanced by solvent-based, fluorescent dye penetrant applied with spray cans. The penetrant distinguishes the crack from ordinary wear marks. Bleeding from the rough area to the left of the crack is caused by penetrant not removed by hand wiping of the surface during cleaning.

The laser system offers wide capability in automated inspection as will be discussed later. For the present it is sufficient to note that the object shown being inspected in Fig. 28-5 easily could be moving as in a production facility. More complex robotic movement of the scanning system and the part holder enables the laser system to very quickly and efficiently inspect complex shapes, as extensively described in Refs. 12 and 13.

28-5 TYPICAL PENETRANT INDICATIONS

Many examples of penetrant inspections are contained in the references and will not be repeated here. Specific note is made of the detailed descriptions and illustrations of penetrant inspections given in Refs. 1, 6, 12–15, 20, 21, 26 and 33–36.

An example of a penetrant indication obtained using visible dye penetrants is shown in Fig. 27-1 [1]. The rather large crack that has been detected with the visible dye penetrants would most likely be otherwise unnoticed. This particular indication was obtained from a pipe hanger lug weld at a power generating station.

Fluorescent dye penetrant frequently is useful in distinguishing fatigue cracks from ordinary wear marks that are visible under ordinary lighting conditions on the surface of a part. For example, Fig. 28-6 shows the upper portion of a pin used in railroad car couplers where the crack has been enhanced by solvent based fluorescent dye penetrant materials applied with spray cans, cleaned by

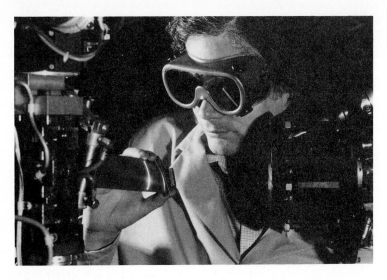

FIGURE 28-7
Laser illumination of fluorescent particle defect indication in turbine blade. (*Courtesy The Leading Edge, GE Aircraft Engines.*)

hand wiping of the surface, and illuminated with a black light. Since the wear marks would not draw penetrant into the material, the crack is clearly distinguishable. In this example, the rough area to the left of the crack is seen to fluoresce indicating that excess penetrant was left there in the cleaning process. Locating a crack in that region might be unreliable under these conditions. More effective cleaning techniques, such as immersion, should be used to obtain adequate inspection of this rough area.

An example of a laser illuminated defect is shown in Fig. 28-7 [33, 34]. A postemulsifiable penetrant process with a hydrophilic remover and a helium/cadmium laser has been used to produce the fluorescent defect indication. The system shown in Fig. 28-7 uses a 40 mW laser with a 0.02 mm (0.008 in) beam directed to a scanner that gives a useful swath of about 25 mm (1 in) of a flying spot laser operating at a scan rate of 325 Hz. The more striking appearance of the laser enhanced fluorescence is notable. In actual inspections, the defect is detected using optical methods incorporating a photomultiplier tube.

28-6 TEMPERATURE EFFECTS ON PENETRANT INSPECTION

The effect of elevated temperatures on penetrant inspection may be encountered in several instances. While slightly elevated temperatures may be used to speed the drying process both after penetrant removal and in the development process, too high temperatures may cause several deleterious effects on the inspection process. Problems that could occur include the loss of brightness in the dyes and slower mobility of the penetrant. While specifications generally restrict penetrant inspections temperatures to 50°C (125°F) or lower, Sherwin and Holden [37] and Mooz [38] describe satisfactory uses of penetrants in higher-temperature applications.

28-7 HARMFUL EFFECTS OF CONTAMINANTS ON HIGH-STRENGTH AND HIGH-TEMPERATURE ALLOYS

In the course of the penetrant inspection, it is possible to introduce harmful chemicals that subsequently can lead to the initiation of defects on the surfaces of some materials. These effects are well known, as described in Ref. 39, and prudent control of penetrant material chemistry can prevent any harm to the material.

The effects of chlorides on various high-strength and high-temperature alloys such as the nickel-based alloys, austenitic stainless steels, and titanium have been found to be particularly serious. Sulfur and halogen are the contaminants that may be present in all penetrant materials and the levels of these elements must be closely monitored. Traces of these chemicals left in crevices on the inspected item are likely to attack the material and initiate cracks as a result of the penetrant inspection.

While the effects of these chemicals on the alloy materials are well documented, the exact levels of contaminant that initiate the attack are not fully established. Before inspecting these materials, the reader is urged to consult an authoritative source, such as Ref. 39.

28-8 SUMMARY

The intent of this section has been to familiarize the reader with the fundamentals of penetrant inspection and the characteristics of the various materials that are used. The relationship of these material characteristics and the physical principles discussed earlier have been further described to show how the penetrant material characteristics affect the inspection process.

Inspection sensitivity has been discussed on several occasions throughout this material. Various means for calculating sensitivity and the effect of the individual processes have been described. An overview of penetrant sensitivity that collectively discusses many of the involved processes is given by Robinson and Schmidt [40].

Penetrant materials may be purchased from a number of commercial suppliers and in a variety of containers ranging from spray cans to bulk 55 gal (208 l) drums. In some cases the materials may be used as-purchased while in other cases the final product must be mixed. Further information on penetrant materials is available from a number of sources, particularly Refs. 13, 14, and

16–18. A general discussion of the inspection characteristics of various penetrant materials is given by Malkes et al [41].

Previous discussion has shown that penetrant performance may be seriously affected by temperature and by the introduction of contaminants into all of the chemicals in the regular inspection processes. To assure satisfactory inspection, an evaluation of the quality of the penetrant materials in use is important. This evaluation may be accomplished through the use of comparators as described by Sherwin et al. [42]. Cracked aluminum blocks and chrome panels are most often used for this evaluation.

From the underlying bases described in this chapter, the material to follow will review penetrant inspection systems and applications.

CHAPTER
29

PENETRANT
INSPECTION
APPLICATIONS

29-1 INTRODUCTION

The type of penetrant inspection system that is used depends on a number of factors. The size and shape of the part may determine whether the materials are applied by immersion or spray or some other process. Whether there is one part to be inspected or several thousand will determine the handling procedure as well as the inspection process. Immersion is more amenable to the inspection of large quantities than are spray techniques. Additionally, any chemical reaction between the penetrant materials and the part being inspected must be considered. Lastly, but of considerable importance, is the sensitivity of the chosen penetrant inspection method. The materials in this section will describe the various inspection systems that use the materials and methods discussed in previous sections.

29-2 PENETRANT INSPECTION SYSTEMS

The fundamental steps in any penetrant inspection are as described in the previous section and shown in Fig. 28-1. Differences in the particular process arise due to specific characteristics of the test as reviewed in the preceding paragraph. Figure 29-1 shows flow diagrams for typical inspections using the

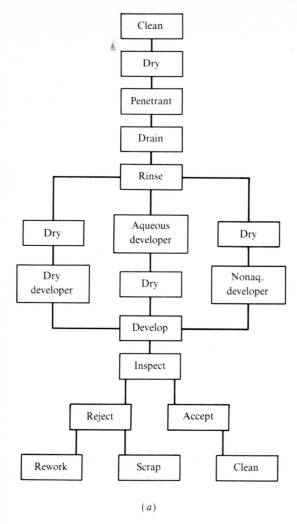

FIGURE 29-1

Flow diagrams for the three principle methods of penetrant inspection. (*Courtesy American Society for Nondestructive Testing.*) (*a*) Water-washable penetrant, ASTM Procedure A-1 or B-1. (*b*) Postemulsifiable penetrant, ASTM Procedure A-2 or B-2. (*c*) Solvent removable penetrant, ASTM Procedure A-3 or B-3.

three most common penetrant methods, as listed in Table 29-1. These diagrams are only for general information and discussion since there are infinite number of variations possible for each. The material presented here is largely summarized from Refs. 13–17 and the interested reader is referred to these sources for additional information.

29-2.1 WATER-WASHABLE METHOD

For the water-washable method shown in Fig. 29-1(*a*), all of the materials used are water soluble. As noted on the figure, this process is identified as ASTM Process A-1 [17].

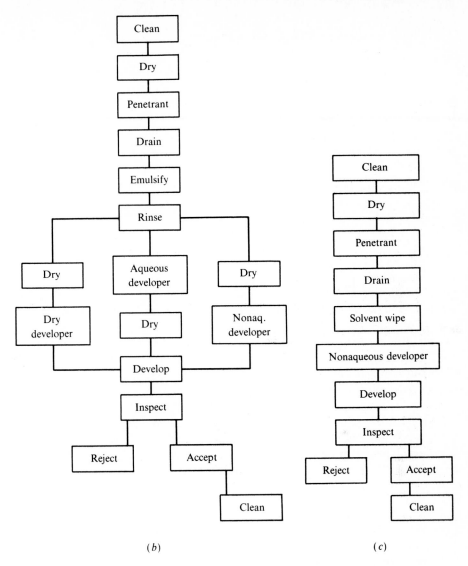

(*b*) (*c*)

FIGURE 29-1
Continued.

The initial cleaning step is generally independent of the test method and will not be discussed further. Drying of the part after cleaning must be thoroughly done since water or other solvent not completely removed could seep into the cracks or pores and either dilute the penetrant or prevent it from entering. Following the drying step, the penetrant application proceeds as described earlier with the requirement that the penetrant be a water-based fluid. Dwell time for the

TABLE 29-1
Classification of liquid penetrant inspection methods.

Method A—Fluorescent, liquid penetrant inspection

Type 1—Water-washable (Procedure A-1)
Type 2—Postemulsifiable (Procedure A-2)
Type 3—Solvent removable (Procedure A-3)

Method B—Visible, liquid penetrant inspection

Type 1—Water-washable (Procedure B-1)
Type 2—Postemulsifiable (Procedure B-2)
Type 3—Solvent removable (Procedure B-3)

Source: Reference 17.

penetrant, of course, depends on the application. Table 29-2 lists typical dwell times for various materials and penetrants.

The importance of the draining step relates to the type of inspection. For example, where the part is immersed or when the penetrant is applied at high volume with a hose, the draining is usually over a tank or sump that enables the recycling of the penetrant. Local application of penetrant to small areas obviously requires minimal consideration of the drain step. Similarly, the rinse or penetrant removal step is different for various applications methods where large volumetric use of water would require adequate draining and disposal of the rinse water.

Developer application follows the rinse step. Where aqueous developers are to be used, there is no need for a drying step prior to the application of the developer. The use of dry or nonaqueous developers not only demands a drying step before developer application but again places great importance on that step since incomplete drying can decrease the brilliance of the penetrant brought out from the crevices. Aqueous developers, of course, must be dried after the application. Often an oven can be used to reduce the drying time although, as previously described, too high temperatures must be avoided since this may reduce the quality of the inspection.

The development step is complete whenever the movement of the developer out of the defect has ceased. While the amount of time required to complete this step is variable, generally it is short. Several typical developer times are listed in Table 29-2. The inspection step following the development is independent of the penetrant method being used and will be discussed separately in a later section. Cleaning of the penetrant chemicals from the inspected part will be required for defect-free parts that are to be returned to service. Depending on deposition, cleaning of the penetrant chemicals may be required from parts being reworked or scrapped.

As previously mentioned, the water-washable systems offer the highest sensitivity levels with minimal detrimental effect from toxic or flammable chem-

TABLE 29-2
Recommended dwell times.

Material	From	Type of discontinuity	Dwell times (min) for Methods A-1 to A-3, and to B-3*	
			Penetrant†	Developer‡
Aluminum, magnesium, steel, brass and bronze, titanium, and high-temperature alloys	*Cast*—castings and welds	Cold shuts, porosity, lack of fusion, cracks (all forms)	5	7
	Wrought—extrusions, forgings, plate	Laps, cracks (all forms)	10	7
Carbide-tipped tools		Lack of fusion, porosity,	5	7
Plastic	All forms	Cracks	5	7
Glass	All forms	Cracks	5	7
Ceramic	All forms	Cracks, porosity	5	7

*For temperature range from 60–125° F(15–50° C). All dwell times given are recommended minimums.
†Maximum penetrant dwell time 60 min, in accordance with 6.4.3.
‡Development time begins directly after application of dry developer and as soon as wet developer coating has dried on surface of parts (recommended minimum).
Source: Reference 17.

icals. Since many of the steps in this technique may be accomplished with immersion or spray, it affords the greatest advantage in the inspection of both very large parts and large quantities of similar parts. When inspecting ferrous-based parts, however, rust inhibitors must be used in the process.

29-2.2   Postemulsifiable Method

Postemulsifiable penetrant inspection systems, also identified as ASTM Process A-2 [17], offer a combination of the advantages of solvent- and water-based inspections. As indicated in Fig. 29-1(b), the differences between the postemulsifiable penetrant process and the previously described water-washable system generally are in the penetrant material that is used and the need for an emulsifier. In the present case, the solvent-based penetrant that is used is followed by an emulsifier application that allows the remainder of the process to follow the water-washable path. Both hydrophilic and lipophilic emulsifiers may be used in this method. The advantage of the postemulsifiable system is that solvent penetrants that may be required for some parts may be removed by water. This lessens the exposure to and the consumption of toxic and flammable fluids while giving a high level of sensitivity. As with the water-washable system, this process is most frequently applied to the inspection of either very large parts or large quantities of similar parts.

29-2.3   Solvent Removable Method

The solvent removable process, ASTM Process A-3, is a fully oil-based inspection process. The steps shown in Fig. 29-1(b) are similar to those previously described except for the method of solvent removal and the type of developer. Penetrant removal must be accomplished by hand wiping of the part with a rag dampened with solvent, since processes such as immersion or rinsing with a hose would most likely remove the penetrant from the defect that is sought. Hand wiping is difficult and time consuming on parts that have many corners, rough surfaces, or inaccessible areas. Of great importance also is the fact that hand removal of the penetrant is time consuming in any case when compared to the large are a rinsing processes available with the water-based and postemulsifiable processes.

The full solvent technique is most often applied for the inspection of either a few small parts or for localized inspection of a small area of a larger part. It also is the method most amenable to portability. The visible dye method, of course, requires no electrical power.

29-2.4   Work Stations for Penetrant Inspection

Considering that penetrant inspection is applied to parts both large and small and both numerous and singular, it is not surprising that the complexity of the work stations for penetrant inspection vary considerably. For individual inspection of a few parts, the work station might be a simple countertop. The inspection

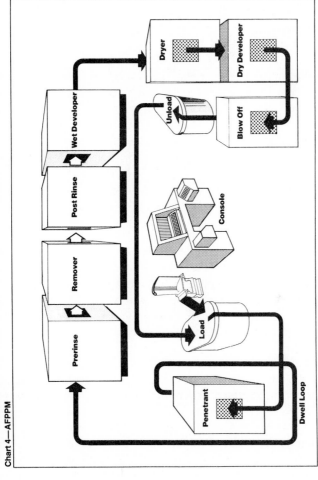

FIGURE 29-2

Flow diagram for automated fluorescent penetrant preprocessing of turbine blades. (*Courtesy General Electric Company, Aircraft Engine Business Group.*)

AFPPM

Automated Fluorescent Penetrant Preprocessing Module

FIGURE 29-3
Work layout for automated fluorescent penetrant preprocessing module used for turbine blade inspection. (*Courtesy General Electric Company, Aircraft Engine Business Group.*)

of a large number of intake and exhaust valves at a diesel engine rebuild assembly might involve a fully automated system where the material handling and the penetrant inspection process were integrated. Further, fully robotic inspection of large and small parts may be used where the demand is sufficiently high. A full description of typical penetrant test equipment and integrated systems is available from a number of sources, particularly Refs. 12–14 and 33–35. A brief discussion of one such system will follow.

The computer-based, automated fluorescent penetrant inspection system developed for the inspection of aircraft turbine blades and vanes provides a useful example of an integrated penetrant inspection system where the material handling and inspection techniques have been developed to an advanced level [13, 33, 34].

A flow diagram for the automated blade preprocessing system is shown in Fig. 29-2 and the work layout is shown in Fig. 29-3. Blades that have been suitably cleaned are loaded into the system at the basket load station shown at the left of the diagrams. After loading into the basket, the blades progress through the various stations. Penetrants are applied using either electrostatic spray or fogging techniques. Following the penetrant application, the basket moves to the hydrophilic remover and water rinse stations, then to the developer area, and finally to the unloading station. The speed of travel for the baskets and the type of penetrant and developer to be applied are selected prior to the start of the preprocessing system. An average of 1500 parts per hour are moved through this preprocessing system.

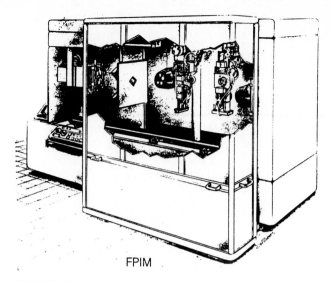

FPIM

FIGURE 29-4
Laser inspection booth for automated fluorescent penetrant inspection of turbine blades. (*Courtesy General Electric Company, Aircraft Engine Business Group.*)

Upon emerging from the preprocessing system, the blades are moved individually through the inspection module that is shown in Fig. 29-4. The laser source and photomultiplier tube scanner assembly are shown at the far right of the cabinet. The scanner is oriented to the left at a blade mounted in one of the automated manipulators. The inspection module processes 500 parts per hour. Although the inspection system operates in a fully automated mode, manual visual inspection is available when needed.

29-3 INSPECTION OF LARGE COMPONENTS AND PIPING

The particular problems presented by the penetrant inspection of either large or lengthy components is one more of movement and manipulation rather than process. The cleaning, penetrant application, developing, and inspection phases are accomplished using techniques and facilities that generally are governed by the large size of the components being inspected.

Application techniques such as electrostatic and conventional spray are most frequently used for large components or structures. Electrostatic penetrant processing has been in use for some years and is described in some detail by Hall [43] and by Borucki et al. [12] This process uses electrostatic potential differences between the part being inspected and the spray source to accelerate the particles onto the application area. Systems for electrostatic application of penetrant are available from commercial vendors.

FIGURE 29-5
Overall view of the chemical cleaning system for DC-10 vertical stabilizer spars. Parts are entering the cleaning system. (*From Douglas Aircraft Company* [*12*], *The Nondestructive Testing Handbook, vol.* 2, *Liquid Penetrant Tests, 2nd ed.*, 1982, *the American Society for Nondestructive Testing.*)

An example of the application of penetrant inspection to large size aircraft components is shown in Fig. 29-5. The vast space and considerable material handling systems required for this inspection are apparent. Besides the cleaning station shown in the photograph, the part must be moved on the overhead trolley system through the entire inspection cycle.

Another example of penetrant inspection of large scale structures is the mechanized scanning system shown in Fig. 29-6, which is for the inspection of welds in piping in nuclear power stations [44]. The system shown performs the entire penetrant operation, i.e., cleaning the surface, applying and removing the penetrant, and applying the developer. Images of penetrant indications are obtained with fiber optics and transmitted to a remote television monitor. A video recorder is used to document position and visual data.

29-4   INSPECTION OF COMPOSITES AND ADHESIVELY BONDED STRUCTURES

Since penetrants will seek entry into any surface opening, they naturally have been applied in research on composite materials where, from the side of the composite sample, disbonds between the layers would provide an opening to the

FIGURE 29-6
Mechanized remote scanner for on-site penetrant inspection of butt welds in nuclear power plant piping. Penetrant indications are displayed on a remote color television monitor and recorded on a videotape recorder. (*Courtesy Southwest Research Institute.*)

penetrant. They also may be used for the inspection of impact damage of composite materials in applications such as aircraft outer skins.

The use of specialized penetrants for the detection of openings in the composite skin that provide access to metallic substructure has been described by Crane [45]. In this case, the opening that leads to the metallic substructure is particularly troublesome since, in service of the aircraft, water penetrating the composite skin would cause corrosion of the aluminum. If undetected, this, of course, could lead to loss of the aircraft. The early detection of these particular openings could minimize the damage that the structure encounters.

A chelating agent added to the carrier fluid provides the unique feature of these penetrants. As described by Crane, the effect of the chelating agent is seen as a shift of the fluorescent band to a longer wavelength, i.e., into the visible light spectrum. Isopropyl alcohol is used as the carrier fluid for this application. The alcohol and chelating agent mixture is applied to the composite skin using conventional techniques. If an opening exists through the skin to a metallic substructure, the chemical action will produce material at the surface that will appear bright yellow when viewed under ultraviolet light. Crane also describes the application of the technique to the inspection of adhesively bonded structures.

29-5 LEAK DETECTION

Leak testing with penetrant material is a very straightforward application of the basic principles of the technique, namely that a fluid introduced at the entrance to a hole or crevice will, by capillary action, follow the boundaries of the anomaly until either it is filled or the fluid exits on the other side. A penetrant material that is discovered on a wall surface opposite to where it was introduced must, then, indicate that a path exists through the wall.

Penetrant inspection techniques are particularly amenable to the detection of leaks in castings, pressure vessels, pipes, welded connections, and a variety of other circumstances. The indications may appear in a number of ways, depending on the type of through-the-wall defect. Among the typical indications are bleeding spots, lines, multiple spots, or rays originating at some single spot.

Both fluorescent and visible dye penetrants may be used for leak detection and the indications on the side to be inspected may be enhanced with developer, if needed, although this may not be required for gross defects. The bleeding procedure may be facilitated with pressure vessels where either internal or external pressure may force the penetrant through the material wall. Examples of the penetrant indications in leak testing are given in a number of sources including Refs. 14, 36, and 46.

29-6   SUMMARY

The material in this chapter has been intended to show a few applications of penetrant testing that indicate, on the one hand, the principles of the basic techniques and, on the other hand, some of the broader capabilities.

Because of the myriad of applications of the liquid penetrant technique, no attempt has been made to cover all of the possible uses. Several of the references previously furnished show individual applications of several penetrant methods. In addition to these sources, Betz [47] presents a most complete description of applications of the liquid penetrant technique for nondestructive testing. These various references should be consulted for more specific details.

CHAPTER
30

SAFETY AND ENVIRONMENTAL CONSIDERATIONS

30-1 INTRODUCTION

As with any inspection system, using penetrant materials involves certain safety and environmental concerns. The risks mostly are from the fumes emitted by the chemicals that may be toxic or flammable or cause skin irritation. Other hazards are the harmful effects to the eye caused by defective black lights and pollution caused by improper disposition of the chemical materials.

Generally, the magnitude of the risk is a direct function of the magnitude of the inspection job. For small scale inspection of a few parts that are small in dimension, the risks are minimized because of the lesser volumes of liquids involved. Large scale inspections, however, involving either large quantities of parts or large sized parts, may pose considerable risk due mostly to the fact that a greater volume of chemicals is used.

The material in this section is intended to acquaint the reader with some of the general risks associated with using penetrant materials and is not intended to be a comprehensive review of the subject. In all cases, the manufacturers of the various chemicals and processes furnish proper safety and use procedures for their products. These should be consulted before becoming involved in a penetrant inspection process. More detailed information also is available from Refs. 12, 39, 48, and 49.

30-2 EFFECTS ON PERSONNEL

The harmful personal effects of penetrant processing chemicals mostly are associated with toxicity and skin irritation. Eye damage may result from either chemical exposure or from short wavelength ultraviolet light that may be emitted by damaged black lights.

Toxicity effects may occur in virtually all steps of the process. Typical harmful conditions caused by material toxicity are headaches, nausea, and chest pains, as well as respiratory effects and skin irritation. All of the cleaning processes have a potentially harmful effect when chemicals are used. Fumes from the acidic, electrical cleaning, and salt bath cleaning processes may cause a number of irritating and harmful effects. Chlorinated hydrocarbons and alcohols may release vapors that are fatal. Also, trichloroethylene and perchloroethylene in high concentrations have a strong narcotic effect. Solvent-based penetrants may be both volatile and toxic. Potential irritation exists in the use of emulsifiers in the postemulsifier method. In the case of aqueous materials, the greatest concern is in the developing chemicals. Dry developers consist of very fine dust particles and the wet developers may be either toxic or flammable.

The relative toxicity of various penetrant process materials has been tabulated by Booth [49] using 1977 Federal Standards. Chemicals in the highest toxicity group include trichlorotrifluoroethane, ethanol (denatured), and acetone. Kerosene, mineral spirits, and high flash naphtha are given to be nontoxic; a variety of other chemicals are between these two groups.

Adequate defenses exist against the harmful effects of toxic chemicals, namely, adequately covering the body with proper clothing, gloves, and protective glasses, and adequate ventilation. In many cases, these safeguards represent normal industrial practice and may be achieved without altering the inspection process.

30-3 FIRE AND EXPLOSION

Fluids that are volatile and flammable are potentially hazardous due to the likelihood of fire and explosion. The flammability of a material is indicated by the flash point. Penetrant materials, as tabulated by Booth [49], range from those having no flash point (the chlorinated hydrocarbons) to the most dangerous materials (acetone, benzene, and toluene), which have flash points at or below 8°C (40°F). Kerosene shows the highest flash point at 65°C (145°F). For small trays or spray applications, the material is normally dispersed rapidly enough so that there is not normally a risk. There may be considerable risk, however, for large, open tanks where there is considerable surface area from which vapors might escape. Citing United States federal regulations, Booth [15] states that the minimum flash point for liquids used in open tanks with no special precautions is 93°C (200°F).

30-4 EFFECT OF ULTRAVIOLET LIGHT ON THE HUMAN EYE

An ultraviolet light correctly filtered for use in penetrant inspection will pose no danger to the human eye since the damaging, short wavelength part of the spectrum has been removed. There is risk, however, from a light that has been damaged so that short wavelength ultraviolet light escapes. Protective goggles that remove the potentially damaging part of the spectrum from the light that strikes the eye are available. Where laser lights are used, the risk is severe. Permanent damage to the eye may result if one looks directly into the laser.

Discomfort and fatigue may result in an inspection environment where a large amount of the surroundings fluoresce. This will occur where the inspector's hands and clothing have become spotted with penetrant material. In addition, some natural products used in clothing will fluoresce. The effect of these conditions can be minimized by washing the hands frequently with soap and water and by wearing a shop or lab smock that can be changed regularly.

30-5 ENVIRONMENTAL EFFECTS OF PENETRANT PROCESS CHEMICALS

The liquids used in the penetrant inspection processes contain a variety of different chemicals. These range from innocuous water to the highly flammable and toxic materials previously discussed. In addition, the emulsifier materials used in the removal of solvent-based materials with water are detergents that carry harmful effects if discarded in an irresponsible manner. It is important, therefore, that the engineer or other individual in charge of the disposition of these materials become thoroughly familiar with the chemicals that are used and the governing regulations and practices for their disposition. Spanner [39] offers specific information on the disposal of penetrant process materials.

30-6 SUMMARY

In many respects, the hazards posed by penetrant inspection are not severe. Often, the safeguards used for assuring safety in penetrant inspection are well established in industrial practice. The most serious hazards typically associated with penetrant inspection are the inhalation of fumes by the operator, the potential of fire or explosion with some of the chemicals, exposure of the eye to damaging ultraviolet and laser light, and the possibility of environmental pollution if the chemicals are improperly discarded.

In general, one can perform penetrant inspections in a safe and responsible manner if proper consideration is given to adequate ventilation and lighting and to proper disposal of the chemicals. Before embarking on extensive penetrant inspections, however, the reader is urged to consult one of several references that are given on safety and environmental matters, particularly Refs. 12, 39, 48, and 49.

CHAPTER
31

SUMMARY

While the penetrant technique is based on some of the earliest scientific observations, it has certainly found circumstances in modern industry and research where it is not only useful, but also is the best technique. With regard to the simplicity of application, it is probably the easiest to use. Realizing, however, that ease of use may not be matched with sensitivity for very small and inaccessible defects, the NDE engineer is cautioned to not choose the penetrant technique simply because of ease in application. It should be chosen on its merits, which are considerable.

There are two general topics related to penetrant inspection that need further comment. They are the repeatability of results and the related need for further automation in the inspection process.

The quality (and hence, repeatability) of a penetrant inspection is very much affected by variations in a multitude of factors, namely, the chemical content and the amount of impurities in the materials, the temperature of the part, the dwell time, the aggressiveness of the cleaning process, and the ambient and inspection lighting conditions, to name a few. In most applications, the control of these factors has been almost totally dependent upon the skill and attentiveness of the human inspector. While the human inspector is well known to be able to find the very smallest flaws, it is also well known that variations in personal attention result in wide fluctuations in the inspection process. With this, then, it often is difficult to repeat a penetrant test and to achieve exactly the same results. For example, slight overwashing can remove penetrant from the very smallest flaws, where, under proper washing conditions, the flaws would have been detected.

From the preceding discussion, it appears that there is considerable benefit to be gained for the penetrant inspection process with the increased adoption of automated precedures where the various parameters previously described could be brought under more uniform control. The design and adoption of more automated processes could assure greater repeatability and reliability in inspection results.

PROBLEMS FOR PART V

Chapter 27

27-1. For a smooth, blind cavity, determine the percentage of the depth that would be filled with water at intervals of 20 and 40 s.

27-2. Discuss the effects of the coefficient of dynamic viscosity, surface tension, and fluid wetting angle on the fill rate for a penetrant fluid in a blind, smooth cavity.

27-3. Discuss the effect of crack depth and crack width on the fill rate for a penetrant fluid in a blind, smooth cavity.

27-4. Describe how an ultraviolet light emitting electromagnetic radiation that is beyond the visible spectrum causes penetrant particles to become visible.

27-5. Describe the role of the filter on the ultraviolet (black) light in increasing the conspicuity of the defect indications.

Chapter 28

28-1. The dye particles and the base (vehicle) fluid contribute in different ways to obtaining a distinct defect indication. Discuss the two roles of these materials.

28-2. Would you expect that there would be a difference in the sensitivity of the visible dye penetrant technique and the fluorescent dye penetrant technique? Justify your answer.

28-3. Describe how the precleaning and cleaning steps in the penetrant process affect the effectiveness of penetrant inspection.

28-4. What is the role of an emulsifier when used with the postemusifiable inspection process? What risks are associated with improper use of the emulsifier?

28-5. An automated, laser excitation and optical scanning system has been described in the text. Do you see any advantages of this system that could justify the extra costs when compared to a system where the part is illuminated with a conventional black light and the indications are detected with a human inspector?

28-6. Compare the expected sensitivity of the water-washable penetrant technique with the solvent removable method.

518

Chapter 29

29-1. Compare the process characteristics of the water-washable and the solvent removable penetrant techniques and comment on the advantages of each.

29-2. Dwell times recommended in Table 29-2 are considerably longer than the cavity fill times described in Chapter 27. Speculate on the reasons for these differences.

29-3. How does improper dwell time affect the effectiveness of a penetrant inspection?

Chapter 30

30-1. What harmful effects to human operators can result from penetrant inspections?

30-2. What single item is important for safe operation of all penetrant systems, whether for large scale inspections (i.e., large size or a large volume of parts) or for small scale systems (i.e., small size or a small quantity of parts)?

30-3. Why would large scale penetrant inspection systems require greater concern for operational safety than would smaller systems that are used to inspect a few small parts?

REFERENCES
FOR
PART V

1. Reinhart, E., "P-11/P-22 Piping Update," *Critical Path*, vol. 1, no. 4, pp. 2–4, July 1986.
2. Prandtl, L., and Tietjens, O. G., *Fundamentals of Hydro- and Aeromechanics*, Dover Publications, New York, 1934.
3. Sears, F. W., and Zemansky, M. W., *College Physics*, 2nd ed., Addison-Wesley, Cambridge, MA, 1952.
4. Sears, F. W., Zemansky, M. W., and Young, H. D., *University Physics*, vol. 1, 7th ed., Addison-Wesley, Reading, MA, 1987.
5. Fox, R. W., and McDonald, A. T., *Introduction to Fluid Mechanics*, 3d ed., John Wiley, New York, 1985.
6. Alburger, J. R., "Dynamic Characteristics of Liquid Penetrants and Processing Materials," in R. C. McMaster (ed.), *Nondestructive Testing Handbook*, vol. 2, 2nd ed., American Society for Metals, Metals Park, OH, 1982, sec. 7, pp. 273–320.
7. Prokhorenko, P.P., Migun, N. P., and Dezhkunov, N. V., "Calculating the Process of Filling of Microcapillaries with Liquids under Ultrasonic Action," *Defectoskopia* (*Soviet Journal of Nondestructive Testing*), vol. 18, no. 4, pp. 326–331, April 1982.
8. Prokhorenko, P. P., Migun, N. P., and Adler, M., "Sensitivity of Penetrant Inspection in the Absorption of Penetrants and by a Sorption Detector from Plane Parallel Cracks," *Defectoskopia* (*Soviet Journal of Nondestructive Testing*), vol. 21, no. 7, pp. 502–513, July 1985.
9. Miller, F. Jr., *College Physics*, 5th ed., Harcourt Brace Jovanovich, San Diego, CA, 1982.
10. Sears, F. W., *Optics*, 3d ed., Addison-Wesley, Reading, MA, 1949.
11. Bowen, E. J., *Luminescence in Chemistry*, Van Nostrand, London, 1968.
12. Borucki, J. S., "Liquid Penetrant Test Equipment," in R. C. McMaster (ed.), *Nondestructive Testing Handbook*, vol. 2, *Liquid Penetrant Tests*, 2nd ed., American Society for Metals, Metals Park, OH, 1982, sec. 5, pp. 153–220.
13. Kaiser, B., and Cable, T., "Automated Fluorescent Penetrant Inspection at San Antonio ALC," REPTECH/0228186/1344(062), General Electric Co., Aircraft Engines Business Group, Evendale, OH.
14. "Liquid Penetrant Inspection," in H. E. Boyer (ed.), *Metals Handbook*, vol. 11, *Nondestructive Inspection and Quality Control*, 8th ed., American Society for Metals, Metals Park, OH, 1976, pp. 20–44.
15. Booth, R. C. et al., "Principle of Liquid Penetrant Inspection," in R. C. McMaster (ed.), *Nondestructive Testing Handbook*, vol. 2, *Liquid Penetrant Tests*, 2nd ed., American Society for Metals, Metals Park, OH, 1982, sec. 2, pp. 17–59.
16. "Back to Basics—Liquid Penetrants," *Materials Evaluation*," vol. 35, no. 11, pp. 24, 26, 28–30, Nov. 1977.

17. "Standard Practice for Liquid Penetrant Inspection Method," *1986 Annual Book of ASTM Standards*, E 165-80, vol. 03.03, American Society for Testing and Materials, Philadelphia, PA, 1986, sec. 3, pp. 212–230.

18. Mlot-Fijalkowski, A., Garcia, V. A., and Robinson, S. J., "MIL-I-25135 Revision D: The Qualified Products List," *Materials Evaluation*, vol. 45, no. 7, pp. 841–844, July 1987.

19. Packman, P. F., Hardy, G., and Malpani, J. K., "Penetrant Inspection Standards," in H. Berger (ed.), *Nondestructive Testing Standards—A Review*, ASTM STP 624, American Society for Testing and Materials, Philadephia, PA, 1977, pp. 194–210.

20. Lomerson, E. O. Jr., "Statistical Method for Evaluating Penetrant Sensitivity and Reproducibility," *Materials Evaluation*, vol. 28, no. 2, pp. 67–70, March 1969.

21. Hyam, N. H., "Quantitative Evaluation of Factors Affecting the Sensitivity of Penetrant Systems," *Materials Evaluation*, vol. 30, no. 2, pp. 31–38, February 1972.

22. Vaerman, J. F., "Fluorescent Penetrant Inspection Process Automatic Method for Sensitivity Quantification," *Proceedings 11th World Conference on Nondestructive Evaluation*, vol. III, Las Vegas, NV, 1985, pp. 1920–1927.

23. Tanner, R. D., Ustruck, R. E., and Packman, P. F., "Adsorption and Hysteresis Behavior of Crack-Detecting Liquid Penetrants on Steel Plates," *Materials Evaluation*, vol. 38, no. 9, pp. 41–46, September 1980.

24. Whitehorn, N. D., "The Effects of Lignocellulose Abrasive Blasting on Subsequent Dye Penetrant Inspection," *British Journal of Nondestructive Testing*, pp. 27–28, January 1985.

25. Hosokawa, T., and Hosoya, M., "The Influence of the Concentration of Hydrophilic Emulsifiers on the Crack-Detectability and Waterwashability," *Proceedings 11th World Conference on Nondestructive Evaluation*, vol. I, Las Vegas, NV, 1985, pp. 286–292.

26. Booth, R. C. et al., "Characteristics and Maintenance of Penetrant Inspection Materials," in R. C. McMaster *Nondestructive Testing Handbook*, vol. 2, *Liquid Penetrant Tests*, 2nd ed., American Society for Metals, Metals Park, OH, 1982, sec. 3, pp. 61–115.

27. Sherwin, A. G., "Overremoval Propensities of the Prewash Hydrophilic Emulsifier Fluorescent Penetrant Process," *Materials Evaluation*, vol. 41, no. 3, pp. 294–299, March 1983.

28. Lovejoy, D. J., Balinsky, Z., and Maya, B., "The Effects of Common Chemical Agents on the Fluorescent Brilliance of Penetrants," *British Journal of Nondestructive Testing*, vol. 27, no. 4, pp. 213–219, July 1985.

29. Glazkov, Yu. A., and Bruevich, E. P., "Determination of the Whiteness of Developers for Penetrant Flaw Inspection," *Defectoskopia* (*Soviet Journal of Nondestructive Testing*), vol. 21, no. 4, pp. 284–288, April 1985.

30. Prokhorenko, P. P., "Effect of Physicochemical Factors on Process of Migration of the Indicator Liquid in Capillary Inspection," *Defectoskopia* (*Soviet Journal of Nondestructive Testing*), vol. 21, no. 3, pp. 185–190, March 1985.

31. Schmidt, J. T., and Robinson, S. J., "Penetrant Fluorescence Measurements, 1982 Model," *Materials Evaluation*, vol. 42, no. 3, pp. 325–332, March 1984.

32. Holden, W. O., "UV/Black Light Measurement, 1982 Model," *Materials Evaluation*, vol. 41, no. 3, pp. 244, 246, 248, 249, March 1983.

33. Sturges, D. J., "Quality—The Critical Challenge" *The Leading Edge*, Fall 1985, pp. 2–7, (General Electric Company, Cincinnati, OH).

34. "Integrated Blade Inspection System," General Electric Company, Evendale, OH (sponsored by US Air Force, US Army, and US Navy and administered by Wright Aeronautical Laboratories).

35. Booth, R. C. et al., "Basic Techniques of Liquid Penetrant Inspection," in R. C. McMaster (ed.), *Nondestructive Testing Handbook*, vol. 2, *Liquid Penetrant Tests*, 2nd ed., American Society for Metals, Metals Park, OH, 1982, sec. 4, pp. 117–151.

36. Sparling, R. H., "Interpretation of Liquid Penetrant Indications," in R. C. McMaster (ed.), *Nondestructive Testing Handbook*, vol. 2, *Liquid Penetrant Tests*, 2nd ed., American Society for Metals, Metals Park, OH, 1982, sec. 8, pp. 321–388.

37. Sherwin, A. G., and Holden, W. O., "Heat Assisted Fluorescent Penetrant Inspection," *Materials Evaluation*, vol. 37, no. 10, pp. 52–56, 61, September 1979.

38. Mooz, W. E., "Precautions in the Use of High Temperature Penetrants," *Materials Evaluation*, vol. 37, no. 12, pp. 26–28, November 1979.

39. Spanner, C., "Control of Penetrant System Chemistry and Effluent Waste Pollution," in R. C. McMaster (ed.), *Nondestructive Testing Handbook*, vol. 2, *Liquid Penetrant Tests*, 2nd ed., American Society for Metals, Metals Park, OH, 1982, Sec. 10, pp. 428–480.

40. Robinson, S. J., and Schmidt, J. T., "Fluorescent Penetrant Sensitivity and Removability—What the Eye Can See, a Fluorometer can Measure" *Materials Evaluation*, vol. 42, no. 8, pp. 1029–1034, July 1984.

41. Malkes, L. Ya., Sukiasova, L. J., Denel', A. K., and Borovikov, A. S., "Sets of Materials for Capillary Flaw Detection," *Defectoskopia* (*Soviet Journal of Nondestructive Testing*), vol. 20, no. 6, pp. 392–395, June 1984.

42. Sherwin, A. G. et al., "Design and Applications of Penetrant Comparators and Reference Panels," in R. C. McMaster (ed.), *Nondestructive Testing Handbook*, vol. 2, *Liquid Penetrant Tests*, 2nd ed., American Society for Metals, Metals Park, OH, 1982, sec. 9, pp. 389–428.

43. Hall, M. A., "Electrostatic Application of Red Dye Penetrants and Dry Powder Developers to Massive Nuclear Components," *Materials Evaluation*, vol. 37, no. 4, pp. 56–58, 1979.

44. Hutchinson, E. G. et al., "Application of Penetrant Testing in the Nuclear Power Industry," in R. C. McMaster (ed.), *Nondestructive Testing Handbook*, vol. 2, *Liquid Penetrant Tests*, 2nd ed., American Society for Metals, Metals Park, OH, 1982, Sec. 13, pp. 551–574.

45. Crane, R. L., "A New Penetrant for Composites and Adhesively Bonded Structures," *Materials Evaluation*, vol. 35, no. 2, pp. 54–55, February 1977.

46. Roehrs, R. J. et al., "Introduction to Leak Testing Technology," in R.C. McMaster (ed.), *Nondestructive Testing Handbook*, vol. 1, *Leak Testing*, 2nd ed., American Society for Metals, Metals Park, OH, 1982, sec. 1, pp. 1–56.

47. Betz, C. E., *Principles of Penetrants*, 2nd ed., Magnaflux Corp, Chicago, IL, 1969.

48. Sparling, R. H., "Management and Personnel for Liquid Penetrant Inspection," in R. C. McMaster (ed), *Nondestructive Testing Handbook*, vol. 2, *Liquid Penetrant Tests*, 2nd ed., American Society for Metals, Metals Park, OH, 1982, sec. 1, pp. 1–60.

49. Booth, R. C. et al., "Techniques for Cleaning Test Object Surfaces," in R. C. McMaster (ed.), *Nondestructive Testing Handbook*, vol. 2, *Liquid Penetrant Tests*, 2nd ed., American Society for Metals, Metals Park, OH, 1982, sec. 2, pp. 222–272.

PART VI

EDDY
CURRENT
TECHNIQUES IN
NONDESTRUCTIVE
EVALUATION

CHAPTER
32

FUNDAMENTAL EDDY CURRENT CONCEPTS

32-1 INTRODUCTION

In eddy current (EC) testing, a fundamental requirement is that the examined material have induced in it a distribution of currents. This restricts the technique to those materials that are electrically conductive. The currents within the inspected materials are formed in response to a changing electromagnetic field that is generated by a changing current in a conductor that is placed in close proximity to the material. They are constrained to flow in closed loops within the inspected material and themselves generate magnetic fields. These fields, in combination with the fields that excite the currents, must be detected either by electromagnetic induction in a coil, or a system of coils, or by sensors such as the Hall element. In many cases the same coil is used both to excite the circulatory (eddy) currents and also to detect their fields.

The parameters that are important in EC testing can be divided into two categories, electromagnetic and physical. The electromagnetic properties that are of importance are:

1. The source frequency or frequencies ($f_1, f_2, \ldots$); these are also the eddy current and magnetic field frequencies
2. The electrical conductivity (σ) of the inspected part and changes therein ($\Delta\sigma$)

3. The relevant magnetic permeability (μ) and changes therein ($\Delta\mu$)

These quantities affect the depth to which eddy currents penetrate into the inspected material. Knowledge of parameters 2 and 3 often enables the frequency to be set optimally when performing tests with single frequency devices or suggests possible combinations of frequencies in the case of multifrequency devices. The physical parameters that must be considered are:

4. The radius of the exciter coil for circular shapes or other relevant dimensions for other shapes
5. The radius of the sensing coil for circular shapes or other relevant dimensions for other shapes

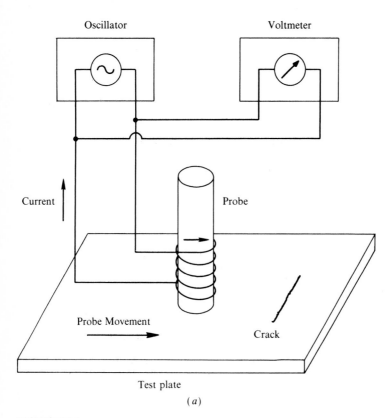

(a)

FIGURE 32-1
(a) Eddy current probe and circuit arrangement where single exciting and sensing coils are combined. (b) Eddy current probe and circuit arrangement for separate exciting and sensing coils. (c) Eddy current probe and circuit arrangement with a Hall element used as the sensor. (d) Eddy current probe arrangement in "cross-axis" configuration. [Reprinted by permission of Atomic Energy of Canada Ltd., from Eddy Current Manual, Volume 1, by V. S. Cecco, G. Van Drunen, and F. L. Sharp (AECL-7523, Rev. 1). Published in the United States by GP Publishing Inc.]

6. The numbers of turns on these coils, which directly affect the strength of the applied field that generates the eddy currents and the induced sense coil voltage

7. The proximity of the exciter coil to the part

8. The proximity of the sensor to the part and to the exciter

9. Part dimensions such as thickness, radius, etc.

10. The proximity of the system to the edges of the part or any other physical change in the part that could alter the path of eddy currents

   The art of EC testing is to select all of the controllable parameters in such a way as to both optimally detect the desired material parameter, for example, the internal or external diameter of a tube, or the presence of corrosion or cracking, or changes in permeability or conductivity or coating thickness, and at the same time suppress possible signals from unwanted or nonrelevant variables. As an example, one might wish to measure the inside diameter of an oil well casing but ignore changes in the magnetic permeability of the material of the casing. Such requirements generally lead to compromise.

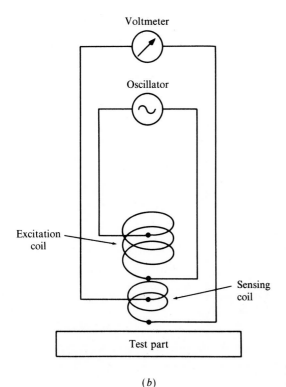

(b)

FIGURE 32-1
Continued.

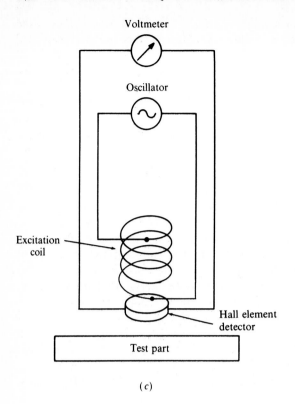

(*c*)

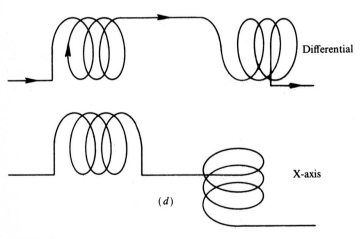

(*d*)

FIGURE 32-1
continued.

The underlying relationships that govern EC testing are derived from Maxwell's equations. Unfortunately, these lead to nonlinear, three-dimensional partial differential equations that often have awkward boundary conditions. Until recently, closed form analytic solutions to such equations were available only for the simplest geometries, such as a tube passing axially through an encircling coil. Now, finite element codes have been developed that provide numerical solutions for a broad range of EC problems. The theorist can at last explain the experimentalist's signals. In this text, however, we will generally follow established principles that have evolved from simple solutions to Maxwell's equations and the trial-and-error technique that has lead EC testing to its present position.

32-2 EQUIPMENT REQUIREMENTS

Generation and detection of ECs requires an oscillator, a means of generating a changing magnetic field close to the part (generally a coil), and a means of measuring voltage in a detector. Figure 32-1 shows various configurations. In Fig. 32-1(a), the exciting coil also serves as the sensing coil, so that the voltmeter detects changes in self-inductance. In Fig. 32-1(b), a separate sense coil is used; this coil can be inside, close to, or remote from the exciter coil, and can be split into differential or "cross-axis" configurations [Fig. 32-1(d)]. Hall sensors can also be used [Fig. 32-1(c)] and configured just as sense coils are. They do, however, require their own drive circuits (not shown).

Oscillation is generally, but not always, sinusoidal, with the frequencies being dependent upon the application. The oscillator itself may provide a range of frequencies, from 5–500 Hz for the remote field EC method (Chap. 34) that is used for the testing of ferromagnetic tubes, to the 1 kHz to 2 MHz range for testing nonferrous materials. Voltage measurements consist of amplitude and phase difference measurements from the exciter coil current.

Clearly, while the oscillator(s) and detection circuitry may be combined into one box and used for a wide variety of EC testing, the sensors must be configured to suit the application. This requires some prior evaluation of the task, with such items as size, orientation, impedance at various frequencies, the presence or absence of ferrite cores, the nature of suspected flaws, and the suppression of nonrelevant indications all being taken into consideration. Here trial-and-error often plays a role, along with various code requirements for the detection of certain standard test flaws prior to beginning the inspection.

32-3 GENERATION AND DETECTION OF EDDY CURRENTS

In the magnetics section of this text (Part III), the relationship between the current flowing through a coil and the resulting field strength $\overline{H}$ was discussed. Close to the coil, this field strength is responsible for a flux density $\overline{B}$ and a flux

Φ, which is defined as

$$\Phi = \int \overline{B} \cdot \hat{n} \, da \qquad (32\text{-}1)$$

i.e., the integral of $\overline{B}$ over the relevant area. In EC testing, the magnetizing coil is generally close to the test sample, so that, as shown in Fig. 32-2, the flux Φ_p that it creates affects the sample. This flux is a function of coil parameters and primary excitation current I_p. If I_p is given by $I_p = I_0 \sin(\omega t)$, where I_0 is the maximum current, then

$$\Phi_p \propto N_p I_0 \sin(\omega t) \qquad (32\text{-}2)$$

where the excitation frequency f and the angular frequency ω are given by $f = 2\pi/\omega$.

The instantaneous magnitude of the current I_p is given by Ohm's Law, i.e., the driving voltage divided by the coil impedance (V_p/Z_p). The oscillating nature of the flux induces circulatory (eddy) currents in the part beneath the coil. Their direction is such that their own magnetic field opposes the field that produces them (Lenz's Law). They spread out into the part, but will naturally be constrained by its boundaries. Being circulating currents, they produce their own secondary flux Φ_s in opposition to Φ_p. The coil now senses an equilibrium flux Φ_e, which is the difference between Φ_p and Φ_s:

$$\Phi_e = \Phi_p - \Phi_s \qquad (32\text{-}3)$$

With the test sample far from the coil, the flux therein is Φ_p, but as the coil approaches the test sample, Φ_s is induced and the net flux in the coil changes. If the sample is not ferromagnetic, then Φ_s is less than Φ_p and the net flux in the coil falls. Since the flux linkage with the coil controls its impedance, then there is a decrease in coil impedance. If the sample is ferromagnetic, Φ_s may far exceed

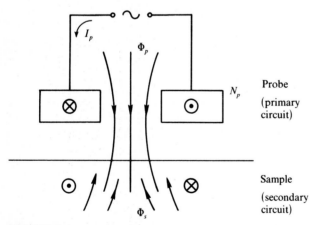

FIGURE 32-2
Excitation of primary and secondary flux with an eddy current probe. [*Reprinted by permission of Atomic Energy of Canada Ltd., from Eddy Current Manual, Volume 1, by V. S. Cecco, G. Van Drunen, and F. L. Sharp (AECL-7523, Rev. 1). Published in the United States by GP Publishing Inc.*]

Φ_p because of the high average permeability, and so may Φ_e. This results in an increase in coil impedance.

If a separate sense coil (N_s turns) is used, the flux linked with it will not generally be the same as the preceding example. Denoting this flux by $\Phi(s')$, the voltage induced in the coil is, from Faraday's Law,

$$V_s = -N_s[d\Phi(s')/dt] \qquad (32\text{-}4)$$

A Hall sensor merely responds to the flux density through it and not to a time derivative, so there will be a 90° phase difference between voltages developed in coils and Hall elements placed at the same location.

32-4 NATURE OF EDDY CURRENTS

As shown in Fig. 32-2, for this excitation situation, the eddy currents that flow in the part in a direction perpendicular to the flux are parallel to the winding and the part surface. Angling the coil would alter the flux and also the currents. In an infinite medium they are unbounded but their density falls off with distance as does the flux. In finite media, they are affected by the requirement that they must remain within the part. In the case of round bars or tubes passing through encircling coils, they flow around the outer surface. The equation that governs their flow is

$$\nabla^2 J = \sigma\mu(\partial J/\partial t) \qquad (32\text{-}5)$$

where J = their current density
σ = the material conductivity
μ = the material permeability

The solution of Eq. (32-5) for a semiinfinite medium is

$$J(x)/J_0 = e^{-(x/d)}\sin(\omega t - x/d) \qquad (32\text{-}6)$$

where $J(x)/J_0$ = the eddy-current density ratio between a depth x and the surface
d = the skin depth given by

$$d = (\pi f\mu\sigma)^{-1/2} \qquad (32\text{-}7)$$

The first term of Eq. (32-6) expresses an exponential decrease of eddy-current density with depth that depends upon excitation frequency f, part permeability μ, and conductivity σ. Increases in all of these parameters serves to limit penetration into the part, as indicated by Eq. (32-7). The second term shows increasing phase, or time lag, between the surface current and that at depth x. The flux density at depth x compared with that at the surface [i.e., $B(x)/B_0$] is also given by the right hand side of Eq. (32-6).

32-5 STANDARD PENETRATION DEPTH AND PHASE LAG

The skin depth d, as defined by Eq. (32-7), is often called the standard depth of penetration. It is the depth at which the eddy-current density (and flux density) is

$1/e$, or 36.8% of its surface value. Equation (32-7) can also be written as

$$d = 50(\rho/f\mu_r)^{1/2} \text{ mm}$$

or

$$d = 2(\rho/f\mu_r)^{1/2} \text{ in}$$

(32-7)

where ρ is the electrical resistivity in $\mu\Omega$-cm (a common way of giving this quantity in scientific tables) and μ_r is the dimensionless relative permeability. For dia- and paramagnetic materials, the value of μ_r can be taken as unity; for ferromagnetics, an average over the local B-H curve scanned by the exciting field may be taken in rough calculations. For many industrial steels around the origin of the B-H curve a value in the range 40–70 can be used. At two skin depths, the current or flux densities fall to $1/e$ of their value at one skin depth, or $1/e^2 = 13.5\%$ of the surface value, and at $5d$, only 0.7% remains.

Since the skin depth relation refers only to plane electromagnetic waves, in commonly encountered EC testing situations (such as shown in Fig. 32-1) or small coils encircling tubes, the relation is not truly valid, but does provide a reasonable estimate of the skin depth.

In more complicated treatments, the skin depth relation is part of the solution of Maxwell's equations for the particular inspection problem. From the same solution arises the fact that the deeper one goes into the material, the greater is the phase difference from the current density at the surface. Again, for a thick material, Eq. (32-6) shows that this phase lag is

$$\beta = x/d \text{ rad (or } 57.3x/d°)$$

(32-8)

Thus, at one standard depth of penetration, this phase lag is 57.3°. In ac theory, often there is a phase difference between the exciting voltage and the resultant current; this is not the same phase lag.

CHAPTER

33

EDDY CURRENT DETECTOR PARAMETERS

33-1 INTRODUCTION

As outlined in Chap. 32, in most eddy current instruments the exciter coil induces eddy currents in a test part and sense coils are used to detect the magnetic effect of these currents. In some instruments, only one coil is used, doubling as exciter and sense coil. In others, coil arrays are used or, where the frequency might be low, Hall sensors might be used. In this chapter, the electrical parameters of such sensors are discussed.

33-2 COIL SENSORS

Coils are generally made of copper wire, wound so as to optimally produce a voltage from a changing magnetic flux linkage by Faraday's Law:

$$V_s = -N_s[d\Phi(s')/dt] \tag{32-4}$$

Obviously, larger voltages for the same flux change occur when larger numbers of turns (N_s) are used. Similarly, larger voltages occur with increasing $d\Phi/dt$, i.e., increasing flux changes or smaller time increments. Thus, induced voltage rises with frequency.

Space limitations often dictate sense coil size, so that to keep the induced voltage relatively high, very fine wire is used, leading to substantial coil resistance.

33-2.1   Coil Resistance

All coils exhibit electrical resistance, which, in the case of direct current, is given by

$$R_s = \rho L / A \ \Omega \tag{33-1}$$

where ρ = the dc resistivity (the reciprocal of the conductivity) in ohm-m
L = the length of the wire in m
A = its cross-sectional area in square meters.

The resistivity of a metal such as copper is the ratio of the electric field (in volts per meter) within the material to the rate of charge transport per unit cross-sectional area. It may be visualized as the effect of the retarding interactions of the material lattice on the conduction electron cloud, i.e., the current within the conductor.

33-2.2   Variation with Temperature

At any temperature, the resistivity is made up of components from interactions with the "impurities" within the lattice (both foreign materials and dislocations) and die out with lattice vibrations so that near to $0°K$, only the impurities contribute. At room temperature (around $300°K$), in relatively pure metals the lattice vibrations contribute the majority of the resistivity and may be written as

$$\rho = \rho_l \left[1 + a(T - T_l) \right] \ \Omega \cdot m \tag{33-2}$$

That is, the resistivity ρ at some temperature T is linearly related to that (ρ_l) at T_l through the quantity a, which is known as the temperature coefficient of resistivity. Equation (33-2) can be rewritten as

$$a = (1/\rho_l)(d\rho/dT) \ \text{per} \ °K \tag{33-3}$$

For copper, $\rho = 1.724 \times 10^{-8} \ \Omega \cdot m$ ($1.724 \ \mu\Omega \cdot cm$) at $20°C$ and $a = 0.00427$ per $°K$.

The positive temperature coefficient, common to most metals, indicates a rise of resistance with temperature. Thus, when measuring the resistive component of a sense coil, as might traditionally be performed with a bridge network, one must be aware of the possible effects of temperature change. Commonly, the leads to the coil are often compensated by dummy leads in a second arm of the bridge.

33-2.3   Variation with Frequency

The passage of dc through a wire creates a magnetic field at points inside the wire. It is zero at the center and rises linearly to its maximum value at the surface. However, when ac is used, the skin effect reduces the field closer to

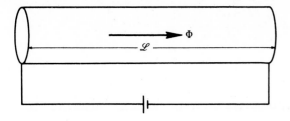

FIGURE 33-1
Long air-cored coil with flux Φ.

the center of the wire, with the result that the conduction electron density is no longer uniform, but larger toward the outer regions of the wire. This leads to increased resistance over the dc case. The effect occurs where the same coil is used as both exciter and sensor. Where a passive independent sense coil is used, if the induced voltage is measured with a bridge circuit, so that no current is allowed to flow from the coil, then the effect is not detected.

33-2.4 Coil Inductance

In addition to resistance, since there is a changing flux in the coil, it also exhibits self-inductance, which is defined as

$$L = d(N\Phi)/dI \tag{33-4}$$

i.e., the flux linkage change per unit current.

In order to see the general form of L for cylindrical coils, consider an infinitely long air-cored coil of length $\mathscr{L}$ as shown Fig. 33-1. Since the flux is given by $\Phi = BA$, and $B = \mu_0 H$, then $\Phi = \mu_0 HA$. But $H = NI/\mathscr{L}$ so that $N\Phi = \mu_0 N^2 IA/\mathscr{L}$, and for this special case,

$$L = \mu_0 N^2 A/\mathscr{L} \tag{33-5}$$

For small, air-cored coils, the difference from Eq. (33-5) is included within a geometrical factor K, so that the self-inductance becomes

$$L = K(\mu_0 N^2 A/\mathscr{L}) \tag{33-6}$$

A concise account of various formulas for K has been given by Ida [1] or it can be found by experimentation with an inductance bridge.

33-2.5 Coil Inductive Reactance

When a coil is excited at angular frequency ω it exhibits another form of resistance that is due to a back emf that is generated in it, which is itself due to the changing flux linkage therein. Known as the inductive reactance (X_L), the magnitude is given by

$$X_L = \omega L = 2\pi f L \tag{33-7}$$

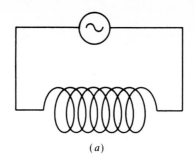

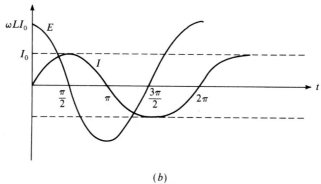

(b)

FIGURE 33-2
(a) Alternating current applied to inductor. (b) Current I and voltage E waveforms, showing that E leads I by 90°.

This quantity is seen to be zero for dc ($f = 0$) and to rise with increasing frequency. A simple way to derive Eq. (33-7) is to note that the back emf that develops when a pure inductance [i.e., no resistance, Fig. 33-2(a)] is connected to an ac source is given by $e = -N\,d\Phi/dt$, or, since $N\,d\Phi = L\,dI$, also equal to

$$e = -L\,dI/dt \tag{33-8}$$

Using $I = I_0 \sin \omega t$ for the current in Eq. (33-8) gives

$$e = -(\omega L)I_0\cos(\omega t) = -(\omega L)I_0\sin(\omega t + 90) \tag{33-9}$$

The applied voltage must be equal and opposite to this, i.e.,

$$E = (\omega L)I_0\sin(\omega t + 90) \tag{33-9a}$$

In terms of rms values, Eq. (33-9a) may be written as $E_{\text{rms}} = (\omega L)I_{\text{rms}}$, which indicates that

$$E_{\text{rms}}/I_{\text{rms}} = \omega L = 2\pi f L \tag{33-10}$$

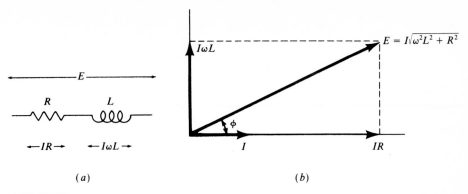

FIGURE 33-3
(a) Generalized coil. (b) Vector relations for coil.

Equations (33-8) and (33-9) are plotted in Fig. 33-2(b) to show that E leads I (or I lags behind E) by 90°. Typically, a 5 mH coil excited at 1000 Hz has an inductive reactance of $X_L = 2\pi(1000)(0.005) = 31.4\ \Omega$.

33-2.6   Total Coil Impedance

Generally, when ac is involved coils possess both resistance and inductive reactance. The phase of the current through the coil is common to both components, so that if it is drawn electrically as shown in Fig. 33-3(a), then the voltages across the resistive and inductive parts are IR and $I(\omega L)$, respectively [Fig. 33-3(b)]. Here the voltage across L is drawn 90° ahead of IR. This creates a right triangle in which the total circuit voltage E is given by

$$E = I\sqrt{\omega^2 L^2 + R^2} \tag{33-11}$$

and analogy with Ohm's Law ($E = IZ$) illustrates that the coil exhibits a total impedance (Z) to the circuit of

$$Z = \sqrt{\omega^2 L^2 + R^2} \tag{33-12}$$

which is the Pythagorean sum of ωL and R. A phase now exists between E and I, which is given by

$$\sin\phi = I\omega L/IR = \omega L/R = 2\pi f L/R \tag{33-13}$$

Now, the current I lags the voltage across the inductor by 90° and lags that in the complete circuit by ϕ. In terms of sinusoids, in which we have taken $E = E_0 \sin\omega t$, then $I = I_0 \sin(\omega t - \phi)$ as shown in Fig. 33-4.

Since the total impedance can often be measured experimentally, the following formulas, which can be seen from inspection of Fig. 33-3(b), are useful

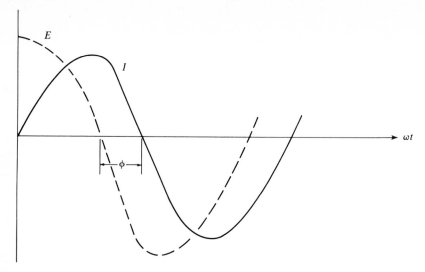

FIGURE 33-4
Relation between total circuit voltage E and current I.

for calculating the resistive and reactive components:

Resistive: $\qquad\qquad\qquad\qquad X_R = |Z|\cos\phi$

Reactive: $\qquad\qquad\qquad\qquad X_L = |Z|\sin\phi$ $\qquad\qquad$ (33-14)

where $|Z|$ is the absolute value of the impedance in ohms.

A second, often negligible, component of the total impedance is that which is due to the coil capacitance. The capacitive reactance (X_c) of a coil is given by $X_c = 1/\omega C = 1/2\pi f C$ (ohms), where C is the capacitance in farads. This can generally be measured with a capacitance bridge. Capacitive reactance may become a factor when one has to consider the use of coil probes attached to the EC instrument by long cables.

With respect to Fig. 33-3(b), the presence of a relatively large capacitive reactance in a circuit introduces a vector IX_c ($= I/\omega C$) that is 180° from the $X_L = I\omega L$ vector, i.e., it is directed downward. In the absence of inductance, Eq. (33-11) becomes

$$E = I\sqrt{1/\omega^2 C^2 + R^2}$$ (33-11a)

and Eq. (33-12) becomes

$$Z = \sqrt{1/\omega^2 C^2 + R^2}$$ (33-12a)

The phase angle is given by

$$\tan^{-1}[(1/\omega C)/R]$$ (33-13a)

When the circuit contains both inductance and reactance, the most general form of the impedance is

$$Z = \sqrt{(\omega L - 1/\omega C)^2 + R^2} \qquad (33\text{-}15)$$

and the phase angle between the current in and voltage across the circuit is given by

$$\tan \phi = (\omega L - 1/\omega C)/R \qquad (33\text{-}16)$$

It can be seen that the possibility arises that, if $\omega L = 1/\omega C$, then Z reduces to just R, and ϕ becomes $0°$. This condition is known as resonance, and many circuits are designed to operate under or close to such a condition.

CHAPTER
34

REMOTE
FIELD
EDDY
CURRENT
TECHNIQUES

34-1 INTRODUCTION

The position of the sensor with respect to the exciter determines whether testing is done in the close or remote field. In close-field inspection, the sensor is placed so as to share at least some of the flux from the primary (exciter) coil. In remote field eddy current (RFEC) testing, it is well removed from the source and is not constrained by the need to remove the effects of the applied signal. The technique may well be limited to tubular products; it has been used in oil well casing inspection for many years.

34-2 SCHMIDT EXPERIMENT

In a classic experiment, Schmidt [2, 3] discovered that if a coil is placed inside a steel tube and excited at low frequency (e.g., 30–120 Hz) as shown in Fig. 34-1, then the phase and amplitude of signals induced in a pickup coil are as shown in Fig. 34-2. The data show that two zones can be clearly identified, one in which the coupling between the coils is direct and that extends roughly two pipe diameters from the exciter, and a remote field region beyond this that extends for many pipe diameters.

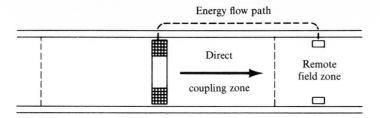

FIGURE 34-1
Remote field representation. The energy flow path is indicated. (*After Schmidt* [2].)

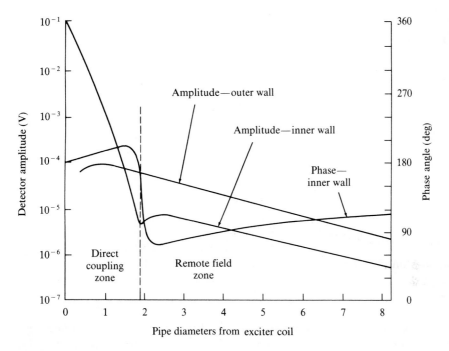

FIGURE 34-2
Amplitude and phase relation for real and remote field eddy currents in a ferromagnetic tube. (*After Schmidt* [2].)

In the direct coupling zone, the detector amplitude falls off rapidly with distance from the exciter, but in the remote field zone, the decay of amplitude is much weaker. If the test is repeated with a sensor at the outer wall, the signal strength is about 10 times higher than at the inner wall. The phase difference between exciter and inner wall pickup shows a large jump when passing through the transition zone. It is also found that placing a conducting shield between the

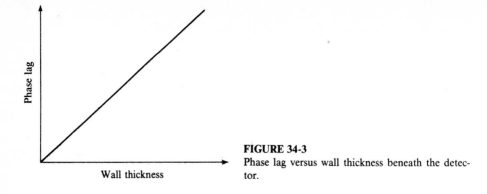

FIGURE 34-3
Phase lag versus wall thickness beneath the detector.

exciter and pickup produces no significant change in the signal strength. This indicates that the energy flow is through two walls, as illustrated in Fig. 34-1.

Testing with tubes of different wall thicknesses reveals a very linear relation between the phase lag and the wall thickness directly beneath the sensor, as shown in Fig. 34-3 In the data shown, this is about $36°$ per millimeter of wall, or $0.92°$ per 0.001 in. The remote field phase lag is therefore a sensitive wall thickness measuring technique. Complex finite element calculations have yielded similar results.

34-3 EXPLANATION OF THE EFFECT

The ac current in the exciter induces substantial circumferential eddy currents in the tube, the pipe wall being effectively a shorted single turn. These diffuse through the wall and spread outward. Eventually, they rediffuse back through the wall, undergoing a second attenuation and time delay.

At the transition zone, the dip is caused by interaction of two signals that have about the same amplitude but different phases. This produces a null effect when vector addition is performed.

A first approximation using the skin depth equations [Eqs. (32-6) and (32-7)] is useful in understanding the technique. These equations show that the skin depth is $1/\sqrt{\pi f \mu \sigma}$ and the phase lag is $x\sqrt{\pi f \mu \sigma}$, i.e., the latter is linear with wall thickness. It is clear, however, that the phase lag is affected by the product of magnetic permeability and electrical conductivity. A knowledge of the phase difference, wall thickness, and electrical conductivity may then provide a means for estimating the average permeability over the region of excitation.

34-4 DETECTED SIGNAL

When the tube passes under the exciter coil, local sections of the tube wall are taken through their *B-H* characteristics. The number of ampere turns of the exciter is generally sufficient to ensure that the tube is magnetized to a high level of flux density. This excitation may be useful in eliminating fields that remain

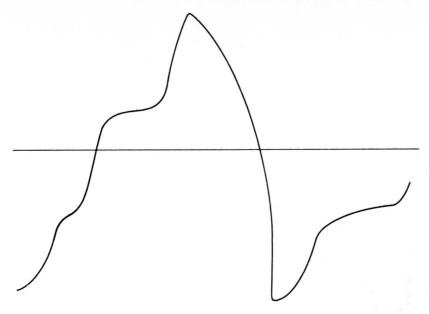

FIGURE 34-4
Detected eddy current signal from nonlinear tube material possibly containing stress.

from prior inspections. However, as the tube passes out of the near field into the remote field region, it experiences ac demagnetization (see the magnetics section of this text). When the sensor arrives at this section of the tube, only very small fields—the remote fields—activate the material. The RFEC therefore scans the permeability of the material around the origin of the *B-H* curve.

Figure 34-4 shows the form of the detected signal. In practice, it is very weak and therefore requires a high degree of amplification. It is that of a nonlinear material and may contain effects from stress within the tube.

From the point of view of practical NDE, the phase of the signal is useful in the determination of the wall thickness directly beneath the sensor and in the detection of pitting. Obviously any flaw that perturbs the flow of eddy currents as the field reenters the tube may be detected.

34-5 EQUIPMENT DESIGN

Figure 34-5 illustrates a (RFEC) system for the assessment of installed oil well tubes [4]. An exciter source remote field eddy current sends a signal down the well logging cable to the exciter coil and determines the reference phase. The pickup coil signal(s) are amplified and sent back up the wire line. It is filtered, amplified, and demodulated before passing to a tuned filter. The phase of the resulting signal is compared with that of the exciter. Laboratory tests do not require modulation–demodulation of the RFEC signal.

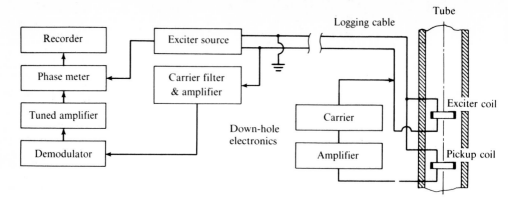

FIGURE 34-5
Remote field system for down-hole tubular applications.

34-6 TECHNIQUE ASSESSMENT

The RFEC technique has been in use since the 1960s for in situ inspection of oil well tubes and is now being applied to ferromagnetic tubes in other industries. As originally used (one exciter, one detector) the system is relatively simple to build, although for down-hole use the engineer must also contend with high temperatures, highly corrosive environments, down-hole electronic layouts no wider than 2.5 cm, and the need to pass signals up and down a wire line at the same time. There is equal sensitivity to external and internal anomalies, and a linear relationship between wall thickness and measured phase lag. Lift-off requirements for the sensor can be virtually eliminated by suitable sensor location, so that dirt and scale problems are minimized. Of great importance is the fact that no couplant is needed. Early applications also showed that the technique gives positive results with two or three frequencies applied to the exciter, so that permeability variations may be reduced.

Unfortunately, the technique is limited to tubular products and works only for ID sensors. The inspection speed is low because of the low excitation frequency, but this should not be a serious problem. The temperature of the inspected metal can seriously affect its electrical conductivity, especially for steel, so that there is a need to correct wall thickness measurements for temperature. (This can easily be accomplished by adding a temperature sensor to the detector system.) The most serious problem, however, is the presence of other nearby metallic objects, since these affect the magnetic fields outside of the tube. Perhaps the worst cases are (*a*) metallic sheets that are perpendicular to the tube under inspection and (*b*) nonconcentric tubes. The former situation occurs with tube sheets in heat exchangers, while the latter occurs with oil well casing and tubing. Such additional metal seriously affects both phase and amplitude.

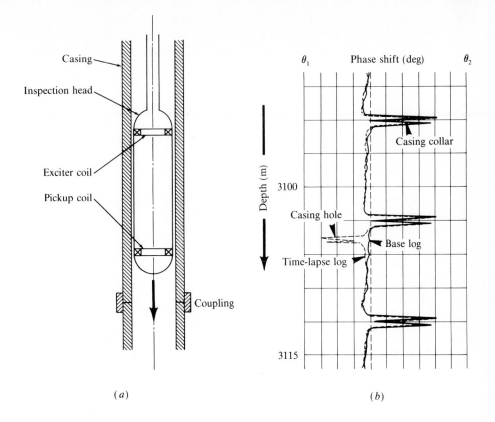

(a)   (b)

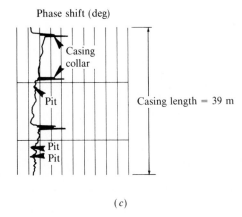

(c)

FIGURE 34-6
(a) Simplest RFEC system in well casing. (b) Typical data from casing over a period of time. (c) Development of pitting.

Example 34-1. Figure 34-6 shows data taken with a relatively unsophisticated tube inspection tool [Fig. 34-6(*a*)] that consists of just two coils. The base log [Fig. 34-6(*b*)] shows large swings due to additional metal and the physical ends of pipes as the exciter and detector pass over couplings (collars) that are about 30 ft (10 m) apart. A later log shows a hole in the tube. Figure 34-6(*c*) shows the presence of pitting. Since the tool does not give an indication of the location of the flaws, such devices generally also include a high-frequency sensor system that discriminates whether the flaw is ID or OD. A typical high-frequency system may operate in the region of 50,000 Hz, since at this frequency, the eddy currents are highly localized by the skin depth equation to the ID. (Here then, the presence of a signal indicates an ID flaw; the lack of one indicates an OD flaw.)

It is clear from the nature of the pickup coil used in this test that the signal must represent an average around the tube. Much better sensitivity is obtained by splitting the detector into an array or connecting sensors so as to detect differentially. These data also indicate that the technique is relatively insensitive to flaws in the vicinity of the coupling.

CHAPTER 35

IMPEDANCE
PLANE
DIAGRAM

35-1 INTRODUCTION

In this chapter, the basic R-L circuit is used to introduce the idea of the impedance plane diagram and normalization of eddy current coil voltages. Förster's ideas of the characteristic frequency and the effective permeability are introduced for a cylindrical bar that completely fills a test coil.

35-2 VOLTAGE PLANE ANALYSIS FOR A CONSTANT VOLTAGE CIRCUIT

Equation (33-12) shows that for constant values of the driving voltage, inductance L and resistance R, and because the vectors $I\omega L$ and IR are at right angles, then the locus of the end of the vector IR is a semicircle. This is traced out by varying ω ($= 2\pi f$) as shown in Fig. 35-1. Here, at $\omega = 0$, the point A is obtained, i.e., the impedance is purely resistive in R. At the point D, $\omega L = R$, or $f = R/2\pi L$, and as 0 is approached, the majority of the impedance is contributed by the inductance.

35-3 NORMALIZATION

When a test coil is connected to an eddy current instrument and held far away from an electrically conductive surface, a convenient reference point—the empty coil impedance—is found. Assume that the empty coil has a fixed resistance (R_0)

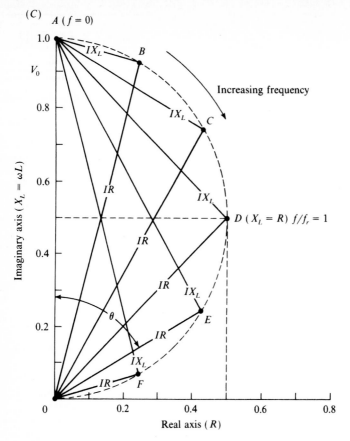

FIGURE 35-1
Impedance plane diagram for a R-L series circuit with constant voltage. (*Courtesy American Society for Nondestructive Testing.*)

and self-inductance (L_0). By introducing the test part into or close to the coil, changes in R and L occur that are related to the part. These changes are caused by the induced eddy current fields and they depend upon whether or not the part is ferromagnetic.

35-3.1   Nonmagnetic Parts

These are materials that are dia- or paramagnetic, i.e., have a relative permeability that is very close to unity. The inductive reactance of the coil falls from ωL_0 to some lower value ωL, while the apparent resistance rises from R_0 to some value R. This latter change may be of little interest. The normalized impedance is given by

$$(Z/Z_0)^2 = \left[(R - R_0)^2 + (\omega L)^2\right]\big/\left[R_0^2 + (\omega L_0)^2\right] \qquad (35\text{-}1)$$

Generally, the empty coil resistance is much less than the impedance, so that R_0

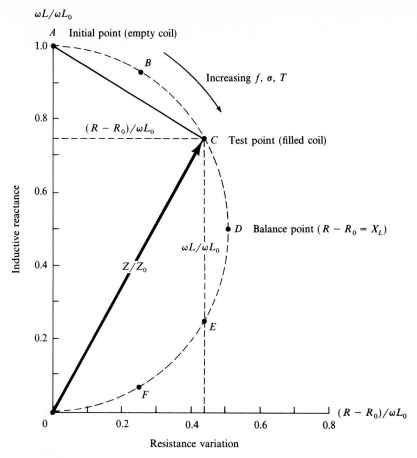

FIGURE 35-2
Normalized impedance plane diagram for a series *R-L* circuit. (*Courtesy American Society for Nondestructive Testing.*)

can be neglected in comparison to $(\omega L_0)^2$ and Eq. (35-1) becomes

$$(Z/Z_0)^2 = \left[(R - R_0)^2 + (\omega L)^2\right]\big/(\omega L_0)^2 \qquad (35\text{-}2)$$

By analogy with Fig. 35-1, the test point (*C*) in Fig. 35-2 lies at the tip of a vector (Z/Z_0), which has components $(R - R_0)/\omega L_0$ and $\omega L/\omega L_0$. The points *A* and *C* represent the empty coil and filled coil at any frequency; thus vector *AC* represents the signal variation between the empty and filled states. Note that the introduction of an object into the coil increases $R - R_0$ and lowers the inductive reactance.

Equal changes in resistive and inductive reactance components occur at *D*; this point is that of maximum sensitivity in EC testing, but may not represent the best operating point. Sometimes it is more essential to separate different effects

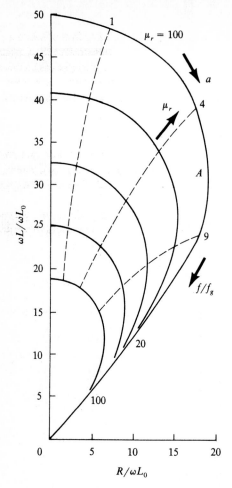

FIGURE 35-3

Normalized impedance plane diagram for solid ferromagnetic parts. (*From Halmshaw* [5]. *Courtesy Edward Arnold.*)

from each other, so that one looks for a point that provides a high phase angle between the effects.

35-3.2 Ferromagnetic Parts

Here, the applied field strength (H) raises the flux density (B) within the coil. This raises signal magnitudes by perhaps factors of 10 (resistive) and 100 (inductive), as shown in Fig. 35-3 [5]. These curves show the effect of increased relative permeability (μ_r) of the magnitudes of the eddy current signals. Both Figs. 35-2 and 35-3 show the effect of increasing the test frequency (f) and part conductivity (σ).

35-4 CHARACTERISTIC FREQUENCY

Consider the case of a solid cylinder with diameter d that is sliding through a test coil (Fig 35-4). Cylindrical symmetry leads to solutions of the relevant Maxwell

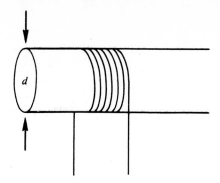

FIGURE 35-4
Cylindrical part sliding through test coil.

equation in terms of Bessel functions of argument A, where

$$A = f d^2 \mu_{\text{rel}}/50.66\rho \qquad (35\text{-}3)$$

where ρ is the resistivity in microhms-cm, d is in centimeters, and μ_{rel} is the relative magnetic permeability for eddy current testing. The characteristic frequency (f_g) arises when $A = 1$, i.e.,

$$f_g = 50.66\rho/d^2\mu_{\text{rel}} \qquad (35\text{-}4)$$

35-5 CONCEPT OF EFFECTIVE PERMEABILITY

When a nonmagnetic cylinder is placed in an ac coil, the field strength from the coil within the cylinder is diminished by the eddy current shielding effect (skin effect), which also includes a phase delay. No eddy current flow occurs at the center of the cylinder. The effect can be described as follows:

1. Assume a uniform field strength H_0 throughout the bar.
2. Assume the magnetization is given by an effective permeability that is: (a) constant in magnitude and phase over the cross-sectional area of the bar and (b) given by a complex number.

The voltage induced in any N-turn coil with sinusoidal flux linkage is

$$V = NA(2\pi f)\mu_0 H_0$$

or, in this case,

$$V = N(2\pi f)(\pi D^2/4)\mu_{\text{eff}}\mu_0 H_0 \qquad (35\text{-}5)$$

This voltage has real and imaginary components given by

$$V_{\text{real}} = N(2\pi f)(\pi D^2/4)\mu_{\text{rel}}\mu_{\text{eff, imag}} H_0$$

$$V_{\text{imag}} = N(2\pi f)(\pi D^2/4)\mu_{\text{rel}}\mu_{\text{eff, real}} H_0 \qquad (35\text{-}6)$$

The switch in real and imaginary components is due to the 90° phase shift that occurs between solid state and coil sensors.

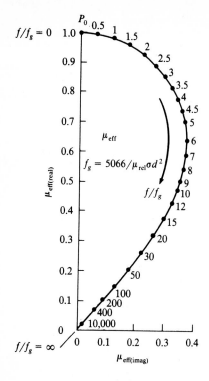

FIGURE 35-5
Effective permeability components for a solid nonferromagnetic bar that fills an encircling test coil. (*Courtesy Inst. Dr. Förster.*)

Figure 35-5 shows the real and imaginary components of a solid, nonferromagnetic bar that completely fills an encircling coil, plotted against the ratio f/f_g [6, after Förster]. To use this figure, first find f_g from Eq. (35-4), then compute f/f_g for the test and obtain the values of $\mu_{eff,real}$ and $\mu_{eff,imag}$ from the figure, and, finally, compute the pickup coil voltage components from Eqs. (35-6). Further, since Eqs. (35-6) show that

$$V_{real}/V_{imag} = \mu_{eff,imag}/\mu_{eff,real}$$

if the normalized real component of the voltage is plotted on the horizontal axis of Fig. 35-5 and the imaginary component on the vertical axis, the normalized coil voltage point lies over the effective permeability point.

In the case of a ferromagnetic bar, μ_{rel} must be obtained from prior experience. Its value may be lowered by performing the test in a high ambient field, i.e., with a high value of B in the material. This is sometimes accomplished by used pulsed eddy current techniques [7].

35-6 MAGNETIC PERMEABILITY USED IN EDDY CURRENT TESTING

Two problems have hampered eddy current testing of steel parts: the ability to lower the relative permeability by using high ambient fields and the effect on the eddy current signal of the variability of the permeability, even at high fields.

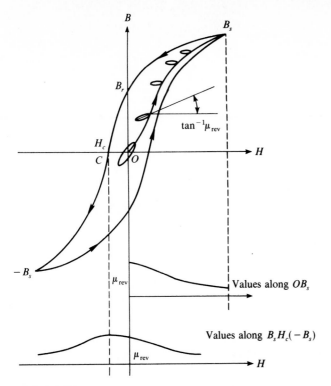

FIGURE 35-6
A *B-H* hysteresis loop showing the permeability μ_{rev} used in eddy current testing of ferromagnetic materials. The lower figures show μ_{rev} versus *H*. (*From Halmshaw [5]. Courtesy of Edward Arnold, London.*)

As Fig. 35-6 shows, the permeability varies considerably both along the virgin curve (OB_s) and around the rest of the loop. The relative permeability used in eddy current testing is that which occurs for small excursions, caused by the exciting field, from the normal (static) permeability. This is known as a "reversible permeability." As shown by the lower plots, it decreases along the virgin curve as B_s is reached and is greatest at the coercivity H_c.

It is clear from these plots that when such a nonlinear material is scanned by an initially purely sinusoidal field, the induced voltage will contain harmonics that carry information about the shape of the reversible permeability loop. These harmonics can be used to check the metallurgical state of the steel, especially those that have an anisotropic permeability tensor, i.e., have different permeabilities in different directions. Further, they can be used to determine differences in the stress state of the steel.

PROBLEMS
FOR
PART VI

Chapter 32

32-1. Find the skin depth for nonferromagnetic stainless steel of resistivity 72 $\mu\Omega$ cm at 150 kHz. What is the phase lag at 2 mm depth?

Chapter 33

33-1. Design an exciter coil that will saturate the tube wall of 30 cm ID, 1.25 cm wall thickness ferromagnetic pipe.

33-2. How would you add signals of 10, 50, and 120 Hz together so as to feed them to the exciter coil of a RFEC system?

33-3. Design a differential pickup probe system for
   (*a*) average wall loss in ferromagnetic tubes
   (*b*) localized wall loss over a 5 cm² area

33-4. How does the electrical conductivity of steel vary with temperature?
   (*a*) Give an equation
   (*b*) Compute the conductivity at 100 and 200°C.
   (*c*) State how the temperature might affect the apparent readings of wall thickness when performing inspection in a 15,000 ft (4.57 km) oil well at 190°C.

33-5. Assuming an applied frequency of 40 Hz, what is the value of $\mu\sigma$ for Fig. 34-3? Look up the value of σ and determine the value of μ.

Chapter 35

35-1. A copper bar of diameter 1 cm and electrical conductivity 50.68 m/Ω mm² fills a 100-turn test coil that is excited at 500 Hz. If the primary field intensity is 1 Oe, calculate (*a*) f/f_g, (*b*) the effective permeability components, and (*c*) the coil voltage components V_{real} and V_{imag}.

35-2. A steel bar of diameter 1 cm, electrical conductivity 10 m/Ω mm², and average eddy current permeability 150 fills a 100-turn test coil, the primary field of which is 1 Oe. If $f = 60$ Hz, calculate (*a*) f/f_g, (*b*) the effective permeability components, and (*c*) the coil voltage components V_{real} and V_{imag}.

REFERENCES
FOR
PART VI

1. Ida, N., "Eddy Current Transducers," *Handbook on Nondestructive Testing*, vol. IV, 2nd ed., American Society for Nondestructive Testing, Columbus, OH, 1987, sec. 3.
2. Schmidt, T. R., "The Remote Field Eddy Current Inspection Technique," *Qual-Test-2 Conference Proceedings, Dallas, TX, Oct. 1983*, American Society for Nondestructive Testing, Columbus, OH, 1983, pp. 9-13–9-30.
3. Schmidt, T. R., "Remote Field Low Frequency Eddy Current Inspection," *Handbook on Nondestructive Testing*, vol. IV, 2nd ed., American Society for Nondestructive Testing, Columbus, OH, 1987, sec. 8, pp. 206–211.
4. Palanisamy, R., "Remote Field Eddy Current Testing: A Review," *Review of Progress in Quantitative Nondestructive Evaluation*, vol. 7-a, pp. 157–164, 1988.
5. Halmshaw, R., *Nondestructive Testing*, Edward Arnold, London, 1987.
6. Förster, F., in R. C. McMaster, P. McIntire, and M. L. Mester, eds., *Nondestructive Testing Handbook* vol. IV, *Electromagnetic Methods*, 2nd ed., American Society for Nondestructive Testing, Columbus, OH, 1986.
7. Dodd, C. V., Deeds, W. E., and Chitwood, D. L., "Eddy Current Inspection of Ferromagnetic Materials using Pulsed Magnetic Saturation," *Materials Evaluation*, to be published.

APPENDIXES

APPENDIX
A

MODELING AND ANALYSIS OF EXPERIMENTAL DATA

A study of manufacturing defects, fatigue failure patterns, and maintenance trends usually requires that data be collected and analyzed. Often, there is a need to compare data sets obtained from the same environment, but at a different time. The results of these analyses are very important in the implementation of an NDE program and statistical analyses techniques offer very useful tools for an engineer. While statistical methods are widely available in computer software, it is proper that the fundamentals be presented at this time in order to allow discussion on the uses of statistical analyses in NDE.

Presentations of statistical techniques applied to the analysis of engineering data are available from a number of sources, including Refs. 1–7 of Part I. There are many techniques used in statistical analysis, each method suited to a particular set of circumstances. The discussion to be given here is merely to show the basic methods of statistical analysis and, hence, will use the most fundamental approach, i.e., the normal distribution. The fundamental assumption of the normal distribution is that all data are symmetrically distributed on either side of some central value. Examples often used to demonstrate the normal distribution are students' quiz scores and people's heights. Under some conditions it is possible to have results where all of the grades lie equally on either side of the average. Similarly, it is possible to have a collection of people whose heights are distributed equally on either side of an average height.

Two parameters are initially required to analyze data with the assumption of a normal distribution, namely the mean $\bar{X}$, and the standard deviation σ. The mean is a measure of the central value of the data and the standard deviation is a

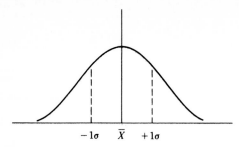

$-1\sigma \quad \overline{X} \quad +1\sigma$

FIGURE A-1
Typical plot of normally distributed data.

measure of the variability. A large standard deviation represents data with a large amount of variation about the mean while a smaller value indicates data that are more closely grouped to the mean. The mean and standard deviation are given by

$$\overline{X} = \sum_{j} \left[\frac{X_j}{n} \right] \tag{A-1}$$

$$\sigma = \sqrt{ \frac{\sum_j \left(X_j - \overline{X} \right)^2}{n-1} } \ , \qquad 0 \le j \le n \tag{A-2}$$

where n is the quantity of data and j is the summation variable. Once the mean and the standard deviation are known for a data set, it is relatively simple to obtain a plot of the curve predicted with the assumption of a normally distributed data set. The expression for this is shown in Eq. (A-3) and a typical plot of this function is shown in Fig. A-1:

$$f(X) = \frac{1}{\sigma\sqrt{2\pi}} \, e^{-(X-\overline{X})^2/2\sigma^2} \tag{A-3}$$

It is important to observe that the mere calculation of a mean and a standard deviation does not force a data set to be normal. For example, the data previously shown in Table 2-2 could be used to develop a normal distribution curve. It is interesting to note, however, that the plot of the data shown in Fig. 2-2 is skewed to the left while the normal distribution curve is symmetric. While this comparison does demonstrate a flaw in the assumption of a normal distribution for all data, it is nevertheless true that this simple assumption can be used in a large number of situations without fear of gross error. Tests exist that permit the adequacy of the normal distribution assumption to be checked. These are described in the references.

Some properties of the normal distribution curve are useful in graphically demonstrating data characteristics. Since all of the data are expected to be found in the realm from $-\infty$ to $+\infty$, the total area under the curve is defined to be equal to 1. This may be expressed by the relationship

$$\int_{-\infty}^{\infty} kf_x(x) \, dx = 1 \tag{A-4}$$

where k is the normalization constant. Further, the standard deviation as

TABLE A-1
Standard probability functions

z	σ	α	$1 - \alpha$
0.99	1	0.32	0.68
1.96	2	0.05	0.95
2.57	3	0.01	0.99

determined by Eq. (A-2) describes lesser areas. For example, 99% of the data would be expected to fall in the region from $\overline{X} \pm 3\sigma$, 95% from $\overline{X} \pm 2\sigma$, and 68% from $\overline{X} \pm 1\sigma$. It also can be shown that the data from -1σ to $+\infty$ represent 84% of the data. Where a normally distributed flaw data set was known to have a mean of $\overline{X}^*$ and a standard deviation of σ^*, it would be expected that 84% of the flaws would be in a range

$$X \leq \overline{X}^* + \sigma^*$$

Tables available in standard references on statistics describe the characteristics of normally distributed data as a function of the distance away from the mean. These functions are often expressed in terms of the standard normal variate z and the standard deviation. Table A-1 gives an abbreviated summary of a table of standard normal probability.

The term $1 - \alpha$ represents the fractional amount of the area of the normal distribution curve within $\pm p\sigma$, where $p = 1, 2, 3, \ldots$.

When evaluating trends, it is often required that a comparison of two or more data sets be made in order to determine if true changes are occurring. Decisions made because of perceived changes may be costly when it is discovered that the true trends were not as perceived. Statistical tests can remove some uncertainty in the decision process.

Where only two data sets are involved and both may be assumed to possess a normal distribution, a comparison of two means test, as described by Duncan (Ref. 7 of Part I), may be performed. The equation used for this comparison is

$$z = \frac{\overline{X}_1 - \overline{X}_2}{\sqrt{(\sigma_1)^2/n_1 + (\sigma_2)^2/n_2}} \tag{A-5}$$

where z is the standard normal variate as given in Table A-1. For $z \geq 1.96$, there is a 95% confidence that the two samples being compared are, in fact, different.

In a situation where a mean value $\overline{X}$ has been obtained from a sample of size n for a particular parameter, the question arises could the $\overline{X}$ from the sample have come from a population whose mean is $\overline{X}'$? This may be evaluated using the expression

$$P\left(-z < \frac{\overline{X} - \overline{X}'}{\sigma/\sqrt{n}} \leq z\right) = 1 - \alpha \tag{A-6}$$

which is the probability that the $(\overline{X} - \overline{X}')/(\sigma/\sqrt{n})$ will be within a given interval that is obtained for a specified confidence level $1 - \alpha$ from tables for the normal variate z, as given in Table A-1. For example, for a 95% confidence level, $\alpha = 0.05$, which yields from the standard tables a value of $z = 1.96$. An increased sample size n will generally produce a smaller standard deviation σ and a mean $\overline{X}$ closer to the population mean $\overline{X}'$.

It is interesting to observe from Eq. (A-6) that where a higher confidence is required, 99% as compared to 68%, for example, a larger spread in values $\overline{X} - \overline{X}'$ is required.

As previously stated, the foregoing discussions on statistical analysis have been furnished merely to give the reader an understanding of the techniques of statistical analysis and the benefits that are possible. There are certainly risks in oversimplification of complex numerical analysis and the reader is directed to one of several references for further understanding.

The use of distributions other than the normal is worthy of brief discussion. There are a myriad of other distributions, e.g., the log normal, Rayleigh, and exponential, that have been shown to be uniquely useful in explaining particular behavior. Additionally, the Weibull analysis has been shown to be useful in investigating data. The Weibull is particularly useful because the numerical results serve as an indicator of the distribution of the data.

At several instances in the material, concern has been expressed about the assurance that experimental data do correctly reveal the parameters about the data set of interest. Statistical confidence intervals are useful in evaluating this assurance. These are discussed in standard references e.g., Ang and Tang (Ref. 3 of Part I).

RADIOGRAPHIC ATTENUATION COEFFICIENTS FOR SELECTED ELEMENTS

Element	Atomic no.	Energy (MeV)	Attenuation coefficient		
			Atomic (10^{-24} cm)	Mass (cm^2/g)	Linear (cm)
Magnesium	12				
		0.02	112	2.27	4.82
		0.05	13.0	0.322	0.561
		0.15	5.62	0.139	0.242
		0.50	3.48	0.0863	0.150
Aluminum	13				
		0.02	156	3.48	9.40
		0.05	16.0	0.357	0.964
		0.15	6.17	0.138	0.373
		0.50	3.78	0.0844	0.228

Element	Atomic no.	Energy (MeV)	Attenuation coefficient		
			Atomic $(10^{-24}$ cm$)$	Mass (cm^2 / g)	Linear (cm)
Titanium	22				
		0.05	94.6	1.19	5.40
		0.10	21.7	0.273	1.24
		0.50	6.50	0.0818	0.371
		1.00	4.67	0.0587	0.266
		2.00	3.31	0.0416	0.189
		4.00	2.52	0.0317	0.144
Iron	26				
		0.05	179	1.93	15.2
		0.10	34.5	0.372	2.93
		0.50	7.79	0.0841	0.662
		1.00	5.55	0.0599	0.417
		2.00	3.94	0.0425	0.334
		4.00	3.07	0.0331	0.260
Nickel	28				
		0.10	43.5	0.447	3.96
		0.50	8.46	0.0869	0.769
		1.00	5.98	0.0614	0.543
		2.00	4.27	0.0439	0.389
		4.00	3.35	0.0344	0.304
Copper	29				
		0.10	48.6	0.461	4.10
		0.50	8.80	0.0834	0.742
		1.00	6.21	0.0589	0.524
		2.00	4.43	0.0420	0.374
		4.00	3.50	0.0332	0.295
Zinc	30				
		0.10	54.2	0.499	3.56
		0.50	9.15	0.0843	0.601
		1.00	6.44	0.0593	0.423
		2.00	4.59	0.0423	0.302
		4.00	3.63	0.0335	0.239
Tungsten	74				
		0.10	1330	4.369	81.5
		0.50	40.1	0.131	2.45
		1.00	20.0	0.0655	1.223
		2.00	13.4	0.0439	0.821
		4.00	12.4	0.0406	0.759
Lead	82				
		0.10	1880	5.47	62.0
		0.50	52.4	0.152	1.72
		1.00	24.2	0.0704	0.798
		2.00	15.9	0.0462	0.524
		4.00	14.7	0.0427	0.484

Source: Reference 18 of Part 4 The *Nondestructive Testing Handbook*, vol. 3, *Radiography and Radiation Testing*, 2nd ed., 1985, American Society for Nondestructive Testing.

APPENDIX
C

NDE CODES, SPECIFICATIONS, AND STANDARDS

Various industrial and professional organizations and companies provide standards and recommended practices that are used in NDE. The lists of these specifications and standards is quite long and is constantly being updated. Rather than reproduce the list here, the reader is directed to sources such as *Materials Evaluation*, *Redi-Reference Guide* of the American Society for Nondestructive Testing, which annually publishes a list of current specifications and standards. Specific organizations may also be consulted about their activities in this area. A list of names and addresses of the various organizations active in the specifications and standards area is supplied:

ABS (American Bureau of Shipping)
45 Eisenhower Dr.
PO Box 910
Paramus, NJ 07653-0910

AEC (RDT) (US Atomic Energy Commission)
Oak Ridge National Laboratory
PO Box 62
Oak Ridge, TN 37830

AIA (Aerospace Industries Association of America)
1725 DeSales St. NW
Washington, DC 20036

AISI (American Iron and Steel Institute)
1000 16th St. NW
Washington, DC 20036

Al. Assoc. (The Aluminum Association Inc.)
818 Connecticut Ave. NW
Washington, DC 20006

ANSI (American National Standards Institute)
1430 Broadway
New York, NY 10018

API (American Petroleum Institute)
211 N Ervey
Dallas, TX 75201-3688

ASME (American Society of Mechanical Engineers)
345 East 47th St.
New York, NY 10017

ASNT (American Society for Nondestructive Testing)
4153 Arlingate Plaza
Caller #28518
Columbus, OH 43228-0518

ASTM (American Society for Testing and Materials)
1916 Race St.
Philadelphia, PA 19103

AWS (American Welding Society)
550 NW LeJeune Rd.
Miami, FL 33126

DOD (US Department of Defense)
c/o Naval Publications and Forms Center
5801 Tabor Ave.
Philadelphia, PA 19120

DOE (US Department of Energy)
c/o Naval Publications and Forms Center
5801 Tabor Ave.
Philadelphia, PA 19120

ICRU (International Commission on Radiation Units
  and Measurements)
7910 Woodmont Ave.
Suite 1016
Washington, DC 20014

NAVSEA (Naval Sea Systems Command)
c/o Naval Publications and Forms Center
5801 Tabor Ave.
Philadelphia, PA 19120

NBS (National Bureau of Standards)
c/o Superintendent of Documents
US Government Printing Office
Washington, DC 20402

NCRP (National Council on Radiation Protection and Measurements)
7910 Woodmont Ave.
Suite 1016
Washington, D.C. 20014

OSHA (Occupational Safety and Health Administration)
200 Constitution Ave. NW
Washington, DC 20001

SAE (Society of Automotive Engineers)
400 Commonwealth Dr.
Warrendale, PA 15096

USAF (US Air Force)
c/o Naval Publications and Forms Center
5801 Tabor Ave.
Philadelphia, PA 19120

USN (US Navy)
c/o Naval Publications and Forms Center
5801 Tabor Ave.
Philadelphia, PA 19120

APPENDIX
D

CLASS
PROJECTS

CHAPTER 11

11-1. Build an experimental test setup that will allow you to measure the B-H properties of ring samples of steel. Items to consider are:
  (a) Ring acquisition, machining, and measuring
  (b) Integrator design and sensitivity and the possibility of drift
  (c) Number of pickup coil turns so as to provide adequate input voltage to integrator
  (d) Number of magnetic field intensity coil turns, so as to provide adequate H from the power supply used to saturate the sample
  (e) Type of output
  (f) Computer control

CHAPTER 13

13-1. Circumferentially magnetize a ring sample of steel with a through slot and investigate how the field strength in the slot varies as the slot width is increased. Use a Hall element gaussmeter.

13-2. Wind a primary coil onto a ring sample of steel, and drill a hole in the steel that is large enough to accept a Hall element probe. Measure the circular field in the hole as the excited coil current is raised.

13-3. Machine slots of various widths and depths in a steel plate and investigate their ability to hold magnetic particles by using a commercial yoke for excitation as shown in Fig. 13-19(b).

13-4. Design and build a system such as that shown in Fig. 13-10 for the measurement of the B-H parameters of relatively short bar samples of steel.

13-5. Use equipment such as that shown in Fig. 13-10 to assess the permeability of various types of magnetic particles.

13-6. Design a flux loop for the internal inspection of small diameter heat exchanger tubing. Use rare earth magnets and determine the best sensor system to use for ID and OD surface flaws.

CHAPTER 14

14-1. Obtain a list of manufacturers of (*a*) applied current circular magnetization equipment and (*b*) capacitor discharge equipment. Compare and contrast manufacturers' specifications for magnetizing ability and safety.

14-2. Measure the surface field strength just outside a tube that is magnetized as shown in Fig. 14-1 with a Hall element gaussmeter. Is the value close to the predicted value?

14-3. Cut slots of width 0.08 mm (0.003 in) and variable depth in a bar or tube sample, and fill them with epoxy. Determine the minimum current through the sample that will cause the slots to hold (*a*) wet and (*b*) dry magnetic powder.

14-4. Magnetize a piece of hollow material longitudinally to saturation. Using a gaussmeter to measure external field strengths, determine the ability of a pulse magnetization system to rotate the flux into the circular direction?

CHAPTER 18

18-1. With a Hall element gaussmeter, digitize the tangential and normal field signals above slots of the same width but differing depths in a steel bar in active and residual fields.

18-2. Write a computer program that will take the first and second derivatives of the fields recorded in Project 18-1.

18-3. Construct a ferrite probe and use it to determine the leakage fields above flaws in a steel plate. At what excitation frequency does the probe work best?

18-4. After obtaining samples of 10 different magnetic powders, construct a device that will measure the bulk permeability. Use an encircling coil around a tube full of the material as a pickup. How do you determine the packing density of the powder?

18-5. Obtain several types of field indicator and place them at the center of a large coil. Determine the ambient field intensity at the center of the coil that causes visible indications for a variety of powders. Use a gaussmeter to measure the field strength at the position of the indicator. Do your rests support the idea that the powder-holding ability of a test flaw varies as the square of the flaw width?

18-6. Design a detector that will respond to the third derivative of the normal component of a MFL signal. Use coils or Hall elements.

BIBLIOGRAPHY

Edwards, J. M., and Stroud, S. G., "New Electronic Casing Caliper Log Introduced for Corrosion Detection," *Journal of Petroleum Technology*, pp. 933–938, August 1966.

MacLean, W. R., "Appts for Magnetically Measuring Thickness of Ferrous Pipes," US Patent 2,573,799, Nov. 1951.

Schmidt, T. R., "The Casing Inspection Tool—An Instrument for the In Situ Detection of External Casing Corrosion in Oil Wells," *Corrosion*, vol. 17, no. 7, pp. 81–85, July 1961.

Stroud, S. G., and Fuller, C. A., "New Electromagnetic Inspection Device Permits Improved Casing Corrosion Evaluation," *Journal of Petroleum Technology*, pp. 257–260, March 1962.

INDEX

ISBN 0-8493-2655-9

x

1997 CRC Press Inc